Current Topics in Microbiology 213/III and Immunology

Springer
Berlin
Heidelberg
New York
Barcelona
Budapest
Hong Kong
London
Milan
Paris
Santa Clara
Singapore
Tokyo

Attempts to Understand Metastasis Formation III

Therapeutic Approaches for Metastasis Treatment

Edited by U. Günthert, P.M. Schlag, and W. Birchmeier

With 14 Figures

Springer

Ursula Günthert, Ph.D., P.D.
Basel Institute for Immunology
Grenzacherstraße 487
CH-4005 Basel, Switzerland

Professor Peter M. Schlag, M.D.
Virchow Klinikum, Chirurgische Abteilung
Robert-Rössle-Klinik am
Max-Delbrück-Centrum Berlin
Lindenberger Weg 80
D-13122 Berlin, Germany

Professor Walter Birchmeier, Ph.D.
Max-Delbrück-Centrum für Molekulare Medizin
Robert-Rössle-Straße 10
D-13122 Berlin, Germany

Cover illustration: Radionuclide scan of liver- and abdominal metastases in a patient with malignant melanoma

Cover design: Künkel+Lopka, Ilvesheim

ISBN-13: 978-3-642-80073-3 e-ISBN-13: 978-3-642-80071-9
DOI: 10.1007/978-3-642-80071-9

Library of Congress Catalog Card Number 15-12910
Softcover reprint of the hardcover 1st edition 1996

Typesetting: Thomson Press (India) Ltd., Madras

SPIN: 10495485 27/3020/SPS – 5 4 3 2 1 0 – Printed on acid-free paper

Preface

In metastasis, tumor cells disseminate from the primary lesion and home to secondary organs where they may remain dormant for a long time. Metastasis formation is still the most feared manifestation for tumor patients and clinicians. Although improvements have been made concerning earlier detection and specific therapy, most of the cancer patients still die of distant metastases. The purpose of these three volumes is to review the recent progress in molecular metastasis research and to attempt to further understand the biology of this multifocal process.

With respect to present day molecular biology, the pioneers of metastasis research established the basic concepts of metastasis formation in the 1970s and 1980s, namely, clonal selection of metastatic cells, heterogeneity of metastatic subpopulations, organ specificity of metastasis and the importance of angiogenesis (Fidler, Kripke, Nicolson, Folkman and others). In the 1980s and 1990s, several of the molecules involved were identified and their network interactions elucidated. These three volumes of *Current Topics in Microbiology and Immunology* compile the most recent developments on these metastasis-related molecules; their interactions, regulation, and ways to interfere with their action. It became evident that metastasis-related molecules are confined to distinct cellular compartments, such as the extracellular space, the cell membrane, the cytoplasmic signalling network, and the nuclear regulatory system.

For the complex metastatic cascade, proteolysis and alterations in adhesive functions are the most obvious and thus one of the most thoroughly investigated processes. Various proteases and precursors (metalloproteinases and serine proteases) and their inhibitors (tissue inhibitors of metalloproteases, plasminogen activator inhibitors and serpins) exhibit a sensitive complex of interplay – we are particularly fascinated by their highly regulated nature. Not only the proteases and their inhibitors are important in all the different stages of metastasis formation, but also to the same extent adhesive and "de-adhesive" interactions: metastatic cells must constantly detach themselves from their old partners and reattach to new ones, as mainly outlined in the first volume and partly in the second volume. Among the widespread members of the adhesion molecule families, certain immunoglobulins, integrins, cadherins, selectins, and hyaluronic acid receptors as well as their ligands are implicated in the spread of metastatic cells. The control of the metastatic cascade by these extracellularly acting molecules is delicately balanced, and slight changes could affect the establishment of the normal cellular organization and consequently promote metastasis formation. Strikingly, some genes of adhesion molecules have recently been identified as tumor suppressor genes in model organisms (e.g. Drosophila) and are in fact mutated in metastasizing human tumors.

Growth of primary tumors and metastases is strictly dependent on angiogenesis, the formation of new blood vessels. How this process is regulated by cytokines is another topic of the second volume. Cytokines are not only important in angiogenesis but are essential for the direct migration of metastatic cells. Cytokines act through specific receptors which mediate signals by different means, e.g., tyrosine phosphorylation. A recent discovery is that cytoplasmic signal transduction components, transcription factors, and cell cycle regulators are also metastasis-related. Many of the presently described genes in metastasis were known as activated oncogenes for several years, but apparently the encoded gene products have a broader spectrum of action than was originally assumed.

We have recently learned that the spread of metastatic cells, especially of micrometastases, is far more extensive than previously expected. A successful antimetastatic therapy therefore requires new strategies: for this reason the third volume comprises novel approaches such as immunotherapy, transfer of tumor-inhibiting genes and anti-sense constructs, as well as interference with signal transduction pathways. Promising new therapeutic approaches also involve the use of anti-angiogenic factors or of recombinant soluble metastasis-related molecules which interfere with ligand interactions.

As the process of metastatic spread is presently regarded as a multifactor event which is yet to be sufficiently understood in the multitude of its aspects, approaches to clinical treatment have to be polypragmatic. Methods of treatment are based on chemotherapy and radiotherapy, refined and adapted to the type of tumor pertaining and the pattern of metastatic spread. Increasingly, therapies which incorporate new insights from immunology and molecular biology are adopted for clinical use. To present a rounded scope of the topic, these current strategies are covered by the third volume in particular. Surgical treatment options are indicated in cases where a curative intervention is feasible e.g. in solitary metastases of colorectal carcinoma, soft tissue, and kidney tumors.

We hope that the reader of these volumes is impressed by the quality of the contents. Metastasis has obviously emerged as a serious discipline of natural sciences due to the fact that the molecular biology of various metastasis-related molecules and their complex interplay became transparent. We are, nevertheless, still in the beginning phase and await further progress from which patients will finally benefit.

Most, if not all of the metastasis-specific processes described are also known to be involved in embryonic development and pattern formation, as well as in leukocyte biology. The disciplines of metastasis research, developmental biology, and immunology can, therefore, profit from and stimulate each other. The genetic analysis of candidate molecules and their interplays in transgenic mice will certainly further broaden our understanding of the molecular basis of metastasis formation.

We would like to thank the authors who have spent their valuable time in writing a chapter for this series. Without their expertise and cooperation, this compilation of newest developments in metastasis research would not have been attainable. Leslie Nicklin (Basel) assisted the edition of this series with her competent skills; we are most grateful for her contribution.

Basel — Ursula Günthert
Berlin — Walter Birchmeier
Berlin — Peter M. Schlag

List of Contents

List of Contents of Companion Volume 213/I

List of Contents of Companion Volume 213/II

List of Contributors

(Their addresses can be found at the beginning of their respective chapters.)

Micrometastasis Detection and Treatment with Monoclonal Antibodies

K. PANTEL and G. RIETHMÜLLER

1 Introduction

In industrialized societies, the most frequent forms of cancers are derived from epithelia of the gastrointestinal and urogenitary tract, mammary ducts, and bronchi. Despite recent progress in early detection and surgical therapy of these tumors, cancer-related mortality has remained high. The major reason for this disturbing discrepancy is that occult dissemination of viable malignant cells can occur at an early stage of tumor growth (RIETHMÜLLER and JOHNSON 1992). This implies that the acquisition of at least some characteristics of metastatic behavior can occur prior to attainment of the unrestrained growth observed in fully developed tumors. In this context, it is important to consider that tumorigenesis and metastasis development are not necessarily the result of the same genetic changes (FIDLER and RADINSKY 1990; LIOTTA et al. 1991).

Institute of Immunology, Ludwig-Maximilians University, Goethestr 31, 80336 Munich, Germany

Although micrometastatic tumor cell spread is thought to determine the prognosis of patients with operable epithelial cancer, it is usually missed by conventional tumor staging. Several groups (including ours) have therefore developed immunocytochemical and molecular assays that allow the specific detection and characterization of individual carcinoma cells disseminated to secondary organs, such as blood, bone marrow (BM), or lymph nodes. Among the various organs investigated, BM is an important indicator organ of micrometastatic spread because it is easily accessible and lacks epithelial cells. Thus disseminated epithelial tumor cells can be detected by use of monoclonal antibodies (mAb) to epithelial-specific marker proteins such as cytokeratins (CK). Moreover, bone or BM is one of the major sites of distant metastasis in several types of epithelial tumors, including cancer of the breast, prostate, or lung.

The present review focuses on new diagnostic and therapeutic approaches to identify and treat patients with minimal residual epithelial cancer.

2 Diagnostic Approaches of Micrometastasis Detection

2.1 Immunocytochemistry

Several methodological studies have been based on surrogate model systems of BM samples spiked with cancer cells cell lines, demonstrating that immunocytologic techniques are superior to conventional histopathologic examinations or flow cytometric analyses (Molino et al. 1991). However, Gross et al. (1995) presented a recent model study in which spiked breast cancer cells were detected in peripheral blood mononuclear cells at frequencies as low as 10^{-7} by using four-color immunofluorescence.

Immunocytochemical methods relying on mAb to a variety of epithelial-specific cytoskeleton and membrane antigens have been applied to detect individual disseminated carcinoma cells in mesenchymal organs (Table 1). Since the specificity of the immunocytochemical screening for single disseminated tumor cells is one of the major concerns, most investigators decided to use mAb rather than polyclonal antisera for tumor cell detection. The most extensive experience stems from immunocytologic analysis of BM, which is an easily accessible and preferential site of metastases for various epithelial tumors, including carcinomas of the breast, lung, and prostate. Previous methodological studies have used surrogate model systems of BM samples spiked with cancer cells from cell lines, demonstrating that the technique can detect two to four cells at a concentration of 10 per 10^6 and, by extrapolation, a 95% chance of detecting one cancer cell at a concentration of 2 per 10^6 (Osborne et al. 1991). The clinical relevance of these evaluations remain disputed, because tumor cells selected in vitro may display different characteristics as compared to cancer cells in vivo.

Table 1. Immunocytochemical detection of isolated tumor cells in secondary organs of patients with various types of epithelial cancer

Secondary organ	Tumor origin	Marker	Prognostic value	References
Bone marrow	Breast	EMA	+	MANSI et al. (1991)
		EMA, TAG12, CK	+	HARBECK et al. (1994)
		CK	+	COTE et al. (1991)
				SCHLIMOK et al. (1994)
				SCHLIMOK et al. (1992)
		TAG12	+	DIEL et al. (1992)
	Colorectum	CK	+	LINDEMANN et al. (1992)
		Ca-19-9	n.d.	JUHL et al. (1994)
	Stomach	CK	+	JAUCH et al. (1995)
				SCHLIMOK et al. (1991)
	Pancreas	CmK Ca-19-9	n.d.	JUHL et al. (1994)
	Lung, SCLC	SM1	n.d.	STAHEL et al. (1985)
		LCA 1-3	n.d.	HUMBLET et al. (1988)
	Lung, NSCLC	CK	+	PANTEL et al. (1996)
	Prostate	CK	n.d.	OBERNEDER et al. (1994a,b)
				RIESENBEREG et al. (1993)
		CK,PSA, EMA		MANSI et al. (1988)
	Bladder	CK	n.d.	OBERNEDER et al. (1994b)
	Kidney	CK	n.d.	OBERNEDER et al. (1994b)
Lymph nodes	Breast	CK	+	DE MASCAREL et al. (1992)
	Lung	BerEp4	+	PASSLICK et al. (1994)
		CK	n.d.	CHEN et al. (1993)
	Colorectum	CK and CEA	n.d.	CUTAIT et al. (1991)
Peripheral blood	Breast	CK and HEA125	n.d.	BRUGGER et al. (1994)

ICC, immunocytochemistry; RT-PCR, reverse-transcriptase polymerase chain reaction; EMA, epithelial membrane antigen; CK, cytokeratin; TAG12, tumor-associated glycoprotein-12; LCA, lung cancer antigens; PSA, prostate-specific antigen; HEA125, human epithelial antigen-125; CEA carcino-embryonic antigen; SCLC, small cell lung cancer; NSCLC, non-small-cell lung cancer; n.d., not determined.

In a more recent investigation, we therefore concentrated on the evaluation of BM samples obtained from cancer patients (PANTEL et al. 1994). Our analysis underscored the remarkable sensitivity and specificity of anticytokeratin (anti-CK) mAb for tumor cell detection in BM (one tumor cell per 2 x 10^6 BM cells). In contrast, mAb to "epithelial-specific" cell membrane proteins are less reliable because of their frequent cross-reactivity with hematopoietic cells present in BM aspirates (SCHLIMOK et al. 1991; PANTEL et al. 1994). Morphological criteria, as invoked by some authors who have used such antibodies (MANSI et al. 1991; DIEL et al. 1992), are rather unreliable because of the heterogeneous size and shape of aspirated carcinoma cells. Previous investigations suggesting the ectopic expression of CK18 in mesenchymal cells were not confirmed by our recent immunocytochemical analyses (SCHLIMOK et al. 1991; PANTEL et al. 1994). In contrast, the epithelial nature of CK18-positive cells in BM was supported by coexpression of histogenetic marker proteins (e.g., prostate-specific antigen PSA) (RIESENBERG et al. 1993) and by negative costaining with mAb to the common leukocyte antigen CD45 (Schlimok et al. 1987; Pantel et al. 1994).

Using anti-CK mAb the detection rate is however affected by blood contamination of the aspirate, the number of aspirates analyzed, and the number of marrow cells screened per aspiration site (Pantel et al. 1994). Thus the result of immunocytochemical screening for single carcinoma cells in BM largely depends on the applied method, which underscores the urgent need for a standardized protocol. At present, we recommend that at least 2×10^6 nucleated cells aspirated from two sites of the iliac crest be screened, using a broad-spectrum CK mAb with proven specificity for epithelial cells. International concerted activities are now required to develop a standardized procedure that may then also serve as a gold standard for other diagnostic approaches.

2.2 Cellular Enzyme Immunoassay

Immunocytochemical analysis is quite cumbersome, which might hamper its introduction into clinical routine unless reliable techniques for the automated staining and scanning of cytospin preparations become available. Attention is therefore focused on the development of alternative methods to process large sample volumes and to assess epithelial antigens with high sensitivity. We have therefore developed a cellular enzyme immunoassay (enzyme-linked immunosorbent assay, ELISA) using mAb against CK19 (Hoechtlen-Vollmar et al. 1995; Pantel et al. 1993b). CK19 was preferred as target antigen over CK18 because it is smaller and more easily soluble. The cutoff point determined by analysis of peripheral blood leukocytes (PBL) from 50 healthy volunteers was set at 250 pg/ml CK. The sensitivity of the assay was determined by mixing HT-29 colon carcinoma cells with PBL; it allowed the detection of ten HT-29 cells in 5×10^6 PBL.

In total, 386 BM aspirates from 242 patients with various types of epithelial tumors were simultaneously analyzed by this ELISA and our immunocytochemical CK assay using mAb CK2 to CK18 for detection of tumor cells in BM. The results were consistent in 62% ($p < 0.001$) of the aspirates, with ELISA being more sensitive than the immunocytologic analysis. Follow-up studies of tumor patients have been initiated to test the prognostic significance of the results obtained with ELISA.

2.3 Molecular Approaches – Polymerase Chain Reaction Technique

Theoretically, the polymerase chain reaction (PCR) method should be the most sensitive technique to detect minimal residual disease. Thus far, this approach has only proven successful in lymphoma patients who showed a prolonged survival after receiving tumor-free marrow transplants as defined by PCR analysis (Gribben et al. 1991). The development of a corresponding approach for screening patients with epithelial cancer is more difficult, because of thepronounced tumor cell hetergeneity.

2.3.1 Detection of Histogenetic Marker Expression

Several groups have developed reverse transcriptase (RT)-PCR assays that screen for epithelial-specific mRNA species (i.e., CK, PSA) in samples from mesenchymal organs such as BM, peripheral blood, and lymph nodes (Table 2). The specificity of this approach might not be absolute but might rather reflect quantitative differences in the expression level of malignant cells and the surrounding autochthonous cells. Thus ectopic expression of small amounts of epithelial mRNA in mesenchymal cells might also be detected if the amplification of the transcribed cDNA is extensive, which can result in false-positive findings.

This problem has been well addressed in the recent study by SCHOENFELD et al. (1994), who used RT-PCR for CK19 mRNA to detect breast cancer micrometastases in axillary lymph nodes. They concluded that the optimal cutoff point to distinguish involved nodes from metastasis-free nodes remained at the level of 40 cycles of PCR. When the sensitivity of the method was increased by a two-stage amplification using nested primers, the CK19 mRNA product was also found in normal lymph nodes from subjects without cancer (SCHOENFELD et al. 1994). In another study, DATTA et al. (1994) recently used an RT-PCR assay for detection of CK19-mRNA in BM and peripheral blood and claimed a sensitivity of ten mammary carcinoma cells per 10^6 hematopoietic cells. However, the specificity of this finding remains obscure, since hematopoietic cells are known to express low levels of CK mRNA that are detectable by PCR technology (TRAWEEK et al. 1993; ZIPPELIUS et al. 1995).

For patients with prostate cancer, the most specific histogenetic marker available at present is PSA. RT-PCR has been successfully applied to detect single prostatic cancer cells expressing PSA mRNA in peripheral blood, BM, and lymph

Table 2. Polymerase chain reaction (PCR)-based detection of isolated epithelial tumor cells in mesenchymal tissues

Tissue	Tumor origin	Marker mRNA	Reference
Bone marrow	Breast	CK19	DATTA et al. (1994)
		CEA	GERHARD et al. (1994)
	Colorectum	CEA	GERHARD et al. (1994)
	Stomach	CEA	GERHARD et al. (1994)
	Pancreas	CEA	GERHARD et al. (1994)
	Prostate	PSA, CK19	WOOD et al. (1994)
			ZIPPELIUS et al. (1995)
Lymph nodes	Breast	CK19	SCHOENFELD et al. (1993)
	Colorectum	Mutant *p53* and Ki-*ras*	HAYASHI et al. (1995)
	Prostate	PSA	DEGUCHI et al. (1993)
Blood	Breast	CK19	DATTA et al. (1994)
	Pancreas	*ras* mutation	TADA et al. (1993)
	Prostate	PSA, PSM	MORENO et al. (1992), ISRAELI et al. (1994), SEIDEN et al. (1994)

CK, cytokeratin; CEA, carcinoembryonic antigen; PSA, prostate-specific antigen; PSM, prostate-specific membrane antigen.

nodes (DEGUCHI et al. 1993; MORENO et al. 1992; WOOD et al. 1994). However, the sensitivity of this approach might be limited by the fact that PSA is a differentiation marker which is only expressed in a subset of primary and micrometastatic prostate carcinoma cells (RIESENBERG et al. 1993; ZIPPELIUS et al. 1995). The expression of another prostatic marker, called prostate-specific membrane antigen (PSM), appears to be less specific than PSA, since RT-PCR analysis of normal bone marrow from non-carcinoma patients shows positive results in 10%–40% of the cases analyzed (ISRAELI et al. 1994; ZIPPELIUS et al. 1995).

2.3.2 Detection of Tumor-Associated Gene Expression

The majority of genes encoding for tumor-associated molecules are not uniquely expressed in tumor cells, but also exert some expression in certain "normal" tissues. Thus the organ in which the disseminated tumor cells are detected needs to be carefully evaluated for such expression to avoid false-positive results. As demonstrated in Table 2, GERHARD et al. (1994) have recently applied RT-PCR to screen for expression of carcinoembryogenic antigen (CEA) in BM samples obtained from patients with various types of epithelial tumors. Surprisingly, these authors did not find CEA mRNA in samples of normal BM and peripheral blood, which is in contrast to our own results, demonstrating CEA mRNA expression in normal BM from non-carcinoma patients (ZIPPELIUS et al. 1995). Although GERHARD et al. claim an increased sensitivity of their PCR assay over immunocytochemistry, they avoided comparison with a true benchmark method, since the number of cells analyzed in their immunocytochemical assay was about ten times smaller than that in our standard assay (2.5×10^5 versus 2×10^6).

More recently, a group of new molecular markers, fibronectin mRNA isoforms, for hepatocellular carcinoma cells have been proposed by TAVIAN et al. (1994). The pattern of alternative splicing of fibronectin mRNA is oncodevelopmentally regulated in liver, as shown by identification of particular isoforms referred to as EDA and EDB in fetal and tumor liver, but not in control adult livers. The usefulness of these fibronectin mRNA isoforms for detection of disseminated hepatocellular carcinoma cells needs to be corroborated in clinical studies.

2.3.3 Detection of Genomic Changes

The successful application of the PCR method for the detection of occult metastatic disease was supported by the fact that lymphoma cells have unique genomic characteristics such as certain chromosomal translocations or idiotypic rearrangements of the immunoglobulin locus. In contrast, the genomic characteristics of epithelial cancer cells are more heterogenous. Among the most common changes are mutations in the K-*ras* oncogene and the *p53* tumor suppressor gene (BOS 1989; HARRIS and HOLLSTEIN 1993). Based on this observation, TADA et al. (1993) have applied PCR to detect *ras* gene mutations in pancreatic juice and peripheral blood of patients with pancreatic adenocarcinoma (Table 2). The sensitivity of their method is three to 30 copies of mutant K-*ras* genes harboring codon 12 single-base changes in the presence of 300 000 normal copies.

More recently, HAYASHI et al. (1995) also described an elegant approach, called the mutant allele-specific amplification (MASA) method, which is capable of detecting one tumor cell in thousands of lymph node cells. The prognostic relevance of such genetically detectable tumor cells was demonstrated in a clinical follow-up study on 71 patients with "node-negative" colorectal cancer (HAYASHI et al. 1995). Primary tumors were screened for K-*ras* and *p53* mutations and corresponding regional lymph nodes were examined at the genetic level by the MASA method. Although the sensitivity of the MASA method needs to be improved to detect micrometastatic tumor cells in BM usually occuring at frequencies of 10^{-5} to 10^{-6}, it appears to be a promising new approach. Nevertheless, screening for genomic changes is very tedious because it requires molecular analysis of the primary tumor from each patient to determine whether the tumor cells of this individual patient carry the detectable alteration.

3 Clinical Significance of Early Tumor Dissemination

At present, information of the prognostic significance of microdisseminated tumor cells is almost exclusively based on immunocytochemical analyses (Table 1). Disseminated tumor cells can be detected using antibodies which distinguish between the tumor cell and the microenvironment under examination. Immunocytochemistry is therefore the benchmark method for the detection of minimal residual disease. The prognostic relevance of results obtained by new molecular approaches, which claim an even higher sensitivity, remain to be corroborated by clinical studies.

3.1 Breast Cancer

Most data on the clinical relevance of micrometastasis is derived from analysis of patients with breast cancer, which is the main cause of cancer-related deaths in women of Western industrial countries (HENDERSON et al. 1991). Although 90%–95% of these patients have no clinicopathologic signs of overt metastases at the time of primary diagnosis, metastatic relapse develops in 50% of them within the following 5 years (BAUM 1977). At first relapse, BM metastases are detectable in 23% of patients by conventional diagnostic techniques, and the rate increases to about 80% in necropsy studies of patients with metastatic breast cancer (KAMBY et al. 1987).

Using conventional histopathologic techniques, the likelihood of identifying isolated tumor cells in BM at the time of primary diagnosis is only about 4% (RIDELL and LANDYS 1979; SCHLIMOK et al. 1987). As one of the first investigators, REDDING et al. used an antiserum against epithelial membrane antigen (EMA) and detected tumor cells in bone marrow at primary surgery in 28.2% of patients without manifest distant metastases (Redding et al. 1983). Although EMA is now known to be a rather nonspecific marker, several groups including ours were able to confirm their results with the incidence of "positive" findings ranging from 20% to about 45%. Most of the mAb used for tumor cell detection are directed at either CK (Cote et al. 1988; Kovarik et al. 1989; Manegold et al. 1988; Pantel et al. 1994; Schlimok et al. 1987) or epithelial cell surface glycoproteins (Cote et al. 1991; Diel et al. 1992; Mansi et al. 1987; Porro et al. 1988).

An important question was whether the incidence of tumor cells in BM was correlated to established risk factors, such as tumor size or lymph node involvement. Diel et al. (1992) found a significant correlation between BM positivity and tumor size ($p < 0.0001$), nodal status ($p < 0.0001$), and histopathologic tumor grading ($p < 0.002$). While an association to the progesterone receptor ($p = 0.008$) was revealed, there was no correlation to the expression of the estrogen receptor or the menopausal status. The London Ludwig Cancer Institute Group found that the presence of EMA-positive cells in BM was significantly related to lymph node involvement, peritumoral vascular invasion, and size of the primary tumor (Berger et al. 1988). Studies applying CK antibodies appeared more frequently to show positive cells in BM of patients with extended primary tumors or with locoregional lymph node involvement (Cote et al. 1988; Schlimok et al. 1987). However, this tendency was not statistically significant.

In order to assess the biological potential of isolated tumor cells in BM, follow-up studies were required. Using a polyclonal EMA antibody, Mansi et al. (1987) found a significantly shorter relapse-free interval in patients with BM positivity after an intermediate follow-up of 28 months. Analysis of the sites of relapse showed that their immunocytochemical assay predicted predominantly the occurrence of bone metastases. After a longer median follow-up time of 76 months (range, 34–108 months), univariate statistical analysis revealed that this immunocytochemical finding predicts an increased rate of relapse in bone ($p < 0.01$) and other distant sites ($p < 0.001$), as well as a decreased overall survival ($p < 0.005$) (Mansi et al. 1991). However, multivariate analysis indicated that this prognostic influence was not independent. Follow-up examinations by Diel et al.(1992) showed that the presence of tumor cells in BM identified by immunostaining for the tumor-associated glycoprotein-12 (TAG12) is a strong and independent predictor for overall distant relapse in breast cancer ($p < 0.0005$, Cox multiple regression analysis). The highest predictive value for distant metastatic relapse resulted from the combination of nodal status, negative progesterone receptor (PR), and presence of tumor cells in bone marrow. Nevertheless, data concerning overall survival were not included.

Initial evidence of an independent impact of early disseminated tumor cells on overall survival came from the recent study by Schlimok et al. (1994), who assessed the prognostic significance of isolated CK-positive cells in BM in a prospective clinical follow-up trial on 349 breast cancer patients. At primary surgery, BM positivity was observed in 24% of the patients without signs of manifest metastases (UICC stages I–III). Following a median observation time of 35 months (range, 12–80 months), the prognostic relevance of this finding with regard to overall survival was demonstrated by univariate (log rank test, $p = 0.02$) and multivariate analysis (Cox proportional hazard model, $p = 0.03$). The power of the prediction based on the immunocytochemical assay was similar to the established prognostic indicators (SCHLIMOK et al. 1994). HARBECK et al. (1994) confirmed these results in another study which examined BM aspirates from 100 patients with primary breast cancer using a cocktail of mAb to EMA, TAG12, and CK.

Previous work by COTE et al. (1991) suggested that the burden of the micrometastatic cells was indicative for an early relapse in breast cancer. Unfortunately, their investigation was based on the evaluation of cell smears, a technique not permitting a reproducible quantitative transfer of cells to the slide surface.

Another series of studies focused on micrometastasis of regional lymph nodes (Table 1). In breast cancer, immunocytochemistry with anti-cytokeratin antibodies was applied to reexamine lymph node sections of node-negative patients (N0 as defined by routine histopathology), and the detection of micrometastatic cells was prognostically significant (DE MASCAREL et al. 1992). However, the clinical value of detection of lymph node micrometastasis is still contentious because the prognosis of patients with breast cancer appears to be determined by the early onset of hematogenous dissemination.

3.2 Gastrointestinal Cancer

The detection and prognostic significance of tumor cells in bone marrow in breast cancer may not be very surprising, because the skeleton is a preferred site of clinical metastases. In contrast, most patients with colorectal cancer develop liver metastases, while overt skeleton metastasis occurs in only 1%–4% of them, as revealed by conventional methods such as bone scans and radiography (BONNHEIM et al. 1986; TALBOT et al. 1989). However, this frequency is higher in necropsy studies. WEISS et al. detected bone metastases in 12% of patients with relapse at other sites (WEISS et al. 1986). In another necropsy study, WELCH et al. found bone metastases in 10% in association with recurrence to other sites, but never as the only site (WELCH AND DONALDSON 1979). By conventional histology, BURKHARDT et al. (1980) detected BM micrometastases in 17% of patients with intestinal carcinoma without systemic manifestations.

Our recent investigations showed that individual CK-positive cells were found on average in about 30% of patients with apparently localized colon cancer at the time of primary surgery (SCHLIMOK et al. 1990; LINDEMANN et al. 1992). The incidence of a positive finding varied considerably depending on the size and location of the primary tumor, the involvement of regional lymph nodes, and the presence of clinically manifest metastases (LINDEMANN et al. 1992; SCHLIMOK et al. 1990). After a median observation time of 35 months (range, 12–58 months), patients with tumor cells in BM showed a significantly shorter disease-free survival than patients without these cells ($p = 0.008$) (LINDEMANN et al. 1992). Using a multivariate logistic regression analysis, the immunocytochemical finding was confirmed as the strongest independent determinant of relapse. Thus the detection of tumor cells in BM in colon cancer may be taken as an indicator of the disseminative capability of an individual tumor.

After the liver, the skeleton is the most frequent site of distant metastasis in gastric cancer (Koga et al. 1987). It is therefore not surprising that tumor cells in bone marrow were detected in 35% of patients (34 of 97) with gastric cancer at primary surgery using our standard CK assay with mAb CK2 (Schlimok et al. 1991). The incidence of tumor cells in BM was positively correlated to established risk factors, such as histologic classification and the locoregional lymph node involvement. A preliminary clinical follow-up analysis of 38 patients demonstrated an increased relapse rate in patients judged as BM positive (Schlimok et al. 1991).

Jauch et al. (1995) recently demonstrated the relevance of the tumor cell burden in predicting disease-free survival in patients with primary gastric cancer. The tumor cell concentration in BM (less than 3×10^{-6} versus more than 3×10^{-6} mononuclear cells, MNC), as determined by immunocytochemical analysis with mAb CK2, predicted disease-free survival in R0-resected patients ($p = 0.007$). Multivariate analysis with stepwise logistic regression revealed that extent of tumor cell dissemination was an independent prognostic indicator for disease-free survival in subgroups of patients with either pT_{1-2} tumors ($p = 0.014$), Laurén's intestinal type carcinomas ($p = 0.008$), or no involvement of lymph nodes ($p = 0.004$).

3.3 Lung Cancer

Small cell lung cancer (SCLC) is a distinct pathologic entity which accounts for 20%–25% of total lung cancer. BM infiltration is common in SCLC and is detected by unilateral routine histologic/cytologic methods in 20% of patients at diagnosis (Bezwoda et al. 1986). Stahel et al. (1985) showed that the level of marrow infiltration is probably much higher. Using an antibody (SM1) reactive to a membrane antigen on SCLC cells, they were able to demonstrate SM1-positive cells in single aspirates of 69% of all patients analyzed. At the initial staging, tumor cells were found by immunocytochemistry in 50% of patients with limited disease and in 77% of patients with extensive disease.

The Brussels Ludwig Cancer Institute Group used a combination of three mAb defining three distinct surface antigens (lung cancer-associated antigens LCA1, LCA2, and LCA3). The antigens are widely expressed on both SCLC and non-small-cell lung cancer (NSCLC) cell lines. In single BM aspirates, LCA-positive cells were present in 26% of patients with limited disease and in 55% of patients with extensive disease (Canon et al. 1988, Humblet et al. 1988, Lebacq-Verheyden et al. 1988). The prognostic relevance of the immunocytochemical approach in SCLC has thus far not been proven in prospective clinical trials.

NSCLC consists of a heterogeneous group of different tumor histologies, with adenocarcinomas and squamous cell carcinomas being the two major groups. About 40% of patients with NSCLC who are staged postoperatively as tumor free (T_{1-2} N_0 M_0 R_0) by routine histopathology and clinical examinations relapse within 24 months after surgery (MOUNTAIN 1986). Thus the incidence of early micrometastases of tumor cells is clearly underestimated by current staging procedures. Applying the immunocytochemical cytokeratin assay to patients with operable NSCLC, tumor cells in BM were found in 83 of 139 patients (59.7%; PANTEL et al. 1996). After a median observation period of 39 months, patients without metastatic lymph node involvement who displayed CK-positive cells in BM at the time of primary surgery relapsed more frequently than patients with a negative marrow finding $p = 0.004$. This prognostic influence was independent from other risk factors (cox analysis: $p = 0.028$).

Nodal micrometastatic tumor cells were revealed by either polyclonal anti-cytokeratin antibodies or mAb BerEp4 against an epithelial cell membrane antigen (CHEN et al. 1993; PASSLICK et al. 1994); the presence of BerEp4-positive cells has recently been shown to be of independent prognostic value (PASSLICK et al. 1994). Interestingly, no correlation between the immunodetection of micrometastases in BM and lymph node was found, supporting the view that the biological processes involved in the two types of metastatic processes may differ considerably. Thus the immunocytologic search for lymph node micrometastases might provide additional information about the extent of occult tumor cell disemination.

3.4 Urogenital Malignancies

Analysis of 427 patients with various carcinomas of the urogenital tract, including cancers of the prostate,bladder,and kidney,showed on overall incidence of tumor cells in BM of 23%, ranging from 19% in renal cell cancer to 37% in prostate cancer (OBERNEDER et al. 1994a,b). There were significant correlations between the presence of tumor cells in BM and established risk factors, such as size and histologic grade of the primary tumor, in the three different tumor entities studied. For example, the incidence of tumor cells in bone marrow in patients with stage C prostate cancer ($T_{3/4}N_0M_0$) was above 50% (PANTEL et al. 1995a). The specificity of the analysis was supported by the presence of CK-positive epithelial cells displaying PSA and the absence of positive staining in control patients with benign prostatic hyperplasia (RIESENBERG et al. 1993; OBERNEDER et al. 1994a). More extensive clinical follow-up is now required to prove the prognostic relevance of tumor cells in BM in malignancies of the urogenital tract.

4 Antibody Therapy Directed at Minimal Residual Cancer

Minimal residual disease offers the advantage of a small and accessible metastatic tumor burden. Considering that most of the carcinoma cells in BM are not proliferating at the time of primary surgery (PANTEL et al. 1993a), adjuvant therapy with conventional chemotherapeutic agents might be rather ineffective. On the other hand, cancer therapy with a variety of different mAbs has thus far failed to significantly change the clinical course of patients with solid tumors at advanced disease stages (Table 3). This might be explained by the fact that solid metastases can develop effective mechanisms which block the delivery of macromolecules from the peripheral blood (JAIN et al. 1990). In contrast, dispersed micrometastatic cells present at low frequencies in secondary organs may represent more suitable targets for therapeutic approaches with mAbs.

4.1 Adjuvant Antibody Therapy in Colorectal Cancer

To prove the postulated benefit of adjuvant antibody therapy, colorectal cancer patients with no distant metastases (Dukes stage C) were treated in a second trial with the murine mAb 17-1A detecting an antigen present on epithelial cells of human colorectum (RIETHMÜLLER et al. 1994). The antibody mediates antibody-dependent cellular cytotoxicity (ADCC) of colorectal carcinoma lines in vitro and suppresses the growth of human tumors grafted into nude mice (HERLYN et al. 1980). Patients in our trial received a total of 900 mg mAb 17-1A over 20 weeks, starting 2–3 weeks after primary surgery. Following a median follow-up of 5 years, therapy with antibody was found to reduce the overall death rate by 30% and decrease the recurrence rate by 27% (RIETHMÜLLER et al. 1994). These data contrast with the results of numerous trials with 17-1A antibody in advanced tumors in which anecdotal remissions were observed in only a few patients and no benefit for survival could be established (for a review, see RIETHMÜLLER et al. 1993). Thus, by carefully selecting the disease stage at which therapy is initiated, the efficacy of antibody therapy is comparable with that of other adjuvant therapies (MOERTEL et al. 1991; KROOK et al. 1991), while the toxicity of 17-1A treatment is considerably lower. As long as the focus of current research is the design of new drugs and not better selection of appropriate target patient groups, we can predict that much energy will be wasted on "magic bullets" directed towards unassailable targets. Nevertheless, the use of antibodies derived entirely from humans and isolated from combinatorial libraries in bacteriophage will most

Table 3. Antibody therapy of patients with advanced solid tumors

Tumor type	Trials (*n*)	Remission (*n*)	Total (*n*)
Acute leukemia	6	9	49
Chronic lymphatic leukemia	4	2	28
B cell lymphoma	10	12	46
T cell lymphoma	5	8	50
Breast, ovarian, and prostate cancer	5	1	47
Lung cancer	3	0	20
Colon cancer	8	4	237
Melanoma	7	7	107

Modified from DILLMAN (1994).

likely soon replace murine antibodies (PERSSON 1993). This technique allows the isolation of high-affinity, antigen-specific Fab or Fv, even from naive human B cells (WILLIAMSON et al. 1993).

4.2 Towards the Development of Micrometastatic Cells as Surrogate Markers of Therapeutic Efficacy

The efficacy of adjuvant therapy can thus far only be assessed retrospectively in large-scale clinical trials following an observation period of at least 5 years. Consequently, progress in this form of therapy is extremely slow and cumbersome and, in addition, therapy is difficult to tailor to the special needs of an individual patient. The importance of a surrogate marker assay that would permit the immediate assessment of therapy-induced cytotoxic effects on residual cancer cells is therefore obvious. The reproducibility of repeated immunocytochemical BM examination was assessed in a preliminary monitoring study on colorectal cancer patients (SCHLIMOK et al. 1990). The majority of these patients exhibited a consistent pattern of BM findings. This observation is in contrast to data published by MANSI et al. (1989) on breast cancer patients. Using a polyclonal EMA antibody, tumor cells were only redetected in two of the 82 patients who were initially judged as BM positive. This discrepancy might be explained by the lack of reliability of their assay.

In a recent pilot study by SCHLIMOK et al. (1995) 40 patients with breast and colorectal cancer were treated in a randomized fashion with either 6 x 100 mg SDZ ABL-364 (Sandoz, Basel, Switzerland) over 2 weeks or human serum albumin as placebo. MAb ABL-364 (murine immunoglobulin IgG_3) is directed at the Lewis Y (LeY) blood group precursor carbohydrate antigen, which is widely expressed on most epithelial tumors (SCHOLZ et al. 1991). Tumor cell lysis is mediated by induction of both ADCC and complement-dependent cytotoxicity (SCHOLZ et al. 1991). CK-positive cells in marrow were monitored on days 15 and 60 after the start of treatment. Even in patients with an extremely low number cells in BM (one to 11/per 4×10^5 MNC), a tendency for reduction of CK-positive tumor cells in BM was seen after antibody therapy. Significant data, however, were only obtained from the ten breast cancer patients who displayed an initial cell count of more than 20 CK-positive cells per 4×10^5 MNC. Of the seven patients treated with antibody, five showed a distinct reduction or eradication of CK-positive/Le^Y-positive cells (96%–100%), while in two patients with CK-positive/Le^Y-negative cells no response was registered. Similarly, in the three patients receiving human serum albumin no significant tumor cell reduction was observed. Because of the marked cytotoxicity that the antibody ABL-364 exhibits in ex vivo experiments with serum of treated patients (SCHOLZ et al. 1991), we postulate that the observed disappearance of disseminated tumor cells is due to the action of the administered antibody.

Despite the preliminary character of this study, it exemplifies a new approach towards a more rational selection of antibodies for adjuvant studies in minimal residual disease. The proposed use of tumor cells in BM as surrogate markers for the prediction of the therapeutic response may benefit from recent improvements in the cytokeratin assay (PANTEL et al. 1994), which now allows a more precise quantitation of the individual tumor load. Clinical studies are now required to evaluate whether the eradication of these cancer cells translates into a longer disease-free and overall survival. Availability of such a surrogate marker would considerably enhance our abilities to rationally design new therapies directed towards minimal residual disease.

5 Concluding Remarks

The outlined current strategies for detection, characterization, and antibody therapy of cancer micrometastasis have the potential to prevent overt metastatic disease in patients with operable primary carcinomas. The remarkable specificity and sensitivity of immunocytochemical cytokeratin assays support their introduction in international tumor staging classifications. Since most CK-positive cells are isolated tumor cells, the term "micrometastases" might be misleading, as it implies the presence of tumor cell clusters, which are found in only about 5%–10% of patients with a positive BM finding. The term "isolated tumor cells" appears to be therefore more accurate. The UICC has recently suggested the denotation $M_{1(i)}$ for the stage of minimal residual disease defined by the presence of isolated tumor cells in secondary organs (HERMANEK et al. 1993). Although we appreciate the inclusion of this critical stage in the UICC tumor classification, we feel that the denotation M_i might be better suited to clearly distinguish between the presence of isolated tumor cells and solid metastases.

The biology of microdisseminated carcinoma cells remains poorly understood. This ignorance is particularly disturbing in diseases such as colorectal cancer, in which the frequent finding of tumor cells in BM revealed by the immunocytochemical method has proved to be a strong and independent predictor of overall metastatic relapse even though manifest skeleton metastasis rarely occurs (LINDEMANN et al. 1992). Thus tumor cells detected in BM at the time of operation of the primary tumor may not necessarily have the potential to form clinically detectable metastases within the remaining life span of the patient; this may be determined by the more aggressively growing liver metastases, while the cells in BM may remain dormant for years (PANTEL et al. 1993a). The extremely low frequency of tumor cells in BM greatly hampers approaches to obtain more specific information on their biological properties. To obtain initial insights, it was therefore important to develop immunocytochemical double-labeling techniques that allow the identification and phenotyping of these cells. Our present results indicate that they represent a selected population of cancer cells, which, however, still express a considerable degree of heterogeneity (PANTEL et al. 1995b; PANTEL and BRAUN 1996). Clinical follow-up analyses have been initiated to evaluate whether the characterization of early disseminated tumor cells will increase the diagnostic precision of the present immunocytochemical assays.

Acknowledgment. The authors gratefully acknowledge the support of the Deutsche Krebshilfe/ Dr. Mildred Scheel Stiftung and the Deutsche Forschungsgemeinschaft, Bonn, as well as the Friedrich-Baur-Stiftung, Munich, Germany.

References

Baum M (1977) The curability of breast cancer. In: Stall BA (ed) Breast cancer management. Heinemann, London, pp 3–13

Berger U, Bettelheim R, Mansi JL, Easton D, Coombes RC, Neville AM (1988) The relationship between micrometastases in the bone marrow, histopathologic features of the primary tumor in breast cancer and prognosis. Am J Clin Pathol 90: 1–6

Bezwoda WR, Lewis D, Livini N (1986) Bone marrow involvement in an aplastic small-cell lung cancer. Cancer 58: 1762–1765

Bonnheim DC, Petrelli NJ, Herrera L, Walsh D, Mittelman A (1986) Osseous metastases from colorectal carcinoma. Am J Surg 151: 457–459

Bos IL (1989) Ras oncogenes in human cancer: a review. Cancer Res 49: 4682–4689

Braun S, Pantel K (1996) Biological characteristics of human micrometastatic carcinoma cells. In: Günthert U, Birchmeier W (eds) Attempts to understand metastasis formation. I Metastasis-related molecules. Springer, Berlin Heidelberg New York (Current topics in microbiology and immunology, vol, 213/I, pp 163–177

Brugger W, Bross KJ, Glatt M et al (1994) Mobilization of tumor cells and hematopoietic progenitor cells into peripheral blood of patients with solid tumors. Blood 83: 636

Burkhardt R, Frisch B, Kettner G (1980) The clinical study of micrometastatic cancer by bone biopsy. Bull Cancer 67: 291–305

Canon JL, Humblet Y, Lebacq-Verheyden, AM, Mannouriez P, Bazin H, Rodhain J, Prignot J, Symann M (1988) Immunodetection of small-cell lung cancer metastases in bone marrow using three monoclonal antibodies. Eur J Cancer Clin Oncol 24: 147–150

Chen ZL, Perez S, Holmes EC, Wang HJ, Coulson WF, Wen DR, Cochran AJ (1993) Frequency and distribution of occult micrometastases in lymph nodes of patients with non-small-cell lung carcinoma. J Natl Cancer Inst 85: 493–498

Cote RJ, Rosen PP, Hakes TB, Sedira M, Bazinet M, Kinne DW, Old LJ, Osborne MP (1988) Monoclonal antibodies detect occult breast carcinoma metastases in the bone marrow of patients with early stage disease. Am J Surg Pathol 12: 333–340

Cote RJ, Rosen PP, Lesser ML, Old LJ, Osborne MP (1991) Prediction of early relapse in patients with operable breast cancer by detection of occult bone marrow micrometastases. J Clin Oncol 9: 1749–1756

Cutait R, Alves VA, Lopes LC, Cutait DE, Borges JL, Singer J, Da Silva JH, Goffi FS (1991) Restaging of colorectal cancer based on the identification of lymph node micrometastases through immunoperoxidase staining of CEA and cytokeratins. Dis Colon Rectum 34: 917–920

Datta YH, Adams PT, Drobyski WR, Ethier SP, Terry VH, Roth MS (1994) Sensitive detection of occult breast cancer by the reverse-transcriptase polymerase chain reaction. J Clin Oncol 12: 475–482

De Mascarel I, Bonichon F, Coindre JM, Trojani M (1992) Prognostic significance of breast cancer axillary lymph node micrometastases assessed by two special techniques: reevaluation with longer follow up. Br J Cancer 66: 523–527

Deguchi T, Doi T, Ehara H, Ito S, Takahashi Y, Nishino Y, Fujihiro S, Kawamura T, Komeda H, Horie M, Kaji H, Shimokawa K, Tanala T, Kawada Y (1993) Detection of micrometastatic prostate cancer cells in lymph nodes by reverse transcriptase-polymerase chain reaction. Cancer Res 53: 5350–5354

Diel IJ, Kaufmann M, Goerner R, Costa SD, Kaul S, Bastert G (1992) Detection of tumor cells in bone marrow of patients with primary breast cancer: a prognostic factor for distant metastasis. J Clin Oncol 10: 1534–1539

Dillman R (1994) Antibodies as cytotoxic therapy. J Clin Oncol 12: 1497–1515

Fidler IJ, Radinsky R (1990) Genetic control of cancer metastasis. J Natl Cancer Inst 82: 166–168

Gerhard M, Juhl H, Kalthoff H, Schreiber HW, Wagener C, Neumaier M (1994) Specific detection of carcinoembryonic antigen-expressing tumor cells in bone marrow aspirates by polymerase chain reaction. J Clin Oncol 12: 725–729

Gribben JG, Freedman AS, Neuberg D, Roy DC, Blake KW, Woo SD, Grossbard ML, Rabinowe SN, Carol F, Freeman GJ, Ritz J, Nadler LM (1991) Immunologic purging of marrow assessed by PCR before autologous bone marrow transplantation of B-cell lymphoma. N Engl J Med 325: 1525–1533

Gross HJ, Verwer B, Houck D, Hoffmann RA, Recktenwald D (1995) Model study detecting breast cancer cells in peripheral blood mononuclear cells at frequencies as low as 10–7. Proc Natl Acad Sci USA 92: 537–541

Harbeck N, Untch M, Pache L, Eiermann W (1994) Tumour cell detection in the bone marrow of breast cancer patients at primary therapy: results of a 3-year median follow-up. Br J Cancer 69: 566–571

Harris CC, Hollstein M (1993) Clinical implications of the p53 tumor-suppressor gene. N Engl J Med 329: 1318–1327

Hayashi N, Ito I, Yanagisawa A, Kato Y, Nakamori S, Imaoka S, Watanabe H, Nakamura K (1995) Genetic diagnosis of lymph-node metastases in colorectal cancer. Lancet 345: 1257–1259

Henderson BE, Ross RK, Pike MC (1991) Toward the primary prevention of cancer. Science 254: 1131–1137

Herlyn DM, Steplewski Z, Herlyn MF, Koprowski H (1980) Inhibition of growth of colorectal carcinoma in nude mice by monoclonal antibody. Cancer Res 40: 717–721

Hermanek P, Henson DE, Hutter RV, Sobin LH (1993) TNM supplement. Springer, Berlin Heidelberg New York

Hoechtlen-Vollmar W, Gruber R, Bodenmüller H, Lindemann F, Schlimok G, Pantel K, Riethmüller G (1995) Occult epithelial tumor cells detected in bone marrow by an enzyme immunoassay specific for cytokeratin 19 (submitted)

Humblet Y, Canon JL, Sekhavat M, Feyens AM, Manouvriez P, Lebacq-Verhyden AM, Bazin H, Prignot J, Symann M (1988) Detection of small-cell lung cancer bone marrow metastases by immunofluorescence. Pathobiology 36: 25–28

Israeli RS, Miller WH Jr., Su SL, Powell CT, Fair WR, Samedi DS, Huryk RF, De Blasio A, Edwards ET, Wise GJ, Heston WDW (1994) Sensitive nested reverse transcription polymerase chain reaction detection of circulating prostatic tumor cells: comparison of prostate-specific membrane antigen and prostate-specific antigen-based assays. Cancer Res 54: 6306–6310

Jain RK (1990) delivery of novel therapeutic agents in tumors: physiological barriers and strategies. J Natl Cancer Inst 81: 570–576

Jauch K, Heiss MM, Grützner U, Funke I, Pantel K, Babic R, Eissner HJ, Riethmüller G, Schildberg FW (1995) The prognostic significance of early disseminated tumor cells in bone marrow of patients with gastric cancer. (submitted)

Juhl H, Stritzel M, Wroblewski A, Henne-Bruns D, Kremer B, Schmiegel W, Neumaier M, Wagener C, Schreiber HW, Kalthoff H (1994) Immunocytological detection of micrometastatic cells: comparative evaluation of findings in the peritoneal cavity and the bone marrow of gastric, colorectal and pancreatic cancer patients. Int J Cancer 57: 330–335

Kamby K, Guldhammer B, Vejborg I, Rossing N, Dirksen H, Daugaard S, Mouridsen HT (1987) The presence of tumor cells in bone marrow at the time of first recurrence of breast cancer. Cancer 60: 1306–1312

Koga S, Takebayashi M, Kaibara N, Nishidoi H, Kimura O, Kawasumi H, Makino M (1987) Pathological characterization of gastric cancer that develop hematogenous recurrence, with special reference to the site of recurrence. J Surg Oncol 36: 239–242

Kovarik J, Lauerová L, Rejthar A, Bártek J (1989) Keratin typing in screening bone marrows for carcinoma cells. Cytotechnology S78:79

Krook JE, Moertel CG, Gunderson LL et al (1991) Effective surgical adjuvant therapy for high-risk rectal carcinoma. N Engl J Med 324: 709–715

Lebacq-Verheyden AM, Neirynck A, Ravoet AM, Humblet Y, Oie H, Linnoila I, Gazdar AF, Minna JD, Symann M (1988) Monoclonal antibodies for the vitro detection of small-cell lung cancer metastases in human bone marrow. Eur J Cancer Clin Oncol 24: 137–145

Lindemann F, Schlimok G, Dirschedl P, Witte J, Riethmüller G (1992) Prognostic significance of micrometastatic tumour cells in bone marrow of colorectal cancer patients. Lancet 340: 685–689

Liotta LA, Steeg PS, Stetler-Stevenson WG (1991) Cancer metastasis and angiogenesis: an imbalance of positive and negative regulation. Cell 64: 327–336

Manegold C, Krempien B, Kaufmann M, Schwechheimer K, Schettler G (1988) The value of bone marrow examination for tumor staging in breast cancer. J Cancer Res Clin Oncol. 114: 425–428
Mansi JL, Berger U, Easton D,McDonnell T, Redding WH, Gazet JC, McKinna A, Powles TJ, Coombes RC (1987) Micrometastases in bone marrow in patients with primary breast cancer: evaluation as an early predictor of bone metastases. Br Med J 295: 1093–1096
Mansi JL, Berger U, Wilson P, Shearer R, Coombes RC (1988) Detection of tumor cells in bone marrow of patients with prostatic carcinoma by immunocytochemical techniques. J Urol 139: 545–548
Mansi JL, Berger V, McDonnell T, Pople A, Rayter Z, Gazet JC, Coombes RC (1989) The fate of bone marrow micrometastases in patients with primary breast cancer. J Clin Oncol 7: 445–449
Mansi JL, Easton D, Berger U, Gazet JC, Ford HT, Dearnaley D, Coombes RC (1991) Bone marrow micrometastases in primary breast cancer: prognostic significance after 6 years' follow-up. Eur J Cancer 27: 1552–1555
Moertel CG, Fleming TR, MacDonald JS et al (1991) Levamisole and fluoroacil for adjuvant therapy of resected colon carcinoma. N Engl J Med 322: 352–358
Molino A, Colombatti M, Bonetti F, Zardini M, Pasini F, Perini A, Pelosi G, Tridente G, Veneri D, Cetto GL (1991) A comparative analysis of three different techniques for the detection of breast cancer cells in bone marrow. Cancer 67: 1033–1036
Moreno JG, Croce CM, Fischer R, Monne M, Vihko P, Mulholland SG, Gomella LG (1992) Detection of hematogenous micrometastasis in patients with prostate cancer. Cancer Res 52: 6110–6112
Mountain CF (1986) A new international staging system for lung cancer. Chest 89 [Suppl]: 225S–233S
Oberneder R, Riesenberg R, Kriegmair M, Bitzer U, Klammert R, Schneede P, Hofstetter A, Riethmüller G, Pantel K (1994a) Immunocytochemical detection and phenotypic characterization of micrometastatic tumour cells in bone marow of patients with prostate cancer. Urol Res 22: 3–8
Oberneder R, Kriegmair M, Pantel K, Staudte S, Hofstetter A, Riethmüller G (1994b) Prognostic relevance of micrometastatic tumor cells in bone marrow of patients with genitourinary tract tumors. J Urol 151: 413A
Osborne MP, Wong GY, Asina Sh, Old LJ, Cote RJ, Rosen PP (1991) Sensitivity of immunocytochemical detection of breast cancer cells in human bone marrow. Cancer Res 51: 2706–2709
Pantel K, Izbicki JR, Passlick B, Angstwurm M, Häussinger K, Thetter O, Riethmüller G (1996) Incidence and prognostic significance of isolated tumour cells detected in bone marrow of non-small cell lung cancer patients without overt metastases. Lancet (in press)
Pantel K, Schlimok G, Braun S, Kutter D, Schaller G, Funke I, Izbicki J, Riethmüller G (1993a) Differential expression of proliferation-associated molecules in individual micrometastatic carcinoma cells. J Natl Cancer Inst 85: 1419–1424
Pantel K, Koprowski H, Riethmüller G (1993b) Conference on cancer micrometastasis: biology, methodology and clinical significance. Int J Oncol 3: 1019–1022
Pantel K, Schlimok G, Angstwurm M, Weckermann C, Schmaus W, Gath H, Passlick B, Izbicki J, Riethmüller G (1994) Methodological analysis of immunocytochemical screening for disseminated epithelial tumor cells in bone marrow. J Hematother 3: 165–173
Pantel K, Aignherr C, Köllermann J, Caprano J, Riethmüller G, Köllermann MW (1995a) Immunocytochemical detection of isolated tumor cells in bone marrow of patients with untreated stage C prostatic cancer. Eur J Cancer 31 A: 1627–1632
Pantel K, Schlimok G, Angstwurm M, Passlick B, Izbicki JR, Johnson JP, Riethmüller G (1995b) Early metastasis of human solid tumors: expression of cell adhesion molecules. In: Ciba Foundation Symposium (eds) Cell adhesion and human disease. Wiley, Chichester, pp 157–173
Passlick B, Izbicki JR, Kubuschok B, Nathrath W, Thetter O, Pichlmeier U, Schweiberer L, Riethmüller G, Pantel K (1994) Immunohistochemical assessment of individual tumor cells in lymph nodes of patients with non-small cell lung cancer. J Clin Oncol 12: 1827–1832
Persson MA (1993) Combinatorial libraries. Int Rev Immunol 10: 153–63
Porro G, Menard S, Tagliabue E, Orefice S, Salvadori B, Squicciarini P, Andreola S, Rilke F, Colnaghi MI (1988) Monoclonal antibody detection of carcinoma cells in bone marrow biopsy specimens from breast cancer patients. Cancer 61: 2407–2411
Redding HW, Coombes RC, Monagham P, Clink HMD, Imrie SF, Dearnaley DP, Ormerod MG, Sloane JP, Gazet JC, Powels TJ, Neville AM (1983) Detection of micrometases in patients with primary breast cancer. Lancet ii: 1271–1274
Ridell B, Landys K (1979) Incidence and histopathology of metastases of mammary carcinoma in biopsies from the posterior iliac creast. Cancer 44: 1782–1788
Riesenberg R, Orberneder R, Kriegmair M, Epp M, Bitzer U, Hofstetter A, Braun S, Riethmüller G, Pantel K (1993) Immunocytochemical double staining of cytokeratin and prostate specific antigen in individual prostatic tumor cells. Histochemistry 99: 61–66

Riethmüller G, Johnson JP (1992) Monoclonal antibodies in the detection and therapy of micrometastatic epithelial cancer. Curr Opin Immunol 4: 647–655

Riethmüller G, Schneider-Gädicke E, Johnson JP (1993) Monoclonal antibodies in cancer therapy. Curr Opin Immunol 5: 732–739

Riethmüller G, Schneider-Gädicke E, Schlimok G, Schmiegel W, Raab R, Höffken K, Gruber R, Pichlmaier H, Hirche H, Pichlmayr R, Buggisch P, Witte J, and the German Cancer Aid 17-1A Study Group (1994) Randomised trial of monoclonal antibody for adjuvant therapy of resected Dukes' C colorectal carcinoma. Lancet 343: 1177–1183

Schlimok G, Funke I, Holzmann B, Göttlinger HG, Schmidt G, Häuser H, Swierkot S, Warnecke HH, Schneider B, Koprowski H, Riethmüller G (1987) Micrometastatic cancer cells in bone marrow: in vitro detection with anti-cytokeratin and in vivo labeling with anti-17-1A monoclonal antibodies. Proc Natl Acad Sci USA 84: 8672–8676

Schlimok G, Funke I, Bock S, Schweiberer B, Witte J, Riethmüller G (1990) Epithelial tumor cells in bone marrow of patients with colorectal cancer: immunocytochemical detection, phenotypic characterization, and prognostic significance. J Clin Oncol 8: 831–837

Schlimok G, Funke I, Pantel K, Strobel F, Lindemann F, Witte J, Riethmüller G (1991) Micrometastatic tumor cells in bone marrow of patients with gastric cancer: methodological aspects of detection and prognostic significance. Eur J Cancer 27: 1461–1465

Schlimok G, Lindemann F, Holzmann K, Witte J, Pantel K, Riethmüller G (1994) Prognostic significance of tumor cells detected in bone marrow of breast cancer patients: a multivariate analysis. Onkologie 17 [Suppl 2]: 128

Schlimok G, Pantel K, Loibner H, Fackler-Schwalbe I, Riethmüller G (1995) Individual metastatic tumor cells in bone marrow of cancer patients reduced by intravenously administered monoclonal antibody: towards a novel surrogate marker for adjuvant therapy. Eur J Cancer 31 A: 1799–1803

Schoenfeld A, Luqmani Y, Smith D, O'Reilly S, Shousha S, Sinnett HD, Coombes RC (1994) Detection of breast cancer micrometastases in axillary lymph nodes by using polymerase chain reaction. Cancer Res 54: 2986–2990

Scholz D, Lubeck M, Loibner H, Steplewski Z, Koprowski H (1991) Biological activity in the human system of isotype variants of oligosaccharide Y-specific murine monoclonal antibodies. Cancer Immunol Immunother 33: 153–157

Seiden MV, Kantoff PW, Krithivas K, Propert K, Bryant M, Haltom E, Gaynes L, Kaplan I, Bubley G, De Wolf W, Sklar J (1994) Detection of circulating tumor cells in men with localized prostate cancer. J Clin Oncol 12: 2634–2639

Stahel RA, Mabry M, Skarin AT, Speak J, Bernal SD (1985) Detection of bone marrow metastasis in small cell lung cancer by monoclonal antibody. J Clin Oncol 3: 455–461

Tada M, Omata M, Kawai Sh, Saisho H, Ohto M, Saiki RK, Sninsky JJ (1993) Detection of ras gene mutations in pancreatic juice peripheral blood of patients with pancreatic adenocarcinoma. Cancer Res 53: 2472–2474

Talbot RW, Irvine B, Jass JR, Dowd GS, Northover JM (1989) Bone metastases carcinoma of the rectum: a clinical and pathological review. Eur J Surg Oncol 15: 449–452

Tavian D, De Petro G, Colombi M, Portolani N, Giulini SM, Gardella R, Barlati S (1994) RT-PCR detection of fibronectin EDA+ and EDB+ mRNA isoforms: molecular markers for hepatocellular carcinoma. Int J Cancer 56: 820–825

Traweek ST, Liu J, Battifora H (1993) Keratin gene expression in non-epithelial tissues. Detection with polymerase chain reaction. Am J Pathol 142: 1111–1118

Weiss L, Grundmann E, Torhorst J, Hartveit F, Moberg I, Eder M, Fenoglio Preiser C, Napier J, Horne CH, Lopez MJ (1986) Haematogenous metastatic patterns in colonic carcinoma: an analysis of 1541 necropsies. J Pathol 150: 195–203

Welch JP, Donaldson GA (1979) The clinical correlation of an autopsy study of recurrent colorectal cancer. Ann Surg 189: 496–502

Williamson RA, Burioni R, Sanna PP et al (1993) Human monoclonal antibodies against a plethora of viral pathogens from single combinatorial libraries. Proc Natl Acad Sci USA 90: 4141–4145

Wood DP, Banks ER, Humphries S, McRoberts JW, Rangnekar M, Lexington KY (1994) Identification of micrometastases in patients with prostate cancer. J Urol 151: 303A

Zippelius A, Honald G, Kufer P, Riethmüller G, Pantel K (1995) Detection of isolated carcinoma cells in bone marrow by RT-PCR. J Cancer Res Clin Oncol 121 [Suppl 2]: A5

Transcriptionally Targeted Gene Therapy

I.R. Hart

1 In Vivo Gene Transfer

Gene therapy approaches for cancer treatment have moved from the laboratory to evaluation in the clinic (Culver and Blaese 1994; Nabel et al. 1993). Many of these protocols are based upon the removal ex vivo of either the neoplastic or immune effector cells from the patient, their genetic manipulation in vitro and then their subsequent return to the patient (Vile and Russell 1994). There are significant limitations to such manipulations including the fact that these rather cumbersome and technically demanding procedures probably will be limited to specialised centres. Their widespread use is unlikely on the grounds of both cost and available technology. Equally it is possible that the selection process of establishing cells in tissue culture, albeit for only a brief period of time, may result in the emergence of clones unrepresentative of the major cellular composition of the tumour – a potential problem for immunotherapy approaches. On a more pragmatic level it may be that, especially in early phase I and II trials, the necessary period of manipulation is greater than the patient's expected life span.

An alternative approach, which has several conceptual advantages, would be the direct in situ modification of tumour cells with novel genetic material. Direct gene transfer recently has been shown to be achievable in humans using cutaneous malignant melanoma as the target tumour (Nabel et al. 1993). The human HLA-B7 gene was injected as a complex of DNA with liposomes into subcutaneously located melanoma deposits of HLA-B7 negative patients (Nabel

Richard Dimbleby Department of Cancer Research, Rayne Institute, St. Thomas' Hospital, Lambeth Palace Road, London SE1 7EH, UK

et al. 1993). In the five patients evaluated not only was the procedure shown to be safe but also as many as 1%–10% of the tumour cells around the injection location expressed protein resulting from expression of the transduced gene. This seminal paper established the fact that direct genetic modification of tumour cells in man is possible using DNA injected into the cancer deposit. Retrovirally mediated gene transfer has also been suggested as being a way to introduce genetic material into human neoplastic cells (OLDFIELD et al. 1993) based upon the demonstration that in situ retrovirally mediated gene transfer has proven feasible in the treatment of brain tumours in rats (RAM et al. 1993). While formal proof of viral integration in the human tumour cells has not yet been published, it seems likely from the rodent experience that this indeed is occuring.

The capacity to transduce and to modify human cells genetically while still in situ therefore already exists, and the refinement of these procedures over the next few years, by the improvement of viral vectors or by the development of improved liposome vehicles, should enhance the efficiency of this phenomenon considerably. What genes, however, are to be introduced into these recipient tumour cells? Currently the genes available appear to fall into one of three general categories:

1. Those which correct the genetic changes which underlie the development and progression of the malignant phenotype, and which can therefore arrest or reverse tumour development.
2. Those which evoke or stimulate an immune response.
3. Those which render cells sensitive to subsequently delivered drugs by coding for enzymes capable of converting inert prodrugs to active cytotoxics.

Utilisation of the genes represented in category 1 is beset by two major difficulties. Firstly, in a process as extended as tumour development and progression, identification of the major underlying genetic aberrations/lesions is not always clearcut; by the time of presentation there are probably numerous gross chromosomal abnormalities. Equally it is not always apparent whether even early changes represent a potential target. In hereditary melanoma, for instance, the p16 (CDKN2) gene located at chromosome 9p21 may be a candidate for introduction of a normal wild-type gene (KAMB et al. 1994a,b). However, the fact that even in melanoma patients mutations in this gene are infrequent, relative to unaffected members of families showing linkage to 9p21 (KAMB et al. 1994b), calls into question the generality of the involvement of p16. That the product of the p16 (CDKN2) gene inhibits an enzyme involved in regulating the cell cycle serves to illustrate the second difficulty associated with attempts at correctional therapy in cancer. Since no 'field effect' is likely to occur, this approach requires the delivery of the corrective gene to every neoplastic cell, or at least every neoplastic stem cell. Failure to correct the deficiency in less than the majority of stem cells might lead to the rapid overgrowth of an "unreversed" clone.

At the present time the vector systems available for the delivery of therapeutic genes do not have the necessary efficiency to achieve anything like the requisite degree of transduction. Accordingly, our efforts have centred upon the utilisation of genes from categories 2 and 3 which, rather than being simply inhibitory in their effects are potentially destructive. Given this destructive capacity it would seem advantageous in the use of genes from either category 2 or 3 to restrict their expression to target tumour cells alone. Again, currently available physical transfer techniques and viral vectors do not possess the necessary degree of selectivity to ensure that this occurs. Rather than relying upon delivery to target gene expression the use of tumour cell specific promoters may be one way of limiting the therapeutic gene to the appropriate cells. Specific expression would occur only in the neoplastic cells, and therefore non-specific delivery systems would be acceptable. This approach was pioneered by Huber and coworkers (1991), using the α-fetoprotein promoter to direct expression to hepatocellular carcinoma and led us to examine similar possibilities in malignant melanoma.

2 Melanin Biosynthesis Pathway and Tyrosinase Regulation

The pigment cell's major function is the synthesis of melanin, and this biochemical pathway frequently functions, and often is elevated, in malignant melanomas. The key regulatory steps in the melanin biosynthesis pathway depend upon the activity of the copper-binding enzyme tyrosinase. Thus the rate-limiting step in the biosynthesis pathway is the oxidation of tyrosine to dopa and that of dopa to dopaquinone (Korner and Pawalek 1982). Several other proteins associated with melanogenesis also have been identified, including proteins related to tyrosinase, for example, TRP-1 (Jackson et al. 1991) and TRP-2 (Tsukamoto et al. 1992). Cell type regulation is due partly to post-transcriptional control of expression, but specificity of melanin synthesis is generally attributable to melanocyte-specific transcription of both TRP-1 (Jackson et al. 1991) and tyrosinase (Kluppel et al. 1991) genes. Tissue-specific gene transcription is a consequence of expression in the appropriate cell type and repression in other cell types (Yavuzer and Goding 1994). Using the reporter gene β-galactosidase, we showed that as little as 769 base pair of the 5' flanking region of the murine tyrosinase and 1.4 kilobase pair of the murine TRP-1 genes directed expression to both human and murine melanocytes and melanoma cells in vitro (Vile and Hart 1993a). Thus high levels of activity were observed in 12 of 14 human and murine melanoma lines tested whereas only basal levels were observed in a panel of 12 non-melanocytic cell types (Vile and Hart 1993a). Direct injection of naked DNA into murine melanoma or a control non-melanoma tumour achieved transduction of up to 10% of the melanoma tumour cells but no detectable expression in the control cancer (Vile and Hart 1993a). Some normal melanocytes manifested reporter gene activity, but expression was not found in other surrounding normal tissues (Vile and Hart 1993a). These results showed that it is possible using the transcriptional specificity of a melanocyte-specific promoter to limit expression of genes to melanocyte-derived cells.

3 Targeting of Immunity-Enhancing Genes

Based upon rather indirect immunological assays and their somewhat capricious clinical behaviour it has always been considered that melanoma is likely to be an immunogenic human cancer (CROWLEY and SEIGLER 1993), (CARRELL and JOHNSON 1993). The recent demonstration that a number of tumour peptide antigens, restricted by class I molecules of the major histocompatibility complex (MHC), can be isolated from this tumour type serves to confirm biochemically these earlier clinical suspicions (PARDOLL 1994). Thus the MZ2-E antigen (product of the MAGE1 gene), for example, is expressed by many melanomas as well as other tumour tupes but not by normal tissues (apart from the testis; VAN DER BRUGGEN et al. 1991) and represents a potential vaccination target. The lack of recognition of such tumour-specific neoantigens by T cells in the patients afflicted with these tumours is puzzling. It has been suggested that tolerance of the neoantigen may relate to the efficiency of epitope processing and presentation in manner which is similar to that of tissue-specific antigens (PARDOLL 1994). Moreover the tumours may lack appropriate levels of MHC class 1 or additional co-stimulatory molecules (see below) and remain functionally "invisible" to cytotoxic T lymphocyte recognition.

It might be possible to manipulate these possible mechanisms for therapy. The co-stimulatory molecule B7 binds to the CD28 and CTLA-4 receptors on T cells and results in optimal activation of these cytotoxic effectors. Introduction of the B7 gene de novo into murine melanoma cells resulted in the rejection of murine melanoma cells after re-introduction into syngeneic animals (TOWNSEND and ALLINSON 1993; CHEN et al. 1992). Using the transcriptional control of tissue-specific promoters to restrict expression of the B7 molecule could result in effective immune stimulation. Recently we have shown (CHONG, HART and VILE, unpublished observations) that expression of the B7 gene driven off the tyrosinase promoter results in B16 and K1735 melanoma cells becoming highly immunogenic.

However, a theoretical advantage to this type of immunostimulatory approach is that there is no necessity to deliver genes to every disseminated cancer deposit; an important point given the lack of efficiency of delivery mentioned above. Activated cytolytic T cells should be capable of dealing with distant growths because of their ability to recirculate. Since such effectors can kill more than a single tumour cell, the need to deliver immune-stimulating genes to a large percentage of target tumour cells is obviated. The finding that established $B7^{-ve}$ micrometastases were eliminated by the immune response generated by the subscutaneous inoculation of $B7^{+ve}$ transfected tumour cells also suggests that once stimulated the effector cells are able to deal with untransfected cells located at a distance (CHEN et al. 1992). Various cytokine cDNAs may be introduced into tumour cells in an effort to amplify the number of responding T cells and other immune effector cells. This approach has often formed the basis for establishing more immunogenic vaccine material (GANSBACHER et al. 1992), but there also is every reason to suppose that the local production of such cytokines in situ as a consequence of direct injection may have a beneficial effect (PLAUTZ et al. 1993). Once again, transcriptional targeting may enhance the efficacy of such in situ modification by stimulating the prolilferating T cells in close vicinity with the cytokine-expressing tumour cells. Clearly the cytokine cDNA chosen to be inserted into the tumour cells should not code for a factor for which the tumour cells possess the cognate receptor to avoid the possibility of autocrine growth stimulation. Using interleukin 2 and 4 as examples of such cytokines in our model systems, we have shown that direct injection of cDNAs expressed from the tyrosinase promoter into established B16 melanoma growths results in gene expression within the melanoma cells (VILE and HART 1994) and a significant number of tumour regressions (VILE and HART, unpublished observations).

4 Targeting of Prodrug-Activating Genes

A number of genes are capable of converting an inert prodrug into an active cytotoxic. These genes, frequently referred to as "suicide" genes, include the bacterial cytosine deaminase gene, which encodes an enzymes that converts 5-fluorocytosine into 5-fluorouracil, and the herpes simplex virus thymidine kinase (HSVtk) gene which encodes a kinase which phosphorylates ganciclovir. This guanosine analogue is phosphorylated to an intermediate by the viral kinase and is then phosphorylated further by cellular kinases to disrupt DNA synthesis. Considerable clinical experience has accumulated with both ganciclovir as an antiviral agent and 5-fluorocytosine as an antifungal agent which makes them attractive compounds for fairly rapid translation to the clinic. We have shown that direct injection of HSVtk cDNA, under the control of the tyrosinase promoter, followed by systemic treatment with ganciclovir, produces a significant reduction in tumour size (VILE and HART 1993b). This reduction in tumour size probably is assisted by the so-called 'bystander' phenomenon in which there appears to be transfer of toxic metabolites between neighbouring cells. In practical terms this means that not every cell requires transduction with DNA so that currently achievable efficiencies of gene transfer may prove effective. We have documented the degree of this bystander effect by mixing a clone of B16 cells, stably transduced with HSVtk, with parental, untransduced B16 cells in varying proportions. When the cell mix was cultured in 1 μg/ml ganciclovir, there was substantial killing of the parent cells; at a 50:50 mix of transfected:untransfected cells over 90% of the total cell population were killed (VILE and HART 1993b). Equally the use of the tissue-specific promoter in the direct injection protocols appears to limit the toxic effects to the melanoma cells since no untoward effects are observed in the treated mice (VILE and HART 1993b).

The direct injection routes of administration of prodrug-activating genes, as distinct from immunity-modulating genes, appears to be limited by the extent of any field effect. At first sight it seems likely that these approaches could be used only for local de-bulking procedures. The central conundrum remains how such genes are to be delivered to viscerally located tumours. We have used recombinant retroviral vectors as systemic delivery agents in murine systems and shown that tissue-specific expression is maintained by incorporating the 5' promoter of the murine tyrosinase gene into these constructs (VILE et al. 1994a). While using these vectors, which are likely to be problematic in human patients, we have obtained some dramatic results that may have profound implications for future gene therapy approaches. The number of recently established lung metastases of murine melanoma in syngeneic mice was reduced substantially in animals treated with ganciclovir relative to controls following multiple intravenous injections of high-titre retroviral supernatant encoding the HSVtk gene (VILE et al. 1994b). The reduction in numbers of lung tumour nodules was greater than anticipated from the extent of transduction even allowing for a marked bystander effect. A comparable reduction in tumour nodules was not seen in athymic immunodeficient mice subjected to identical protocols, suggesting that an immune component is somehow involved in this response. We then demonstrated that, whereas the parental cells are only poorly immunogenic, a partial but effective anti-tumour immune response is generated following killing of neoplastic cells in vivo as a result of treatment with ganciclovir (VILE et al. 1994b). This combination of direct killing, via a prodrug activation effect, and an augmented immune response, possibly as a consequence of the liberation of tumour neoantigens, might mean that prodrug activation by gene therapy could play a part in the therapy of disseminated disease.

As stated, it is unlikely that murine retroviruses will be used in humans. Apart from safety considerations, such as the risk of insertional mutagenesis, human complement inactivates these vectors (WELSH et al. 1975). Therefore a duplication of our approach in mice is not likely to succeed in man. It may prove possible, however, to use direct injection of DNA, coding for a prodrug-activating enzyme, into a local tumour deposit to stimulate a systemic immune response. Experiments to test this possibility currently are under way.

References

Carrell S, Johnson JP (1993) Immunologic recognition of malignant melanoma by autologous T lymphocytes. Curr Opin Oncol 5: 383–389

Chen L, Ashe S, Brady WA, Hellström I, Hellström KE, Ledbetter JA, McGowan P, Linsley PS (1992) Costimulation of antitumour immunity by the B7 counter-receptor for the T lymphocyte molecules CD28 and CTLA-4. Cell 71: 1093–1102

Crowley NJ, Seigler HF (1993) Possibilities of immunotherapy and gene therapy for malignant melonoma. Semin Surg Oncol 9: 273–278

Culver KW, Blaese RM (1994) Gene therapy for cancer. Trends 10: 174–178

Gansbacher B, Houghton A, Livingston P, Minasian L, Rosenthal F, Gilboa E, Golde D, Oettgen H, Steffens T, Yang SY, Wong G (1992) Clinical protocol: a pilot study of immunization with HLA-A2 matched allogeneic melanoma cells that secrete interleukin-2 in patients with metastatic melonoma. Hum Gene Ther 3: 677–690

Huber BE, Richards CA, Krenitsky TA (1991) Retroviral-mediated gene therapy for the treatment of hepatocellular carcinoma: an innovative approach for cancer therapy. Proc Natl Acad Sci USA 88: 8039–8043

Jackson IJ, Chambers DM, Budd PS, Johnson R (1991) The tyrosinase related protein-1 gene has a structure and promoter sequence very different from tyrosinase. Nucleic Acids Res 19: 3798–3804

Kamb A, Gruis NA, Weaver-Feldhaus J, Liu Q, Harshman K, Tavtigian SV, Stockert E, Day RS III, Johnson BE, Skolnick MH (1994a) A cell-cycle regulator potentially involved in genesis of many tumor types. Science 264: 436–440

Kamb A, Shattuck-Eidens D, Eeles R, Liu Q, Gruis NA, Ding W, Hussey C, Tran T, Miki Y, Weaver-Feldhaus J, McClure M, Aitken JF, Anderson DE, Bergman W, Frants R, Goldgar DE, Green A, MacLennan R, Martin NG, Meyer LJ, Youl P, Zone JJ, Skolnick MH, Cannon-Albright LA (1994b) Analysis of the p16 gene (cdkn2) as a candidate for the chromosome 9p melanoma susceptibility locus. Nature Genet 8: 22–26

Kluppel M, Beermann F, Puppert S, Schmid E, Hummler E, Schutz G (1991) The mouse tyrosinase promoter is sufficient for expression in melanocytes and in the pigmented epithelium of the retina. Proc Natl Acad Sci USA 88: 3777–3781

Korner A, Pawalek J (1982) Mammalian tyrosinase catalyses three reactions in the biosynthesis of melanin. Science 217: 1163–1165

Nabel GJ, Nabel EG, Yang Z-Y, Fox BA, Plautz GE, Gao X, Huang L, Shu S, Gordon D, Chang AE (1993) Direct gene transfer with DNA-liposome complexes in melanoma: expression, biologic activity, and lack of toxicity in humans, Proc Natl Acad Sci USA 90: 11307–11311

Oldfield EH, Ram Z, Culver KW, Blaese RM, De Vroom HL, Anderson WF (1993) Clinical protocol: gene therapy for the treatment of brain kinase gene and intravenous ganciclovir. Hum Gene Ther 4: 39–69

Pardoll DM (1994) Tumour antigens: a new look for the 1990s. News and views. Nature 369: 357–358

Plautz GE, Yang ZY, Wu BY, Gao X, Huong L, Nabel GJ (1993) Immunotherapy of malignancy by in vivo gene transfer into tumors. Proc Natl Acad Sci USA 90: 4645–4649

Ram Z, Culver KW, Walbridge S, Blaese RM, Oldfield EH (1993) In situ retroviral mediated gene transfer for the treatment of brain tumors in rats. Cancer Res 53: 83–88

Townsend SE, Allinson JP (1993) Tumor rejection after direct costimulation of $CD8^+$ cells by B7-transfected melanoma cells. Science 259: 368–370

Tsukamoto K, Jackson IJ, Urabe K, Montague PM, Hearing VJA (1992) Second tyrosinase related protein, TRP-2, is a melanogenic enzyme termed DOPAchrome tautomerase. EMBO J 11: 519–526

Van der Bruggen P, Traversari C, Chomez P, Lurquin C, De Plaen E, Van den Eynde B, Knuth A, Boon T (1991) A gene encoding an antigen recognized by cytolytic T lymphocytes on a human melanoma. Science 254: 1643–1647

Vile RG, Hart IR(1993a) In vitro and in vivo targeting of gene expression to melanoma cells. Cancer Res 53: 962–967

Vile RG, Hart IR (1993b) Use of tissue-specific expression of the herpes simplex virus thymidine kinase gene to inhibit growth of established murine melanomas following direct intratumoural injection of DNA. Cancer Res 53: 3860–3864

Vile RG, Hart IR (1994) Targeting of cytokine gene expression to malignant melanoma cells using tissue specific promoter sequences. Ann Oncol 5 Suppl 4: S59–S65

Vile RG, Russell SJ (1994) Gene transfer technologies for the gene therapy of cancer. Gene Ther 1: 88–98

Vile RG, Miller N, Hart IR(1994a) Retroviral vectors for the gene therapy of malignant melanoma. Gene Ther 1: 307–316

Vile RG, Nelson JA, Castleden S, Chong H, Hart IR (1994b) Systemic gene therapy of murine melanoma using tissue specific expression of the HSVtk gene involves an immune component. Cancer Res 54: 6228–6234

Welsh RM, Cooper NR, Jensen FC, Oldstone MBA (1975) Human serum lyses RNA tumour viruses. Nature 257: 612–614

Yavuzer U, Goding CR (1994) Melanocyte-specific gene expression: role of repression and identification of a melanocyte-specific factor, MSF. Mol Cell Biol 14: 3494–3508

Recombinant Antibody Fusion Proteins for Cancer Immunotherapy

R.A. Reisfeld[1] and S.D. Gillies[2]

1 Introduction

The last decade has seen the extensive development of monoclonal antibodies (mAb) combined with rapid advances in recombinant DNA technologies. These developments have greatly accelerated and expanded research efforts to generate new approaches for cancer therapy. The focus of many such efforts has been on immunotherapy, because of the unique specificity, diversity, and biological activity of antibodies that make them potentially ideal reagents in the laboratory and the clinic. Initial emphasis was primarily placed on mAb directed against tumor-associated antigen, including growth factor receptors, expressed to a greater extent on the cell surface of tumor cells than on normal cells and tissues. Two basic strategies were applied to substantially reduce tumor cell dissemination and growth in preclinical models as well as in clinical applications. The first of these made use of antitumor antibodies simply as vehicles to deliver to tumor

[1]Scripps Research Institute, 10666 N. Torrey Pines Rd, La Jolla, CA 92037, USA
[2]Fuji Immuno Pharmaceuticals Corporation, 125 Hartwell Avenue, Lexington, MA 02173, USA

cells either radionuclides, chemotherapeutic drugs, or toxins conjugated to these mAb by various means. Although several of these approaches resulted in some modest successes in the clinic, these were mainly confined to mAb conjugates with either radionuclides or toxins when applied to certain sensitive tumors, such as non-Hodgkins lymphoma. The second strategy applied to suppress tumor dissemination and growth made use of the natural effector mechanisms of antibodies to destroy tumor cells. These include triggering the complement cascade at the tumor cell surface, with consequent lysis or binding to Fc receptors on the surface of specialized effector cells, such as phagocytes or natural killer (NK) cells, thereby triggering phagocytosis or antibody-dependent cell-mediated cytolysis (ADCC). This approach has thus far also met only with modest success in the clinic; however, in contrast to the first strategy, it has been in part successful in patients with solid tumors, such as colon carcinoma (RIETHMÜLLER et al. 1994), melanoma (SALEH et al. 1992; MURRAY et al. 1994), and neuroblastoma (HANDGRETINGER et al. 1992, 1995), which usually are quite resistant to standard therapies.

A series of observations made with a variety of murine antihuman tumor mAb in either preclinical experiments or in clinical applications strongly suggested that conventional murine mAb needed to be improved in a variety of ways to be optimally effective for cancer therapy. These included: (a) rendering them less immunogenic in humans by generating human/mouse chimeric mAb or humanized mAb through complementarity-determining region (CDR) grafting and (b) structuring mAb to make optimal use of their natural effector mechanisms to kill tumor cells. In this regard, the unique domain structure of antibody molecules made them particularly accessible to genetic engineering, since functional domains with antigen-binding activities (Fc, Fab fragments) or effector functions (Fc fragments) lend themselves to be used separately as fragments or for interchange between antibodies (for reviews, see WINTER and MILSTEIN 1991; WRIGHT et al. 1992). The transfer of binding sites from one antibody to another is aided by the rigid β-pleated sheet framework structure of the variable domains that are surmounted with antigen-binding loops (WINTER and MILSTEIN 1991). Structural features such as these helped to generate a whole range of designer antibodies, ranging from chimeric or humanized mAb to a whole variety of recombinant mAb. These include the following: antigen-binding fragments, including domain-deletion mutants that can be used for targeting attached toxins or radionuclides (for reviews, see PAI and PASTAN 1994; BRINKMANN and PASTAN 1994; SCHLOM et al. 1991; SCHLOM 1995); antibody-like molecules that combine a binding site derived from a cell surface receptor or cell adhesion molecule, with antibody constant domains, i.e., immunoadhesions (for a review, see ASHKENAZI et al. 1993); hybrid antibodies assembled by using fragments from different antibodies to achieve novel antibody-related effector functions (for a review, see SMITH and MORRISIÖN 1994); and recombinant antibody fusion proteins with growth factors or cytokines.

The ability of cytokines to elicit cellular immune responses led to the application of cytokine gene therapy to create tumor cell vaccines that induce protective immunity. Since the pioneering work of TEPPER et al. (1989), numerous papers have been published on the potential of transfecting tumor cells with cytokine genes as an antitumor therapy primarily in weakly immunogenic murine tumor systems (for reviews, see PARDOLL 1993; COLOMBO and FORNI 1994). However, the clinical application of such an approach would be both time consuming and costly, since it would involve the isolation, transfection, and readministration of an individual patient's tumor cells. Another approach was applied by constructing a recombinant fusion protein between granulocyte-macrophage colony-stimulating factor (GM-CSF) and a murine B cell lymphoma-derived idiotypic protein. This construct could be converted into a strong immunogen capable of inducing idiotype-specific antibodies that protected recipient animals from challenge with an otherwise lethal dose of tumor cells (TAO and LEVY 1993).

An alternative approach would be to use the binding specificity of antitumor antibodies to target cytokines to tumor sites. Such an approach would not suffer from the limitation of being patient specific, since it would involve a simple injection or infusion of a recombinant antibody–cytokine fusion protein to any cancer patient suited for this type of therapy. In fact, successful therapy might be achieved with only a small percentage of the tumor mass being targeted by the fusion protein. This would be of considerable advantage and serve a dual purpose. First, it would circumvent the problem of tumor heterogeneity and tumor penetration that has severely limited the efficacy of antibodies targeting chemotherapeutic drugs, radionucleotides, and toxins. Second, the tumor cells targeted by the recombinant antibody–cytokine fusion protein could serve as an immunogen that would generate a specific and hopefully long-lasting cellular immune response that would result in successful cancer therapy.

This article will deal with recombinant antibody–cytokine fusion proteins and summarize and discuss some of the results obtained by us and other investigators with these constructs that have the potential to substantially suppress dissemination and growth of human tumor cells.

2 Targeting of Human Cytotoxic T Lymphocytes to Kill Heterologous Epidermal Growth Factor Receptor-Bearing Tumor Cells

Cytotoxic T lymphocytes (CTL) have been successfully targeted to effectively lyse tumor cells for which they lack specificity. This was achieved by heterobifunctional reagents that bridge a specific marker on the tumor cell to CD3, a component of the T cell receptor (TCR) (LIU et al. 1985; JUNG et al. 1986). Alternatively, preferential killing of tumor cells by nonspecific CTL could also be obtained by conjugating an analogue of α-melanocyte-stimulating hormone (α-MSH) to an anti-CD3 antibody by chemical means (LIU et al. 1988). A genetic approach was applied in another attempt to optimize killing of heterologous epidermal growth factor (EGF) receptor (EGFR)-bearing human tumor cells by

CTL. Thus, a recombinant antibody fusion protein was constructed between a chimeric mouse/human antihuman CD3 antibody and EGF and tested for its ability to mediate lysis of EGFR-bearing melanoma cells by human CTL (GILLIES et al. 1991a). This construct was made by fusion of an EGF-coding sequence to the 3'-end of the human γ_1-heavy (γ_1-H) chain gene sequence and subsequently expressing the modified gene in transfected cells together with the variable regions of a murine antihuman CD3 antibody. This fusion protein could compete with EGF for its receptor and was able to effectively mediate lysis of EGFR-bearing epidermoid carcinoma, melanoma, and neuroblastoma cells by a tumor-infiltrating lymphocyte (TIL) line or by a CTL line established from peripheral blood. Lysis could be achieved at concentrations as low as 10^{-12}–10^{-11} *M* that were significantly less than the k_D for EGF binding to EGFR (2×10^{-10} *M*). This reaction was specific, since in the absence of the fusion protein, the CTL had little or no cytolytic activity against the tumor targets that were tested. The use of recombinant fusion proteins between anti-T cell antibodies and growth factors, such as EGF, may be useful in evaluating adoptive immunotherapy regimens where a patient's peripheral blood-derived CTL are provided with an additional target specificity. Certainly, such recombinant fusion proteins offer a more direct approach to obtain heterobifunctional reagents of uniform high quality than that provided by heterobispecific antibodies or chemical conjugates that require additional in vitro manipulations.

3 Antibody–Superantigen Fusion Proteins for T Cell-Based Tumor Therapy

In an attempt to produce tumor-specific agents for T cell-based tumor therapy, DOHLSTEN et al. (1994) made a recombinant fusion protein of staphylococcal enterotoxin A (SEA) and the Fab region of the C215 murine monoclonal antibody reacting with human colon carcinoma cells. The rationale for this approach is based on the finding that strong immune responses against allografts usually involve T lymphocytes that are able to recognize peptides via the TCR presented in the context of major histocompatibility complex (MHC) molecules. Since the T cell response to a tumor is usually insufficient to interfere with its growth, an alternative approach to immunotherapy is to target tumor cells with a large number of T cells activated by SEA to release cytokines and produce cell-mediated immunity (SCHERER et al. 1993).

The recombinant fusion protein Fab C215-SEA produced a more than tenfold reduction in MHC class II binding compared to native SEA, and the affinity of this fusion protein for the C215 tumor antigen was approximately 100-fold greater than for MHC class II human colon carcinoma cells in vitro, demonstrating a functional substitution of MHC class II-dependent presentation of SEA with tumor specificity. Treatment with Fab C215-SEA resulted in 85%–99% inhibition of tumor growth and increased survival of Balb/c mice carrying B16 murine melanoma cells expressing a transfected C215 antigen. In this syngeneic, experimental metastasis model, the therapeutic effect depended on antigen-specific targeting of the fusion protein, since native SEA and an irrelevant SEA fusion protein failed to inhibit tumor growth (DOHLSTEN et al. 1994). These results suggest that a Fab–SEA fusion protein can convey superantigenicity to tumor cells that evoke T cell to suppress tumor growth in a syngeneic tumor metastasis model. However, it remains to be determined whether the use of a highly immunogenic bacterial superantigen that elicits tumor targeting of pseudospecific T cells will also be successful for the clinical treatment of solid human tumors.

4 Recombinant Antibody–Cytokine Fusion Proteins

The rationale for constructing recombinant antibody–cytokine fusion proteins is to achieve optimal biological effectiveness by combining the unique targeting ability of antibodies with the multifunctional activities of cytokines (for a review, see BURKE et al. 1993). Specifically, the hypotheses that needed to be tested with this approach were: (a) that these fusion proteins can target cytokines to tumor sites and thereby stimulate and expand immune effector cells sufficiently to achieve efficient tumor cell lysis and (b) that low doses of the antibody–cytokine fusion protein will be more effective than equivalent amounts of cytokine per se, in suppressing tumor growth and prolonging the life span of experimental animals. Should these hypotheses prove correct, one might then anticipate that the potentially lower effective dose levels of the antibody–cytokine fusion protein may be less toxic than the relatively high-dose levels of cytokines used thus far in clinical applications and that this will lead to a more effective immunotherapy of cancer.

GILLIES et al. (1991b, 1992) chose the antiganglioside GD2 antibody 14.18 (MUJOO et al. 1987) to construct recombinant antibody–cytokine fusion protein based on several experimental findings. First, murine anti-GD2 mAb 14.18 (IgG_3) and its isotype switch variant 14. G_{2a} (IgG_{2a}) as well as the chimeric human/mouse mAb ch14.18 (human IgG_1) served as specific probes for structural and functional studies of this disialoganglioside antigen that is associated with neuroblastoma, melanoma, sarcoma, and small cell lung carcinoma. The mAb 14.18 effectively lysed neuroblastoma cells in vitro by either complement-dependent cytotoxicity (CDC) or ADCC. This mAb (k_D, 2.2×10^{-8} *M*) detected as many as 1.2×10^7 antigenic sites on human neuroblastoma cells and markedly suppressed the growth of tumor xenografts in athymic (nu/nu) mice (MUJOO et al. 1987; REISFELD and CHERESH 1987). The isotype switch variant 14.G_{2a} isolated from mAb 14.18 (IgG_3) by fluorescence-activated cell sorter analysis had a k_D of 3.4×10^{-8} M and bound to the same chemically defined GD2 on neuroblastoma and melanoma cells. The mAb 14.G_{2b} and 14.G_{2a} effectively lysed these tumor cells by CDC and ADCC (MUJOO et al. 1989; REISFELD 1992). Effector cells isolated from the peripheral blood of either normal individuals or neuroblastoma patients were equally suitable for ADCC (Bruchelt et al. 1989; Reisfeld 1992). The chimeric mAb ch14.18, which is of human IgG_1 isotype, exhibited these same properties (Reisfeld 1992). Importantly, mAb $14.G_{2a}$ (Handgretinger et al. 1992) and ch14.18 (Handgretinger et al. 1995) were effective in phase I clinical trials of patients with neuroblastoma, resulting in a number of prolonged partial remissions and several long-term complete remissions of this disease. Both antibodies were somewhat less effective in phase I clinical trials of melanoma patients, where treatment resulted in several partial remissions (Saleh et al. 1992; Murray et al. 1994). For the reasons summarized above, the anti-GD2 mAb 14.18 was considered a potentially effective fusion partner for recombinant antibody–cytokine fusion proteins.

4.1 Recombinant Antibody–Tumor Necrosis Factor-β Fusion Proteins

The biological functions of tumor necrosis factor (TNF-α) and lymphotoxin (LT or TNF-β) provided the rationale for constructing recombinant antibody fusion proteins with one of these cytokines, i.e., LT. Thus, while both TNF-α and LT were reported to kill tumor cells, they exert a number of effects on a variety of cell types. The antitumor function of TNF was first studied in terms of its cytotoxic activity for neoplastic cells in vitro (Havell et al. 1988). It was initially assumed that the in vivo therapeutic activity of TNF is based on its ability to bind to its two cell receptors and to directly kill tumor cells. However, it became evident that the in vivo activity of TNF may not be based entirely on its ability to cause tumor regression directly, but is also aided by its capacity to cause hemorrhagic necrosis at the centers of established tumors, thereby destroying the tumor's vasculature. In this way, destruction of much of the tumor's center results from ischemia (Havell et al. 1988).

The rationale for using mAb fusion proteins with TNF-β is to make use of some of its multiple activities. Specifically, one of these will be the activation of inflammatory mechanisms involving granulocytes and macrophages. A further activity of this cytokine is to increase interleukin (IL)-2 receptors on T cells and thereby augment IL-2-mediated proliferation of T cell effectors. Thus once individual modalities have been optimized, there is a rationale to attempt combination therapies with IL-2 and TNF fusion proteins. An additional reason for making use of antibody–TNF fusion proteins is that ligation of TNF to its TR60 and TR80 receptors can induce apoptosis via induction of distinct signal pathways (Grell et al. 1994). This finding may well lead to the investigation of additional basic mechanisms involved in the antitumor effects of antibody–TNF fusion proteins. Another rationale for trying to achieve an increased concentration of TNF in the tumor microenvironment is to more effectively induce destruction of tumor vasculature as a consequence of the hemorrhagic necrosis caused by TNF in the center of established tumors. In this regard, Havell et al. (1988) have clearly shown that treatment of established immunogenic murine sarcoma with TNF often causes hemorrhagic necrosis of the centers of established tumors, but only rarely results in complete regression of the ring of living tumor tissue that survives central hemorrhagic necrosis. In fact, the therapeutic action of TNF against this tumor does not solely depend on its ability to directly destroy tumor cells in vivo, but also on the ability of this cytokine to directly or indirectly destroy the tumor's vasculature. In this way, destruction of most of the tumor center results from ischemia. For these reasons, a combination of mAb–TNF-β and mAb–IL-2 fusion proteins may have an optimal antitumor effect.

Human LT was genetically conjugated to the constant region of a human γ_1 immunoglobulin gene at the end of either the second (CH2–LT) or third (CH3–LT) constant region domain (Fig. 1) (Gillies et al. 1991b). The altered H chain constant regions were combined in a plasmid vector together with the variable regions of mouse antiganglioside GD2 antibody 14.18 and the human k-constant region. The resulting immunoconjugate constructs were expressed in transfected hybridoma cells and tested for both their antibody and LT activities. The two constructs were assembled to varying degrees depending on whether the third H chain constant region was present. However, both forms retained their ability to bind antigen and mediate ADCC, but only CH3–LT was able to mediate the lysis of melanoma target cells in the presence of human complement. In fact, as depicted in Fig. 2A, the CH3–LT conjugate mediates lysis at approximately the same concentrations as ch14.18, suggesting that the fusion of LT to the carboxy terminus of the H chain did not affect complement fixation. The CH2–LT conjugate, on the other hand, did not mediate lysis at concentrations as high as 10 µg/ml. The ability of the CH3–LT conjugate to mediate ADCC was also not significantly impaired by the addition of LT to the carboxy terminus (Fig. 2B). In contrast, the CH2–LT conjugate required higher concentrations to achieve a comparable degree of lysis of the

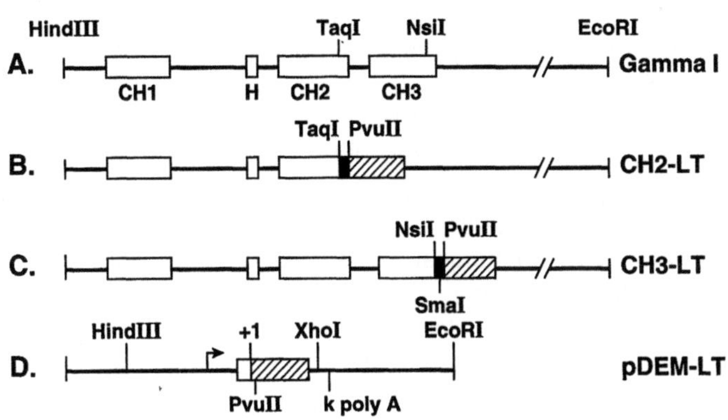

Fig. 1A–D. Construction of fusion proteins between lymphotoxin (*LT*) and the human immunoglobulin (Ig) H chain. **A** Restriction map of a human Cγ1 gene fragment cloned in plasmid pBR322. **B** The Cγ1 gene fused to LT at the end of the CH2 domain. **C** The Cγ1 gene fused to LT at the end of the CH3 domain. **D** The cDNA encoding LT cloned in expression vector pDEM including promotor (*arrow*), the natural leader peptide of LT (*open box*), the first residue of the mature protein (*+1*), and mouse *k*L chain polyA and 3' untranslated sequence. *Open boxes* represent protein-coding regions of Cγ1 in **A–C**; *black boxes* represent synthetic linkers used to join the protein-coding sequences; *striped boxes* represent LT-coding sequences. (From Gillies et al. 1991b)

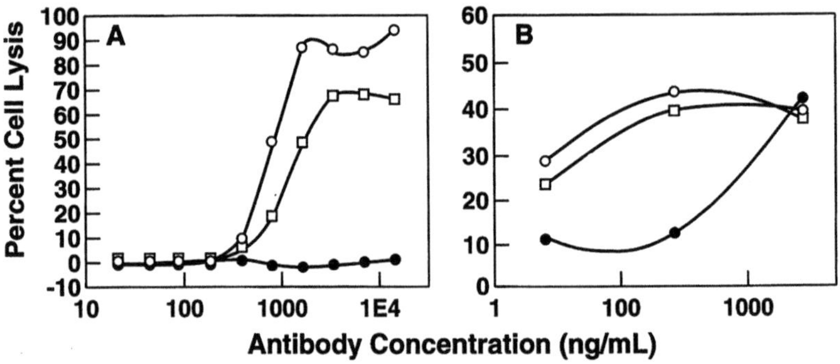

Fig. 2A,B. Complement-dependent cytoxicity (CDC) and antibody-dependent cellular cytoxicity (ADCC) activities of ch14.18 and immunoglobulin (Ig)/ lymphotoxin (LT) immunoconjugates. **A** The complement-dependent lysis of ^{51}Cr-labeled GD2-bearing M21 cells was assayed with the indicated amounts of ch14.18 antibody (*open circles*), CH2–LT (*closed circles*), or CH3–LT (*open squares*) together with human complement. **B** The antibody-dependent lysis of ^{51}Cr-labeled M21 cells was assayed with the antibody or conjugates as above and freshly prepared peripheral blood leukocytes (PBL) at an effector to target ratio of 200:1. Both assays were carried out for 4h at 37°C. The percentage of specific lysis was calculated by subtracting the spontaneously released counts and dividing the corrected experimental values by the total counts obtained by detergent lysis and multiplying by 100. (From GILLIES et al. 1991b)

antigen-bearing melanoma target cells. Since only a fraction of the CH2–LT preparation was assembled into complete antibody molecules, it is possible that the assembled molecules are fully active in mediating ADCC, but that higher concentrations are required to compensate for the fraction of inactive molecules. The presence of the CH3 exon in the fusion protein resulted in complete H chain assembly and thus in higher LT and effector activities (GILLIES et al. 1991b).

Although the assembly of H chains may likely result in LT dimerization, this possibility requires further examination by cross-linking studies. Consequently, the degree of H chain assembly is not the only factor affecting the biological activity of the conjugates. In fact, CH2–LT and CH3–LT preparations differed by as much as 100-fold in activity, but only two- to three fold in terms of assembly. Thus, the conformation of the LT portions of the dimerized conjugates must differ significantly to account for this discrepancy.

The data of GILLIES et al. (1991b) suggest that a free amino terminus is not necessary for LT binding to its receptor, since in the highly active CH3–LT construct the amino terminus of the LT domain is peptide bonded to the Ig H chain. This would agree with the results of FENG et al. (1988), who found that the fusion of LT to the carboxy-terminal end of interferon-γ resulted in demonstrable LT activity. However, one cannot rule out the possibility that the conjugate is proteolytically cleaved during the assay period and releases a free LT protein that binds to the receptor. In fact, plasmin treatment of the two conjugates was found to cleave the Ig and LT molecules at, or very near, their junction and to release active LT, while leaving the chimeric antibody intact (S.D. GILLIES et al., unpublished observations). In summary, the data of GILLIES et al. (1991b) clearly demonstrate that an anti-GD2 antibody can be genetically fused to LT without the loss of antigen-binding activity or receptor-binding and biologic activity of this cytokine. The ability of this fusion protein to target to LT-bearing tumors in vivo remains to be determined. Although the 14.18–LT fusion protein may be useful in delivering LT to the tumor site and may elicit inflammatory responses and other biological activities, it may not be useful as a direct cytotoxic molecule. Thus, even though 14.18–LT was specific for the abundantly expressed and internalized cell surface molecule GD2, it did not render an antigen-positive, LT-resistant tumor cell sensitive to LT-mediated killing (GILLIES et al. 1991b). However, this result is not surprising, since TNF (and presumably LT) does not have to enter a cell in order to kill it. Thus, the mechanism of TNFα and/or LT-mediated cell killing may be due to specific receptor signaling rather than general uptake into cells. The GD2-positive, LT-resistant tumor cells that were not sensitive to LT-mediated killing most likely did not express sufficient TNF receptors. In this regard, ligation of TNF to molecularly cloned TR60 (SCHALL et al. 1990) and TR80 (LOETSCHER et al. 1990) TNF receptors was recently reported to trigger distinct signaling pathways that induce apoptotic cell death (GRELL et al. 1994). It should be pointed out that both TNFα and TNFβ (LT) share the same specific TNF receptors, TR60 and TR80 (LEWIS et al. 1991).

4.2 Recombinant Antibody–Interleukin-2 Fusion Proteins

The rationale for constructing recombinant antibody–IL-2 fusion proteins is based on a large body of in vitro and in vivo data obtained with this cytokine in attempts to improve the efficacy of cancer immunotherapy. Initially, the ready availability of large amounts of molecularly cloned and defined cytokines catalyzed the widespread use of IL-2 for cancer immunotherapy. This was due largely to this cytokine's stimulatory effects on a broad range of immune cell types, including both T and B lymphocytes, monocytes, macrophages, and natural killer (NK) cells. Stimulation of peripheral blood lymphocytes by IL-2 in vitro and in vivo produced lymphokine-activated killer (LAK) cells, derived mainly from NK cells, that were applied in clinical trials, which produced antitumor responses in some patients with melanoma and renal carcinoma (ROSENBERG 1988; ROSENBERG et al. 1994). Specifically, treatment with high doses of IL-2 in patients with advanced melanoma proved to be somewhat less effective than the combined IL-2–LAK therapy resulting in response rates (complete and partial remission) (CR+PR) of 21% versus 24% (ROSENBERG 1988). However, because of a short half-life in the circulation, relatively large doses of IL-2 had to be administered to patients, which lead to toxicities, including fluid and colloid loss into visceral organs and soft tissues. Treatment-related mortality with high doses of IL-2 alone or in combined LAK–IL2 therapies was 1.5% among 652 patients who received 1039 treatment courses (ROSENBERG et al. 1989). Clinical results of LAK–IL-2 therapy were ultimately improved by using TIL, a class of T lymphocytes infiltrating into tumors. Those cells with IL-2 receptors could be propagated in vitro in large numbers by culturing with IL-2. The best clinical results obtained with this type of effector cells, including cyclophosphamide to eliminate T suppressor cells, was a response rate (CR+PR) of 39% obtained with 28 melanoma patients (ROSENBERG 1992). Recently, a 5-year clinical trial of 86 patients with metastatic melanoma indicated that treatment with autologous TIL plus high-dose bolus IL-2 resulted in objective responses in about one third of the patients (ROSENBERG et al. 1994).

In addition to these clinical data, the rationale for constructing recombinant antibody–IL-2 fusion proteins was further strengthened by several in vitro data. Thus treatment of T cells with anti-CD3 antibody prior to IL-2 exposure greatly increased T cell cytolytic activity (WEIL-HILLMAN et al. 1991). In addition, expansion of TIL by culture in the presence of high concentrations of IL-2, together with periodic target cell stimulation, caused substantial increases in cytolytic activity (ROSENBERG 1988). Both of these approaches involve costimulation of IL-2 and T cell antigen receptors for expansion and maintenance of T cell cytolytic activity. Consequently, GILLIES et al. (1992) proposed that an optimal strategy could combine IL-2 activation and tumor antigen presentation together with a tumor-specific antibody, such as antiganglioside GD2 mAb ch14.18, which mediates both CDC and ADCC activities.

The aim was to target IL-2 to tumors expressing ganglioside GD2 to ensure that local T cells would simultaneously recognize their cognate antigens in the context of MHC, while receiving an IL-2 activation signal. In addition, the anti-GD2 antibody would also be available to target Fc receptor-bearing cells, such as NK cells, that were activated by the targeted IL-2. Finally, it was anticipated that the targeting of cytokines by recombinant antibody fusion proteins to tumor sites will obviate the requirement to target every tumor cell, as is the case for antibody conjugates with drugs, toxins, and to some extent radionuclides. The difficulty in achieving this objective with highly heterogenous tumor cell populations has been at least one of the reasons why such conjugates have thus far had only very limited success in cancer therapy.

4.2.1 Construction and Characterization

FELL et al. (1991) initially reported the genetic construction of a fusion protein consisting of a human/mouse chimeric F(ab') of anticarcinoma mAb L6 and human IL-2. In this case, genomic DNA fragments encoding the L6 Ig light (L) chain and H chain variable (V) region gene segments were cloned and used to create chimeric L and H fusion gene constructs. The chimeric L chain construct consisted of L6V_L and human C_k gene segments, and the H chain fusion construct contained the L6V_H exon plus the C_{H1} exon of human γ_1. This was followed by a synthetic exon encoding a mutated version of the human IgG_1 hinge region, i.e., a 13-amino acid linker and the mature sequence of human IL-2. In an attempt to prevent inappropriate intrachain disulfide formation while synthesizing the hinge sequence, the two cysteines that normally form interchain disulfide bonds between H chains were replaced with serine and proline residues, respectively. After cotransfection of the two vectors into the Ig-deficient murine myeloma cell line Ag8.653, screening of cell lines for the assembled fusion protein with a pair of L6 anti-idiotypic antibodies indicated that an IL-2 fusion protein was produced at a concentration of 10μg/ml. Specific binding activity of the fusion protein for the L6 antigen was maintained on L6-positive carcinoma cells, and phytohemagglutinin (PHA)-activated human peripheral blood T cells demonstrated direct binding of the fusion protein to the human IL-2 receptor (IL-2R). In fact, the L6–IL-2 fusion protein also delivered an early signal via interaction with IL-2R that was at least qualitatively the same as that delivered by recombinant human IL-2 (Fell et al. 1991).

However, while the L6–IL-2 fusion protein supported the growth of the IL–2-dependent murine CTLL2 cell line in a dose-dependent manner, it was, unfortunately, far less active in this regard than recombinant human IL-2 (rhIL-2) on a molar basis. Thus, the L6–IL-2 fusion protein in culture supernatants was 30- to 100-fold less active than rhIL-2, and the purified L6–IL-2 fusion protein was 200-fold less active than rhIL2. However, the L6–IL-2 fusion protein exhibited biological activity, as tumor cells coated with this fusion protein stimulated T cell proliferation. The amount of IL-2 activity transferred by 10×10^3 tumor cells over the course of the cell proliferation assay was approximately that of 3 U rIL2/ml. Although the L6–IL-2 fusion stimulated the destruction of carcinoma cells by CD3 blasts equally well as rIL-2, it was far less efficient than rIL-2 at potentiating the effect of resting PBL. Fell et al. (1991) postulate that, although their L6–IL-2 fusion protein is approximately 200-fold less potent than equivalent amounts of rhIL, this reduced activity may allow doses of protein to be administered that accumulate within the tumor with less toxicity than rIL-2. They base this contention on unpublished data suggesting that the molar amount of mAb L6 required to achieve appreciable tumor localization in metastatic lesions is approximately 100 times the amount of rIL that can be safely administered in vivo. These authors expect that the putative biological potency of L6–IL-2 in vivo will also be aided by a longer serum half-life than that of rIL-2; however, this fact has not been established as yet for this particular fusion protein.

Gillies et al. (1992) reported on a recombinant fusion protein consisting of chimeric antiganglioside GD2 antibody (ch14.18) and rhIL-2, designated ch14.18–IL-2. In contrast to the findings reported by Fell et al. (1991) with their Fab L6–IL-2 fusion protein, Gillies et al. (1992) found that direct fusion of the IL-2 to the carboxyl terminus of the immunoglobulin H chain of intact ch14.18 did not reduce rIL-2 activity, as measured in a standard proliferation assay with either mouse or human T cell lines. The approach of Gillies et al. (1992) differed from that of Fell et al. (1991) in that antibody and IL-2 sequences were fused directly without the introduction of an artificial linker consisting of 13-amino acid residues.

In constructing the ch14.18–IL-2 fusion protein, Gillies et al. (1992) used plasmid constructs where the Ig–IL-2 fusion protein expression vector was constructed by fusing a synthetic human IL-2 sequence to the carboxyl end of the human $C_\gamma 1$ gene. In order to join the NH_2-terminal codon of mature IL-2 to the exact end of the CH3 exon (CH3–IL-2), a synthetic DNA linker was used, extending from the *Sma*I site near the end of the antibody-coding sequence to the single *Pvu*II site in the IL-2 sequence. The fused gene was inserted into the vector pdHL2-14.18, as described previously, for an antibody–LT fusion protein (Gillies

1991b). Two additional constructs were made in which the IL-2 sequence was fused to the *Sac*1 site in the hinge region of the human $C_\gamma 3$ gene (Fab–IL-2) or to the end of the CH2 exon at a *Taq*1 site (CH2–IL^{-2}). In each of these constructs, synthetic DNA linkers were used to fuse the antibody and IL-2 sequences without the introduction of any additional amino acid residues.

The expression plasmids were introduced into the Ig-deficient murine myeloma cell line Sp2/O-Ag14 by protoplast fusion and selected in medium containing 100 n*M* methotrexate (MTX). The clones secreting the largest amount of ch14.18–IL-2 were propagated in increasing concentrations of MTX and subcloned in medium containing 5 μ*M* MTX. Purification of the CH3–IL-2 fusion protein was done on protein A sepharose, while the Fab–IL-2 and CH2–IL-2 fusion proteins were purified on an antihuman k-chain affinity column.

Characterization of the various ch14.18–IL-2 fusion constructs revealed that the H chain fusion proteins were expressed together with the V regions of mAb 14.18 and the human C_γ gene. Although secreted H chains did associate with the chimeric L chain to form Fab–IL-2 or CH2–IL-2 fusion proteins, the latter failed to assemble into a whole antibody. This was the case even though it had an intact hinge region, presumably containing covalent disulfide bonds normally involved in inter-H chain binding. As far as antigen-binding activities were concerned, the Fab–IL-2 protein containing the 14.18 V regions showed no antigen-binding activity, while the CH2–IL-2 fusion protein was strongly positive. The genetically engineered 14.18 Fab revealed greatly reduced antigen-binding activity, suggesting that this antibody required bivalency for full activity. However, Fab 14.18–IL-2 and 14.18 CH2–IL-2 retained IL-2-specific activity ranging from 5 to 6.5×10^6 U/mg when normalized for IL-2 content. The lack of 14.18 Fab–IL-2 antigen-binding activity did not indicate that this is generally the case for such constructs, since a Fab–IL-2 fusion protein could be constructed with anti-TAG72 mAb that retained both antigen-binding and IL-2 activities (Gillies et al. 1992).

Construction of a whole antibody–IL-2 fusion protein by fusing the coding sequence of IL-2 to the end of the H chain CH3 exon (CH3–IL-2) resulted in a fully assembled antibody fusion protein with full IL-2 activity. Interestingly, this construct had enhanced antigen-binding activity when compared to mAb ch14.18. Furthermore, cleavage with plasmin reduced this increased antigen-binding activity of CH3–IL-2 back to that of ch14.18, as IL-2 was cleaved from the fusion protein at the carboxyl lysine residue of the antibody H chain, just before the +1 residue of mature IL-2 (Gillies et al. 1992). This finding suggests that the fusion of cytokines may actively alter the structure of antibody domains, resulting in changes in antigen–antibody interactions.

Evaluation of biological activities of CH3–IL-2 indicated that fusion of IL-2 to the carboxyl terminal of the immunoglobulin H chain did not reduce IL-2 activity when measured in proliferation assays with either mouse or human T cell lines. In addition, the CH3–IL-2 fusion was found to be remarkably stable throughout its purification and during subsequent storage for up to 4 years, thus far, at −20 °C. The effector functions of CH3–IL-2, i.e., the ability to mediate CDC or ADCC, were found to be maintained, although somewhat decreased when compared with ch14.18, similar to results reported for the CH3–LT fusion protein (Gillies et al. 1991b).

The CH3–IL-2 fusion protein exhibited enhanced TIL cytotoxic activity of autologous tumor targets. This was observed when the human 660 TIL line, which is CD3+, CD8+, antigen specific, and MHC class I restricted, was used to test the ability of the CH3–IL-2 (14.18) fusion protein to stimulate the killing of GD2-positive autologous melanoma tumor cells (660 mel). Figure 3A provides an example of a killing assay performed with 660 TIL shortly after antigen stimulation with 660 mel cells. In this experiment, the tumor target cells were first coated with CH3–IL-2 fusion protein or with mAb ch14.18 and then used as targets in a 7-h ^{51}Cr release asay. It is evident from Fig. 3A that at the higher effector to target (E/T) ratio of 50:1, mAb ch14.18 alone stimulated killing; however, this occurred to a much lesser extent than with CH3–IL-2. The effect of CH3–IL-2 was even more pronounced at the lower E/T ratio (10:1) or with TIL that had been deprived of IL-2 stimulation for 4 days. Figure 3B depicts results from a similar experiment, comparing CH3–IL-2 and exogenous IL-2, performed 1 week later when autologous killing activity had declined. In these 16 h cytotoxicity assays, the addition of IL-2 (100 U/ml) to the assay mixture produced little effect. However, the stimulatory effect of CH3–IL-2 was striking, especially at the lower E/T ratio (10:1) and when IL-2-depleted effector cells were used. In fact, in all cases, the amount of stimulation obtained by coating tumor cells with the CH3–IL-2 fusion protein exceeded that obtained by equivalent amounts of IL-2.

4.3 Recombinant Single-Chain Antibody–Interleukin-2 Fusion Proteins

SAVAGE et al. (1993) reported the construction, expression, and characterization of a recombinant single-chain antibody–IL-2 fusion protein (SCA–IL-2) that retains

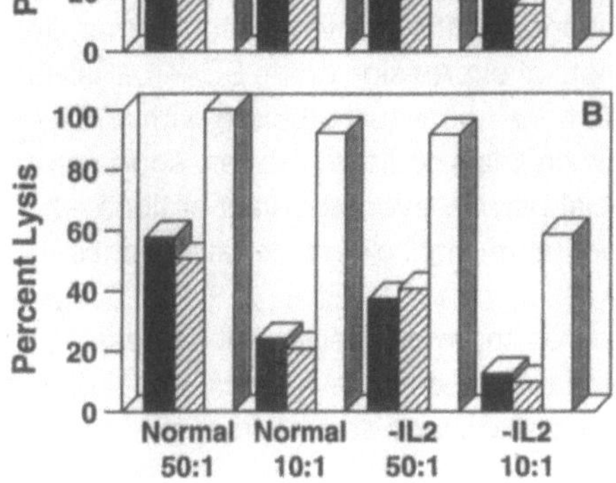

Fig. 3A,B. Stimulation of autologous cytolytic activity by CH3–interleukin (*IL*)-2-coated tumor cells. **A** Freshly stimulated 660 tumor-infiltrating lymphocyte (TIL) cells were used as effectors against 660 mel targets in a 7-h ^{51}Cr release assay. Before the assay, the target cells were either untreated (*filled bars*) or coated with ch14.18 (*hatched bars*) or CH3–IL-2 (*white bars*). The effector cells (660 TIL) were taken from growing cultures (*Normal*) or were cultured for 4 days without IL-2 (*-IL2*) and were used at an effector to target ratio of 50:1 or 10:1. **B** Normal and IL-2-depleted 660 TIL cells, 1 week after antigen stimulation, were used as effectors against untreated (*filled bars*) or CH3–IL-2 coated (*white bars*) 660 ml targets at the indicated effector to target ratio in a 16-h release assay. One set was incubated with added IL-2 (100 units [U]/ml; *hatched bars*). (From GILLIES et al. 1992)

both the antigen-binding characteristics of the parent SCA and the immunostimulatory actions of IL-2. These investigators chose to use an SCA comprised of variable heavy (V_H) and variable light (V_L) chain domains of the antilysozyme antibody D1.3 simply to test the feasibility of constructing it. The rationale advanced for making such a construct was to utilize such well-known characteristics of SCA as good tissue penetration (Yokota et al. 1992), rapid renal clearance of non-localized protein, and potentially low immunogenicity (Colcher et al. 1990). The reason given by Savage et al. (1993) for using the antilysozyme SCA D1.3 was its thorough characterization, well-documented secretion, ease in affinity purification, and detection by enzyme-linked immunosorbent assay (ELISA). Because of the unsuitability of an SCA for cytotoxicity assays, Savage et al. (1993) could not supply any evidence indicating concentration of IL-2 at tumor sites. However, these investigators were able to demonstrate that this fusion protein retains antigen-binding activity and also possesses the immunostimulatory actions of IL-2 when tested with lymphocytes bearing either high or low-affinity IL-2 receptors. Thus, fluorescence-activated cell sorter (FACS) analyses indicated that the SCA–IL-2 fusion protein was produced as an intact protein, rather than a degraded one, consisting of the two parent molecules. The fusion protein produced a significant increase in fluorescence compared with the negative control, while free SCA or free IL-2 alone showed no increase over background values. Specificity of the fusion protein's interaction with murine CTLL-2 cells was indicated by reduction of fluorescence to background levels in the presence of excess free rIL-2. This competitive inhibition of SCA–IL-2 binding also indicates that the fusion protein binds to the cells through specific receptor–ligand interactions. Result from proliferation assays with IL-2-dependent CTLL-2 cells revealed that half-maximal stimulation occurred equally well at equivalent concentrations of free rIL-2 (0.4 U/ml) and SCA–IL-2 fusion protein. Similar dose–response curves were also obtained when the respective effects of rIL-2 were compared with those of the SCA–IL-2 fusion protein in these experiments. Native mAb D1.3–SCA had no proliferative effect in these assays.

Savage et al. (1993) also pointed to a few potential problems with the SCA–IL-2 fusion protein. Thus, after expression in cultures of *Escherichia coli*, K12 KS476, the fusion protein, when probed with anti-SCA antiserum, revealed a 46-kDa component, but also at least five other components of lower molecular mass that apparently are due to proteolytic degradation of the carboxy-terminal section of the fusion protein. Another problem mentioned by these investigators is that, due to proteolytic degradation in *E. coli*, the yield of expression of the SCA–IL-2 is very low and will have to be improved. An additional, potential problem with an SCA–IL-2 fusion protein that may be seen in clinical trials could be the very short serum half-life time of such a construct, especially since even an intact antibody–IL-2 fusion protein was reported to have a very rapid initial clearance rate (α-phase) in normal Balb/c mice (Gillies et al. 1993).

Recently, Dorai et al. (1994) reported on the mammalian cell expression of single-chain Fv (sFv) antibody proteins and their C-terminal fusions with IL-2 and other effector domains. In this case, the model sFv was derived from murine mAb

741F8, which binds to the extracellular domain of the c-*erbB2*/HER2 antigen, a putative growth factor receptor in the tyrosine kinase family of oncogenes with an EGF-like ligand (HOLMES et al. 1992). The c-*erbB2* receptor is overexpressed in ovarian and breast carcinomas and therefore represents an important target antigen for antibody-based therapy. DORAI et al. (1994) described methods for the stable expression of sFv, sFv', and sFv fusion proteins in Chinese hamster ovary (CHO) cells and myeloma cells, as well as for transient expression in the primary human embryonic kidney cell line 293, to permit evaluation of plasmid construction. DORAI et al. (1994) mentioned briefly that they also constructed a fusion of IL-2 attached to the C-terminus of 741F8 sFv using a glycine-rich linker of 14 residues. This fusion protein was expressed in 293 and CHO cells, detected in western blots with anti-741F8 sFv antibodies, as well as by immunoaffinity chromatography on a c-*erbB2* resin. DORAI et al. (1994) state that this purified sFv–IL-2 fusion protein had a specific activity equivalent to the free IL-2 control. However, they did not show any experimental data to back up this statement, nor did they provide information on the antigen-binding activity of this sFv–IL-2 fusion protein.

4.4 Biological Activity and In Vivo Clearance of Antitumor Antibody-Cytokine Fusion Proteins

In a first step to evaluate recombinant antibody–cytokine fusion proteins as potential antitumor agents, GILLIES et al. (1993) examined their pharmacokinetic properties in normal Balb/c and athymic (nu/nu) mice. This included the evaluation of already constructed and characterized recombinant fusion proteins between the chimeric mouse/human antiganglioside GD2 mAb ch14.18 and the human cytokines LT (TNF-β) (GILLIES et al. 1991b) and IL-2 (GILLIES et al. 1992), respectively. Two additional Ig–cytokine fusion proteins were also constructed and tested for both biologic function and blood clearance. These included the fusion of ch14.18 Ig to either human GM-CSF or TNF-α. The mature TNF-β or IL-2 protein sequences were directly fused to the carboxyl terminal of the CH3 exon of the human $C_\gamma 1$ gene. The proteins secreted by transfectants, obtained with these constructs, assembled into complete antibody molecules with the cytokine peptide-linked to the Ig H chains. In both cases, these cytokines retained their biologic activities (GILLIES et al. 1991b, 1992).

The TNF-α fusion protein was made so that it could be compared to a TNF-β fusion protein (GILLIES et al. 1991b). These two cytokines have similar biological properties, although TNF-α has a more potent inflammatory effect (GOEDDEL et al. 1986). A similar fusion protein was also reported by another group (HOOGENBOOM et al. 1991). The second construct was made with GM-CSF, a protein that is structurally different from IL-2 and TNF and that was reported to enhance effector activity (ADCC) mediated by antitumor antibodies, including ch14.18 (BARKER et al. 1991). Both H chain gene fusions were expressed together with the 14.18 V regions and human C_γ in plasmid pdHL2 transfected into Sp2/0Ag14 murine hybridoma cells and selected in increasing concentrations of methotrexate (0.1-1.0 μ*M*) (GILLIES et al. 1989).

Analyses of the purified fusion proteins for chain composition and assembly by sodium dodecyl sulfate polyacrylamide gel electrophoresis (SDS-PAGE) showed that when the GM-CSF–ch14.18 fusion protein was tested under nonreducing conditions, it migrated as a single molecular species with a higher molecular mass than mAb ch14.18. Under reducing conditions, this structure dissociated into the normal L chain and a fusion H chain with a relative molecular mass consistent with the fusion of GM-CSF (approximately 75 kDa). A similar result was obtained with the TNF-α construct (GILLIES et al. 1993). The antigen-binding activity of the antibody fusion proteins was compared to that of ch14.18 in direct binding assays using GD2-coated plates. These data indicated that, as mentioned above, the fusion of IL-2 to ch14.18 resulted in a marked increase in GD2 antigen binding, while the fusion of TNF-β caused a much smaller increase, as determined by both direct and and competitive binding assays (GILLIES et al. 1991b, 1992). The fusion of TNF-α to ch14.18 also caused a slight increase in antigen binding, while the fusion of GM-CSF to this same mAb slightly reduced its antigen-binding activity.

The biological activity of the Ig–GM-CSF construct was found to be reduced when analyzed for proliferation with a GM-CSF-dependent acute myelogenous leukemic cell line. Thus, when compared to recombinant GM-CSF (produced in bacteria) on a molar basis, the fusion protein required approximately five times the concentration to achieve an equal level of cell proliferation when tested at 100–800 pg/ml. However, at a concentration of 1 ng/ml, the activity of the ch14.18–GM-CSF fusion protein approached that of rhGM-CSF. The TNF-α activity of the ch14.18–TNF-α fusion protein was found to be equal to the TNF-α standard (GILLIES et al. 1993).

Examination of the serum clearance rate of the fusion proteins following intravenous injection into athymic (nu/nu) mice indicated that the ch14.18–IL-2 fusion protein, despite its large molecular mass (approximately 200 kDa) was rapidly removed from the serum. In contrast, the ch14.18–GM-CSF fusion protein had an intermediate clearance rate. Specifically, a comparison of the half-life of the α- and β-phases indicated that the β- or elimination-phase half-life values of free ch14.18 (59 h) as well as of ch14.18–IL (30 h) and ch14.18–GM-CSF (28 h) fusion proteins were similar; the real difference in their catabolic rates is mainly due to the α-phase (distribution phase) half-life. These values are 0.3 h for the IL 2 fusion protein and 1.4 h for the GM-CSF protein, as compared with 7.5 h for free ch14.18. These findings suggest an intravascular event to be responsible for the rapid clearance, removing most of the fusion proteins into the extravascular space. Simple aggregation does not seem to be the cause for this rapid clearance, since this was not observed upon long-term storage.

Another possible explanation for the rapid clearance of the cytokine fusion proteins, other than the Ig–GM-CSF construct, is that the overall structure of these constructs is changed by the added domains. This may result in the recognition of the fusion protein by some type of clearance receptor. The observation that fusion of the same cytokines (IL-2, TNF-α, TNF-β) resulted in both enhanced antigen binding and rapid clearance suggests that these altered characteristics are due to structural changes that are not present in the GM-CSF construct. Another finding which indicates that fusion of cytokines may cause changes in antibody conformation is the susceptibility to cleavage by plasmin shown by the IL-2, TNF-α and TNF-β fusion proteins. This cleavage occurs between the carboxyl terminal of the ch14.18 antibody and IL-2 and reduces the increased antigen binding of the fusion protein to that of ch14.18 antibody (GILLIES et al. 1992). Interestingly, probing of the plasmin sensitivity of the ch14.18–GM-CSF fusion protein indicated that the junction was resistant to cleavage. This finding suggests a different protein structure for this construct which exhibits no increase in antigen-binding activity over free antibody, but also has a longer half-life than the IL-2, TNF-α, and TNF-β fusion proteins (GILLIES et al. 1993).

A common link between these two properties of antibody–cytokine fusion proteins may be the antibody's CH2 domain. Thus, this domain's interaction with Fab may reduce antigen binding in the 14.18 anti-GD2 system by decreasing the flexibility of the Fab arm. In fact, removal of the CH2 domain of mAb ch14.18 greatly enhanced its GD2 binding, possibility by increasing the flexibility of the Fab arm (GILLIES and WESLOWSKI 1990). It is possible that cytokine–CH2 interactions may be sufficiently strong to disrupt CH2–Fab interactions and thus free the Fab arm for antigen binding. This same CH2–cytokine interaction could cause a distortion in the Ig structure that results in rapid clearance, especially since this domain was found to be critical in determining Ig half-life (MUELLER et al. 1990).

Interactions between the carboxyl-terminal cytokine and the CH2 domain may vary with differences in surface charge interactions; however, such interactions may involve a looping back arround the CH3 domain. This type of structure is likely to be susceptible to proteolytic cleavage. In contrast, a fusion protein would be more resistant whenever the cytokine does not interact with the CH2 domain. This may explain the differences observed in antigen-binding and serum half-life between ch14.18–GM-CSF and the other fusion proteins (GILLIES et al. 1993). The various antibody–cytokine fusion proteins constructed and characterized by GILLIES et al. (1991b, 1992, 1993) were all biologically active with little or no loss in either antigen-binding or cytokine activity. An expression to this finding was the ch14.18–CM-CSF fusion protein, which was approximately 20% active; however, the GM-CSF standard used for comparison was produced in bacteria, whereas the fusion protein was expressed in mammalian cells. In this regard, it is well known that recombinant non-glycosylated GM-CSF produced in bacteria or treated enzymatically to remove carbohydrates when produced in mammalian cells has an approximately 20-fold greater specific activity than the fully glycosylated form of GM-CSF (MOONEN et al. 1987). Consequently, the glycosylated ch14.18–GM-CSF fusion protein produced in mammalian cells may well be as active as native glycosylated GM-CSF.

Although the antigen-binding activities, biologic cytokine activities, and the in vivo clearance of antibody–cytokine fusion proteins are clearly important properties to be established, one of the key questions to be answered is whether or not these fusion proteins are useful for targeting the various cytokines to tumors in vivo. In this regard, a potential problem, aside from rapid clearance, is that the fusion proteins may bind effector cells prior to reaching the tumor. One can only speculate as to the effect cytokine bivalency of fusion proteins may have on receptor internalization and signal transduction. There is also the possibility that the fusion protein may serve a bridging function between effector and target cells as an additional adhesion molecule. Thus, GD2-positive human melanoma cells coated with ch14.18–IL-2 fusion protein were already shown to be much more readily lysed by their autologous CTL than were uncoated cells in the presence of free IL-2 (GILLIES et al. 1992). The question whether the antibody–cytokines' relatively short half-lives in the circulation present a serious problem that will impact on their effectiveness in suppressing tumor dissemination and growth in vivo can only be answered by appropriate in vivo experiments. This will be the subject of the last section of this article, which describes the second step in the biological evaluation of antibody–cytokine fusion proteins, namely a determination of their efficacy in a preclinical model of tumor metastasis.

NARAMURA et al.(1994) determined some of the mechanisms by which cellular cytotoxicity was mediated by a recombinant antibody–IL-2 fusion protein directed against human melanoma cells. In this case, functional characteristics were established for a fusion protein between rhIL-2 and the mouse/human chimeric mAb 225 (ch225), directed against the human EGFR. The ch225–IL-2 fusion protein was constructed as already reported for the ch14.18–IL-2 fusion protein by fusing a synthetic human IL-2 sequence to the carboxyl end of the human Cγ1 gene (GILLIES et al. 1992). Antigen-binding activity of the ch225–IL-2 fusion protein was essentially the same as that of ch225. This was evident from dissociation constants (k_D) for radiolabeled ch225 ($2.62 \times 10^{-10} M$) and ch224–IL-2 ($3.3 \times 10^{-10} M$) calculated from saturation binding curves obtained with the human melanoma cell line M24met. The biologic activity of ch14.18–IL-2, in terms of IL-2 proliferative activity tested with the IL-2-dependent cell line CTTL-2, was the same as that of equivalent amounts of rhIL-2 and was calculated as two molar equivalent of IL-2 per mole of ch225–IL-2 fusion protein (NARAMURA et al. 1994).

Determination of the cytotoxicity mediated by fresh human peripheral blood mononuclear cells (PBMC) against melanoma 24met and C8161 target cells coated with either ch225 or ch225–IL-2 indicated that the fusion protein was superior to ch225 in this regard. As far as different effector cell populations are concerned, ch225 and ch225–IL-2 did mediate NK cell cytotoxicity, while resting T cells were not stimulated by either ch225 or the ch225–IL-2 fusion protein. However, unlike fresh PBMC, purified NK cells coated with ch225 or ch225-IL-2 killed both melanoma target cells with the same efficacy (NARAMURA et al. 1994).

Experiments were designed to assess the relative involvement of FcγRIII and IL-2R in NK cell-mediated cytotoxicity, since these cells constitutively express both receptors. NK cells interacted with the ch225–IL-2 fusion protein mainly through FcγIII, while the involvement of IL-2R was secondary. When human T cells were activated through CD3 and stimulated with exogenous IL-2, they themselves were found to be capable of killing melanoma cells, an effect that was further enhanced by the presence of the ch225–IL-2 fusion protein. Since $CD8^+$ cell populations were much more potent effectors than $CD4^+$ T cells, NARAMURA et al. (1994) tested the cytotoxicity of activated $CD8^+$ T cells mediated by the ch225-IL-2 fusion protein against human melanoma cells. In this case, an equivalent amount of rhIL-2 was also found to be capable of stimulating activated T cells to lyse melanoma cells. These data imply that the effect of the ch225–IL-2 fusion protein on activated T cells was most likely mediated by IL-2R. This is in contrast to observations made with NK cells, where as little as 0.1 μg ch225–IL-2/ml was enough to stimulate these effector cells, while equivalent amounts of ch225–IL-2 did not potentiate T cells. Taken together, these data suggest that ch225–IL-2 may be a potent immunotherapeutic agent that is able to recruit effector cells expressing FcγR and that it also may be capable of delivering relatively large amounts of IL-2 to tumor sites, to locally activate TIL/CTL.

4.5 Immunotherapy with a ch14.18-Interleukin-2 Fusion Protein in a Preclinical Neuroblastoma Metastasis Model

The rationale for evaluating the efficacy of the ch14.18–IL-2 fusion protein in a preclinical model of human neuroblastoma metastasis is based on preclinical data obtained with this construct that are already summarized in a prior section of this article. Encouraging results obtained with mAb ch14.18 per se in several phase I clinical trials that are also summarized in this article further strengthened the basis of this rationale. Although systemic rhIL-2 therapy in children with progressive neuroblastoma after high-dose chemotherapy and bone marrow transplantation resulted in only modest regressions of metastatic tumors (FAVROT et al. 1989), a further rationale for the potential use of ch14.18–IL-2 fusion protein was provided by additional clinical data. Thus, ADCC-mediated killing of neuroblastoma cells was augmented by $CD56^+$ NK cells, which were produced in increased amounts by IL-2 treatment of neuroblastoma patients (HANK et al. 1990). Furthermore, therapy of neuroblastoma patients with rhIL-2 and antiganglioside GD2 mAb $14.G_{2a}$ generated conditions within the peripheral blood of these patients that enabled their own lymphocytes to mediate ADCC sufficiently to kill human neuroblastoma cells in vitro (HANK et al. 1994).

The primary aim of SABZEVARI et al. (1994) was to test the hypothesis that the ch14.18–IL-2 fusion protein can specifically target rhIL-2 to tumor sites and is more effective than equivalent amounts of free rhIL-2 in achieving efficient tumor cell lysis. This was demonstrated in an adoptive immunotherapy model for experimental hepatic metastasis of human neuroblastoma in severe combined immunodeficient (SCID) mice reconstituted with human lymphokine-activated killer (LAK) cells. Specifically, 1 day after intrasplenic injection of 5×10^5 SK-N-AS human neuroblastoma cells and induction of hepatic metastasis, a control group received daily i.p. injections of 0.2 ml phosphate-buffered saline (PBS). The other animals were each injected i.p. with 4×10^7 human LAK cells and then randomized into experimental groups, each being injected i.p. for 7 days with one of the following: 0.2 ml PBS; ch14.18; ch14.18 plus rhIL-2 or ch14.18–IL-2 fusion protein containing an equivalent amount of rhIL-2. When animals were killed their livers examined for neuroblastoma metastases 1 month after tumor cell injection, there was no statistically significant difference from controls in the number of hepatic metastases found in SCID mice treated with either PBS or LAK cells. Most of these animals presented with a very large number (more than 500) of metastatic foci in their livers and exhibited up to three fold higher liver weights than untreated animals (Table 1). Although SCID mice treated with LAK cells plus ch14.18 showed a statistically significant decrease (p=0.01) in the number of metastatic foci, all of these animals presented with metastases ranging from five to 100 liver foci and had increased liver weights. In contrast, all SCID mice treated with the ch14.18–IL-2 fusion protein revealed a complete absence of macroscopic metastatic liver foci (p=0.003) and also presented with normal liver weights (Table 1, Fig. 4). However, when SCID mice were treated with a mixture of ch14.18 and rhIL-2 at dose levels equivalent to the ch14.18–IL-2 fusion protein, they also showed a complete lack of macroscopic metastatic liver foci. Identical results were obtained when SCID mice were treated with high doses (7.5×10^5 IU/day) of rhIL-2 per se (Table 1). Since these same effects were observed with both ch14.18–IL-2 and a mixture of ch14.18 and rhIL-2 at these relatively high doses, i.e., 250µg ch 14.18–IL-2, a comparison was made with their antitumor effect achieved at lower doses. Surprisingly, as little as 1 µg ch14.18–IL-2 per injection proved effective in suppressing dissemination and growth of metastasis. Indeed, doses of 1, 8, and 16 µg ch14.18–IL-2 fusion protein were far more capable of suppressing growth of tumor metastasis than equivalent doses of rhIL-2 (Table 2). Moreover, another set of experiments indicated that relatively low doses of ch14.18–IL-2 (1 and 16 µg) were also more proficient than equivalent amounts of rhIL-2 (3000 and 48000 IU) in prolonging the life span of tumor-bearing SCID mice reconstituted with human LAK cells. Specifically, the median survival of control animals was 45 days, while animals injected with the ch14.18–IL-2 fusion protein at either 1 µg/day or 16 µg/day essentially doubled their median survival when compared to controls for animals injected with equivalent amounts of free rhIL-2 (Table 3).

Table 1. Effect of a ch14.18–interleukin (IL)-2 fusion protein on experimental liver metastases of SK-N-AS neuroblastoma cells in severe combined immunodeficient (SCID) mice reconstituted with human lymphokine-activated killer (LAK) cells

Treatment[a]	Foci[b] (n)	Liver Weight[c] (g)
Control (PBS)	500 (250, 300, 500, 500, 500, 500, 500, 500, 500, 500, 500, 500, 500)	3.06 ± 1.3
LAK*	500 (2, 500, 500, 500, 500, 500, 500)	3.6 ± 2.7
LAK + ch14.18**	43 (5, 10, 26, 36, 50, 55, 70, 100)	1.63 ± 0.2
LAK + ch14.18+IL-2***	0 (0, 0, 0, 0, 0, 0, 0)	1.5 ± 0.1
LAK + ch14.18–IL-2****	0 (0, 0, 0, 0, 0, 0, 0, 0)	1.4 ± 0.05

From Sabzevari et al. (1994). PBS, phosphate-buffered saline.
[a]Daily dose x7 of ch14.18 and ch14.18–IL-2 was 250 µg and recombinant human IL-2 (rhu IL-2) 7.5×10^5 IU.
[b]Median (individual values are given in parentheses).
[c]Mean ± standard deviation.
*P=0.8; **P=0.01; ***P=0.008; ****P=0.003.

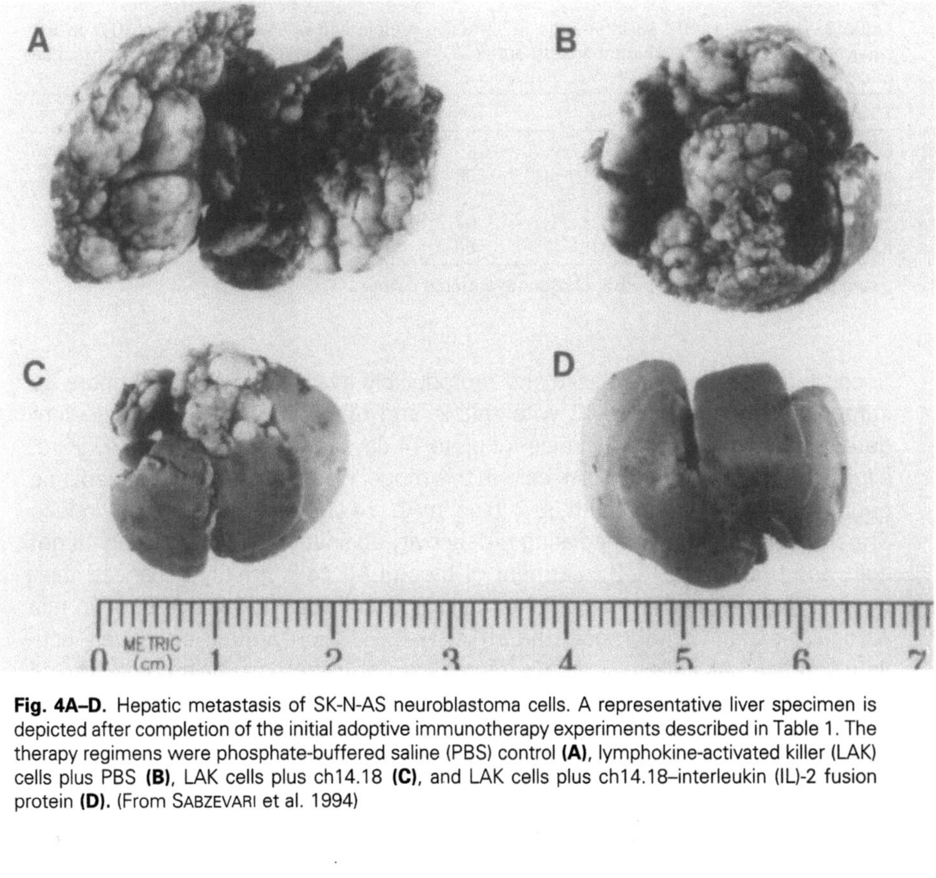

Fig. 4A–D. Hepatic metastasis of SK-N-AS neuroblastoma cells. A representative liver specimen is depicted after completion of the initial adoptive immunotherapy experiments described in Table 1. The therapy regimens were phosphate-buffered saline (PBS) control **(A)**, lymphokine-activated killer (LAK) cells plus PBS **(B)**, LAK cells plus ch14.18 **(C)**, and LAK cells plus ch14.18–interleukin (IL)-2 fusion protein **(D)**. (From SABZEVARI et al. 1994)

Table 2. Effect of a ch14.18–interleukin (IL)-2 fusion protein and equivalent doses of IL-2 on experimental metastases of human neuroblastoma cells in severe combined immunodeficient (SCID) mice reconstituted with human lymphokine-activated killer (LAK) cells

Treatment	Dose	Foci in the liver (*n*)
Control (PBS)	–	500, 500, 500, 500
ch14.18–IL-2*	1 μg	0, 0, 0, 0, 1, 1
IL-2**	3000 IU	0, 0, 76, 80, 500, 500
ch14.18–IL-2*	8 μg	0, 0, 0, 0, 0, 2
IL-2***	24 000 IU	0, 0, 12, 30, 72, 80, 500
ch14.18–IL-2****	16 μg	0, 0, 0, 0, 0, 0, 0
IL-2*	48 000 IU	0, 1, 4, 17, 18, 86

(From SABZEVARI et al. (1994). Tested by the nonparametric Wilcoxon rank sum test. PBS, phosphate-buffered saline. *P=0.0095; **P=0.114; ***P=0.024; ****P=0.0061.

The choice of the adoptive immunotherapy model in SCID mice reconstituted with human LAK cells was based on the following considerations. First, the model was highly reproducible, as hepatic metastases were routinely found in 100% of the SCID mice 4 weeks after intrasplenic injection of human neuroblastoma cells.

Table 3. Effect of a ch14.18–interleukin (IL)-2 fusion protein and equivalent doses of IL-2 on the survival of severe combined immunodeficient (SCID) mice with experimental metastases of human neuroblastoma

Treatment	Dose	Median survival (days)	Range (days)
Control (PBS)	–	45	40 – 78
ch14.18–IL-2	1 µg	98	61 – 128
IL-2	3000 IU	53	40 – 60
ch14.18–IL-2	16 µg	97	78 – > 140
IL-2	48 000 IU	56	43 – > 140

(From Sabzevari et al. 1994). PBS, phosphate-buffered saline.

Second, LAK cells were easily and reproducibly available in large numbers by stimulation of human PBMC with rhIL-2, and up to 20% of LAK cells were detected in the liver of SCID mice-for up to 14 days (Takahashi et al. 1993). Third, a further rationale for using LAK cells in this model was that treatment of pediatric neuroblastoma patients with IL-2 plus mAb $14.G_{2a}$ was reported to induce effector cells capable of mediating LAK activity against NK-resistant Daudi target cells and that further IL-2 treatment of these LAK cells greatly enhanced their cytolytic activity against neuroblastoma cells (Hank et al. 1994). Although in this particular experimental model the ch14.18–IL-2 fusion protein effectively activated human LAK cells to suppress neuroblastoma dissemination and growth, the ability of recombinant antibody–IL-2 fusion proteins to activate effector cell is by no means limited to LAK cells. Thus, treatment of neuroblastoma patients with anti-GD2 antibody plus rhIL-2 induced ADCC of the patients PBMC sufficiently to effectively kill neuroblastoma cells in vitro (Hank et al. 1994). In preclinical studies of human melanoma, it was also found that antibody–IL-2 fusion proteins can activate TIL (Gillies et al. 1992), as well as other leukocytes bearing FcγRIII and/or high-affinity IL-2 receptors, including NK cells and $CD8^+$ activated T cells (Naramura et al. 1994).

Sabzevari et al. (1994) clearly established that a high dose (250 µg/injection) of ch14.18–IL-2 is more effective than either LAK cells or LAK cells plus ch14.18 in suppressing growth of human neuroblastoma metastasis in SCID mice. However, it was particularly impressive that the fusion protein can also achieve this effect at very low doses (1 µg/injection) and more effectively than equivalent amounts of rhIL-2. These findings are encouraging for two reasons. First, they strongly support the hypothesis that recombinant antibody–cytokine fusion proteins can specifically target cytokines to tumor sites and stimulate immune effector cells sufficiently to achieve efficient tumor cell lysis. Second, the fact that very low doses of the ch14.18–IL-2 fusion protein proved more effective than equivalent amounts of rhIL-2 in suppressing tumor growth in SCID mice and in prolonging their life span suggests that it may also be feasible to apply an optimal biologic dose of ch14.18–IL-2 for future treatment of pediatric neuroblastoma patients. Third, based on these data, one might anticipate that the potentially lower effective doses of ch14.18–IL-2 may produce less toxicity than the relatively high doses of rhIL-2 found necessary thus far to achieve antitumor effects in clinical applications.

5 Conclusions

The extensive development of recombinant DNA technologies during the last 15 years has led to many advances in molecular concepts of biology. Among these is the highly successful application of these technologies to the genetic engineering of mAb, which greatly extended the usefulness of these unique molecules as biological and molecular probes. Concurrent with these developments came the realization that the relatively slow progress in the clinical treatment of cancer, particularly of solid tumors, required the development of novel approaches to tumor therapy. The intensive development and rapid advances of novel biological and molecular concepts in immunology during the last 30 years provided much of the impetus for applying such new concepts and technologies to cancer immunotherapy.

The subject of this article is just one of these concepts, namely the development of recombinant antibody fusion proteins for cancer immunotherapy. This is a rather recent event that has occured primarily during the last five years. Consequently, there is presently the kind of optimism for this approach that is characteristic of many new scientific concepts. This is especially true, since the "unforgiving yardstick of clinical trials" has not yet been applied to this new approach to cancer immunotherapy. In fact, the primary emphasis thus far has been on developing the technologies of constructing novel antibody fusion proteins and to characterize their biochemical and biological properties. Most of these evaluations have been done in vitro. Only one attempt has been made quite recently to assess the efficacy of a recombinant antibody–cytokine fusion protein in suppressing dissemination anf growth of a solid human tumor in vivo. The results of this study were encouraging, as they indicated than an antibody–IL-2 fusion protein was more effective in suppressing tumor growth and prolonging the life span of experimental animals than equivalent amounts of free rhIL-2 (Sabzevari et al. 1994). However, it has to be pointed out that the experimental model applied was a highly favorable one, which involved experimental hepatic metastasis of human neuroblastoma in SCID mice reconstituted with human LAK cells. At the same time,it also needs to be emphasized that in spite of these limitations, the data obtained in these experiments clearly provided a proof of principle. Thus, these findings support the hypothesis that a recombinant antibody–IL-2 fusion protein can specifically target this cytokine to tumor sites and stimulate immune effector cells sufficiently to achieve efficient tumor cell lysis, resulting in effective suppression of tumor dissemination and growth. It is also encouraging that the antibody–IL-2 fusion protein was able to achieve this suppression of tumor growth concurrent with a doubling in life span of experimental animals at doses considerably lower than those of free IL-2 per se. On the basis of these results, one can be cautiously optimistic and hope that it may prove feasible in the future to apply optimal biological doses of antibody–IL-2 fusion proteins to pediatric neuroblastoma patients and hopefully other cancer patients as well. Clinical experience indicates that only very high and often toxic doses of rhIL-2 provide effective treatment; however, on the basis of the preclinical data of SABZEVARI et al. (1994) one can hope that much lower effective doses of antibody–IL-2 fusion proteins might achieve effective treatment results in patients concurrent with far less toxicity.

What are the future perspectives for recombinant antibody fusion proteins? Obviously, as in all applications of new biological response modifiers, much of these perspectives will depend to some extent on the initial results obtained in phase I clinical trials with a ch14.18–IL-2 fusion protein that will be started this year with pediatric neuroblastoma patients (R. HANDGRETINGER, personal communication). Needless to say, the successful development of a second generation of more effective recombinant antibody fusion proteins is essential for the optimal realization of future perspectives. In this regard, it is important to realize that this field of research is in an early phase of development and that additional and novel research is required for its advance, including, of course, extensive, critical, preclinical and clinical evaluations. If these developments take place, one can anticipate that recombinant antibody fusion proteins will aid in providing a much needed impetus for more effective immunotherapy of human cancer.

References

Ashkenazi A, Capon DI, Ward RHR (1993) Immunoadhesins. Int Rev Immunol 10: 219–227
Barker E, Mueller BM, Handgretinger R, Herter M, Yu AL, Reisfeld RA (1991) Effect of a chimeric anti-ganglioside GD2 antibody on cell-mediated lysis of human neuroblastoma cells. Cancer Res 51: 144–149
Brinkmann U, Pastan I (1994) Immunotoxins against cancer. Biochim Biophys Acta 1198: 27–45
Bruchelt G, Handgretinger R, Kimmig A, Goeke B, Siedner R, Reisfeld RA, Niethammer D, Treuner J (1989) Effects of granulocytes on human neuroblastoma cells measured by chemiluminescence and chromium-51 release assay. J Biolumin Chemolumin 3: 93–103
Burke F, Naylor NS, Davies B, Balkwill F (1993) The cytokine wall chart. Immunol Today 14: 165–170
Colcher D, Bird R, Roselli M, Hardman KD, Johnson S, Pope S, Dodd S, Pantoliano MW, Milenic DE, Schlom J (1990) in vivo tumor targeting of a recombinant single-chain antigen-binding protein. J Natl Cancer Inst 82: 1191–1197
Colombo MP, Forni G (1994) Cytokine gene transfer in tumor inhibition and tumor therapy: where are we now? Immunol Today 15: 48–51
Dohlsten M, Abrahmsen L, Björk P, Lando PA, Hedlund G, Forsberg G, Brodin T, Gascoigne NRJ, Förberg C, Lind P, Kalland T (1994) Monoclonal antibody super antigen fusion proteins: tumor-specific agents for T cell based tumor therapy. Proc Natl Acad Sci USA 91: 8945–8949
Dorai H, McCartney JE, Hudziak RM, Tai M-S, Laminet AA, Houston LL, Huston JS, Oppermann H (1994) Mammalian cell expression of single-chain Fv (sFv) antibody proteins and their C-terminal fusions with interleukin-2 and other effector domains. Biotechnology 12: 890–897
Favort MC, Floret D, Negrier S, Cochat P, Bouffet E, Dacheng Z, Franks CR, Bijman T, Brunat-Mentigny M, Philip I, Philip T (1989) Systemic interleukin-2 therapy in children with progressive neuroblastoma after high dose chemotherapy and bone marrow transplantation. Bone Marrow Transplant 4: 499–504

Fell HP, Gayle MA, Grossmaire L, Ledbetter JA (1991) Genetic construction and characterization of a fusion protein consisting of a chimeric F (ab') with specificity for carcinomas and human IL2. J Immunol 146: 2446–2452

Feng GS, Gray PW, Shepard HM, Taylor MW (1988) Antiproliferative activity of a hybrid protein between interferon-γ and tumor necrosis factor-β. Science 241: 1501–1503

Gillies SD, Wesolowski JS (1990) Antigen binding and biological activities of engineered mutant chimeric antibodies with human tumor specificities. Hum Antibody Hybridomas 1: 47–54

Gillies SD, Lo K-M, Wesolowski JS (1989) High-level expression of chimeric antibodies using adapted cDNA variable region cassettes. J Immunol Methods 125: 191–202

Gillies SD, Wesolowski JS, Lo K-M (1991a) Targeting human cytotoxic T lymphocytes to kill heterologous epidermal growth factor receptor-bearing cells. J Immunol 146: 1067–1071

Gillies SD, Young D, Lo K-M, Foley SF, Reisfeld RA (1991b) Expression of genetically engineered immunoconjugates of lymphotoxin and a chimeric anti-ganglioside GD2 antibody. Hybridoma 3: 347–356

Gillies SD, Reilly EB, Lo K-M, Reisfeld RA (1992) Antibody-targeted interleukin 2 stimulates T-cell killing of autologous tumor cells. Proc Natl Acad Sci USA 89: 1428–1432

Gillies SD, Young D, Lo K-M, Roberts S (1993) Biological activity and in vivo clearance of anti-tumor antibody/cytokine fusion proteins. Bioconjugate Chem 4: 230–235

Goeddel DV, Aggarwal DD, Gray PW, Leung DW, Nedwin GE, Palladino MA, Patton JS, Pennica D, Shepard HM, Sugarman BJ, Wong GH (1986) Tumor necrosis factors: gene structure and biological activities. Cold Spring Harbor Symp Quant Biol 51: 597–609

Grell M, Zimmermann G, Hülser D, Pfizenmayer K, Schewrich P (1994) TNF receptors TR60 and TR80 can mediate apoptosis via induction of distinct signal pathways. J Immunol 153: 1963–1972

Handgretinger R, Baader P, Dopfer R, Klingbiel T, Reuland P, Reisfeld RA, Treuner J, Niethammer D (1992) A phase I study of neuroblastoma with the anti-ganglioside GD2 study of neuroblastoma with the anti-ganglioside GD2 antibody 14. G2a. Cancer Immunol Immunother 35: 199–204

Handgretinger R, Anderson K, Lang P, Dopfer R, Klingbiel T, Schrappe M, Reuland P, Gillies SD, Reisfeld RA, Niethammer D (1995) A phase I study of human/mouse chimeric anti-ganglioside GD2 antibody ch14.18 in patients with neuroblastoma. Eur J Cancer 31A: 261–267

Hank JA, Robinson RR, Surfus J, Mueller BM, Reisfeld RA, Cheung NK, Sondel PM (1990) Augmentation of antibody-dependent cell mediated cytotoxicity following in vivo therapy with interleukin 2. Cancer Res 50: 5234–5239

Hank JA, Surfus J, Gan F, Chew T-L, Hong R, Tans K, Reisfeld RA, Seeger RC, Reynolds CP, Bauer M, Wiersma S, Hammond D, Sondel PM (1994) Treatment of neuroblastoma patients with anti-ganglioside GD2 antibody plus interleukin-2 induces antibody-dependent cellular cytotoxicity against neuroblastoma detected in vitro. J Immunother 15: 29–37

Havell EA, Fiers W, North JR (1988) The anti-tumor function of tumor necrosis factor (TNF). J Exp Med 167: 1067–1085

Holmes WE, Sliwkowski MX, Akita RW, Henzel WJ, Goeddel DV, Vandlen RL (1992) Identification of heregulin, a specific activator of p185erb2. Science 256: 1205–1210

Hoogenboom HR, Raus JC, Volckaert G (1991) Targeting of tumor necrosis factor to tumor cells: secretion by myeloma cells of a genetically engineered antibody-tumor necrosis factor hybrid molecule. Biochim Biophys Acta 1096: 345–354

Jung G, Honsik CJ, Reisfeld RA, Muller-Eberhard HJ (1986) Activaton of human peripheral blood mononuclear cells by anti-T3: killing of tumor target cells coated with anti-target-anti-T3 conjugates. Proc Natl Acad Sci USA 83: 4479–4483

Lewis M, Tarhaglia GA, Lee A, Bennett GL, Rice GC, Wang GHW, Chen EY, Goeddel DV (1991) cloning and expression of cDNAs for two distinct murine tumor necrosis factor receptors demonstrate one receptor is species-specific. Proc Natl Acad Sci USA 88: 2830–2834

Liu MA, Kranz DM, Kurnick JT, Boyle LA, Levy R, Eisen HN (1985) Heteroantibody duplexes target cells for lysis by cytotoxic T lymphocytes. Proc Natl Acad Sci USA 82: 8648–8652

Liu MA, Nussbaum SR, Eisen HN (1988) Hormone conjugated with antibody to CD3 mediates cytotoxic T-cell lysis of human melanoma cells. Science 239: 395–398

Loetscher H, Pan YC, Lahn HW, Gentz R, Kufe DW (1990) Molecular cloning and expression of the human 55 kDa tumor necrosis factor receptor. Cell 61: 351–359

Moonen P, Mermod J-J, Ernst JF, Herschi M, DeLemarter JF (1987) Increased biological activity of deglycosylated recombinant human granulocyte/macrophage colony-stimulating factor produced by yeast animal cells. Proc Natl Acad Sci USA 84: 4428–4433

Mueller BM, Reisfeld RA, Gillies SD (1990) Serum half-life and tumor localization of a chimeric antibody deleted of the CH2 domain and directed against the disialoganglioside GD2. Proc Natl Acad Sci USA 87: 5702–5705

Mujoo K, Cheresh DA, Yang HM, Reisfeld RA (1987) Disialoganglioside GD2 on human neuroblastoma: target antigen for monoclonal antibody mediated cytolysis and suppression of tumor growth. Cancer Res 47: 1098–1104
Mujoo K, Kipps TJ, Yang HM, Cheresh DA, Wargalla U, Sander DJ, Reisfeld RA (1989) Functional properties and effect on growth suppression of human neuroblastoma tumors by isotype switch variants of monoclonal anti-ganglioside GD2 antibody 14.18. Cancer Res 49: 2857–2861
Murray JL, Cunningham JE, Brewer H, Mujoo K, Zukiwski AA, Podoloff DA, Kasi LP, Bhadkamkar V, Fritsche HA, Benjamin RS, Legha SS, Ater JL, Jaffe N, Itoh K, Ross MI, Bucana CD, Thompson L, Cheung L, Rosenblum MG (1994) Phase I trial of murine monoclonal antibody 14.G2a administered by prolonged intravenous infusion in patients with neuroectodermal tumors. J Clin Oncol 12: 184–193
Naramura M, Gillies SD, Mendelsohn J, Reisfeld RA, Mueller BM (1994) Mechanisms of cellular cytotoxicity mediated by a recombinant antibody-IL2 fusion protein against human melanoma cells. Immunol Lett 39: 91–99
Pai LH, Pastan I (1994) Immunotoxins and recombinant toxins for cancer treatment In: De Vita VT, Hellman S, Rosenberg SA (eds) Important advances in oncology. Lippincott, Philadelphia, pp 3–19
Pardoll DM (1993) Cancer vaccines. Immunol Today 14: 310–316
Reisfeld RA (1992) Potential of genetically engineered monoclonal antibodies for cancer therapy. Pigment Cell Res Suppl 2: 109–112
Reisfeld RA, Cheresh DA (1987) Human tumor antigens. In: Dixon FJ (ed) Advances in immunology, vol 40. Academic, New York, pp 323–377.
Riethmüller G, Schneider-Gädicke E, Schlimok G, Schmiegel W, Raab R, Höffken G, Gruber R, Pichlmaier H, Hirche H, Pichelmair R (1994) Randomized trial of monoclonal antibody for adjuvant therapy of resected Dukes' C colorectal carcinoma. Lancet 343: 1172–1183
Rosenberg SA (1988) Immunotherapy of cancer using interleukin-2: current status and future prospects. Immunol Today 9: 58–62
Rosenberg SA (1992) The immunotherapy and gene therapy of cancer. J Clin Oncol 10: 180–199
Rosenberg SA, Lotze MT, Yang JC, Aebersold PM, Linehan WM, Seipp CA, White DE (1989) Experience with the use of high-dose interleukin-2 in the treatment of 652 cancer patients. Ann Surg 210: 474–485
Rosenberg SA, Yanneli JR, Yang JC, Topalian SL, Schwartzentraber DS, Weber JS, Parkinson DR, Seipp CA, Einhorn JD, White DE (1994) Treatment of patients with metastatic melanoma with autologous tumor-infiltrating lymphocytes and interleukin-2. J Natl Cancer Inst 86: 1159–1166
Sabzevari H, Gillies SD, Mueller BM, Pancook JD, Reisfeld RA (1994) A recombinant antibody interleukin-2 fusion protein suppresses growth of hepatic human neuroblastoma metastases in severe combined immunodeficiency mice. Proc Natl Acad Sci USA 91: 9626–9630
Saleh MN, Khazaeli MB, Wheeler RH, Dropcho E, Liu T, Urist M, Miller DM, Lawson S, Dixon P, Russell CH, LoBuglio AF (1992) Phase I trial of the murine monoclonal anti-GD2 antibody 14.G2a in metastatic melanoma. Cancer Res 52: 4342–4347
Savage P, So A, Spooner RA, Epenetos AA (1993) A recombinant single chain antibody interleukin-2 fusion protein. Br J Cancer 67: 304–310
Schall TJ, Lewis M, Koller KJ, Rice GC, Wong GH, Goeddel DV (1990) Molecular cloning and expression of a receptor for human tumor necrosis factor. Cell 61: 361–370
Scherer MT, Ignatowicz L, Winslow G, Kappler J, Marrack P (1993) Superantigens: bacterial and viral proteins that manipulate the immune system. Annu Rev Cell Biol 9: 101–128
Schlom J (1995) Monoclonal antibodies in cancer therapy. In: De Vita VT, Hellman S, Rosenberg SA (eds) Biologic therapy of cancer, 2nd edn. Lippincott, Philadelphia, pp 507–521
Schlom J, Milenic DE, Roselli M, Colcher D, Bird R, Johnson S, Hardman KD, Guadagui F, Greiner JW (1991) New concepts in monoclonal antibody based radioimmuno-diagnosis and radioimmunotherapy of carcinoma. Nucl Med Biol 18: 425–435
Smith RIF, Morrison SL (1994) Recombinant polymeric IgG: an approach to engineering more potent antibodies. Biotechnology 12: 683–688
Takahashi H, Nakada T, Puisieus I (1993) Inhibition of human colon cancer growth by antibody-directed human LAK cells in SCID mice. Science 259: 1460–1463
Tao M-H, Levy R (1993) Idiotype/granulocyte-macrophage colony-stimulating factor fusion protein as a vaccine for B-cell lymphoma. Nature 362: 755–758
Tepper RI, Pattengale PK, Leder P (1989) Murine interleukin-4 displays potent anti-tumor activity in vivo. Cell 57: 503–510
Weil-Hillman G, Schell K, Segal DM, Hank JA, Sosman JA, Sondel PM (1991) Activation of human T cells obtained pre- and post-interleukin-2 (IL2) therapy by anti-CD3 monoclonal antibody plus IL2: implications for combined in vivo treatment. J Immunother 10: 267–277

Winter G, Milstein C (1991) Man-made antibodies. Nature 349: 293–299
Wright A, Shin S-U, Morrison SL (1992) Genetically engineered antibodies: progress and prospects. Crit Rev Immunol 12: 125–168
Yokota T, Milenic DE, Whitlow M, Schlom J (1992) Rapid tumor penetration of a single-chain Fv and comparison with other immunoglobulin forms. Cancer Res 52: 3402–3408

Influence of Local Cytokines on Tumor Metastasis: Using Cytokine Gene-Transfected Tumor Cells As Experimental Models

Z. QIN and T. BLANKENSTEIN

1 Introduction

The molecular mechanisms and, in particular, the role of cytokines involved in the process of metastasis, though one of the most important aspects of cancer research, are largely unknown. In order to metastasize, a malignant cell has to detach from the primary tumor, invade the extracellular matrix and the blood or lymphatic vessels, arrest in the circulatory system, extravasate into target organs, and initiate tumor cell growth. This multistep process may involve interactions between metastatic tumor cells and the local environment, where cytokines, chemokines, angiogenetic factors, and other factors produced by tumor cells or tumor-infiltrating cells will influence or control the metastatic process in different ways.

Cytokines regulate immune responses in a variety of ways. Because of their immunostimulating properties, such as induction of tumoricidal activity of immune cells, enhancement of tumor antigen presentation, and direct antitumor activity, cytokines, e.g., interleukin (IL)-2, tumor necrosis factor (TNF), interferons (IFN), have been widely used to treat cancer patients by systemic application

Max Delbrück Center for Molecular Medicine, Robert-Rössle-Str. 10, 13122 Berlin, Germany

(Rosenberg et al. 1989). Although they show some (usually transient) therapeutic effects, systemically applied cytokines have a major drawback, namely more or less severe toxicity resulting from the need to apply unphysiologically large amounts.

Recently, an alternative strategy, tumor cell-targeted cytokine gene therapy, has been proposed for cancer treatment. This strategy aims at increasing the concentration of certain cytokines at the site of tumor growth. At least in experimental tumor models, this can easily be achieved by gene transfer and expression of the cytokine in vitro which usually does not change the growth characteristics of the cultured cells. Upon injection into mice, the gene-modified tumor cells start to proliferate and to secrete the transfected cytokine. Depending on the transfected cytokine and the particular tumor cell line, this can lead to induction of an inflammatory response, activation of immune cells, and eventually rejection of the tumor cells (for a review, see Blankenstein 1994a). First of all, these experiments have shown that local cytokines may exert potent antitumor activity without any toxicity, presumably because the source of the cytokine, the tumor cells, is eradicated before the cytokine concentration increases to a toxic level. Furthermore, we are currently left with the puzzling situation that more than ten cytokines are active in this experimental system, even though the underlying immunological mechanisms seem at least partially to differ from each other. Tumor cells transfected with cytokine gene (or other genes with immunostimulating activity) are of therapeutic use provided a systemic immunity is induced and distantly growing, nontransfected tumor cells can be destroyed. Indeed, this has been shown in several tumor models and gives these gene-modified tumor cells the character of a vaccine. However, neither a general consensus exists on whether any particular cytokine is superior to the others–perhaps granulocyte-macrophage colony-stimulating factor (GM-CSF), IL-2, or IL-7 – nor has it been established that cytokine gene -modified tumor cells represent a qualitative improvement compared to classical tumor vaccines. In any case, cytokine gene-transfected tumor cells are relevant to the subject of metastasis, because they are thought to induce host immunity against its own micro-metastasis. On the other hand, several experiments have shown that the transfected cytokine may have opposite effects, namely to augment metastasis of the tumor cells. In the following, we will restrict ourselves to models with metastasizing mouse tumor cells transfected with cytokine genes and will illustrate the two potential sides of the coin. More detailed reviews on this subject have recently been published (Pardoll 1993; Colombo et al. 1994; Blankenstein 1994a; Porgador et al. 1994).

2 Experimental Systems

Several aspects of effects of transfected cytokines have been addressed: (a) Does the secretion of the cytokine alter the metastatic properties of the tumor cells in the form of either spontaneous metastasis of subcutaneously (s.c.) injected cells or experimental metastasis of intravenously (i.v.) injected cells? (b) Does immunization with cytokine gene-modified tumor cells protect the mice from a challenge with the parental, metastasizing cells? (c) Can mice with existing metastasis be cured by injection of cytokine gene-modified tumor cells?

Mouse tumor cell lines which have been used as models for analysis of metastasis include the Lewis lung carcinoma 3LL, the melanoma B16, the Friend leukemia 3Cl-8, the mammary adenocarcinoma TS/A, and the T lymphoma ESb (Table 1). A variety of cytokines have been expressed in these tumor cell lines, and their tumorigenicity, metastatic behavior, and properties as antitumor vaccines have been analyzed.

Table 1. The influence of local cytokine production on tumor metastasis

Transfected cytokine	Tumor cells	Cytokine production	Metastasis	Improved vaccine properties	Reference
IL-2	3LL	7 U/ml	↓	+	Porgador (1993a)
	3LL	1 U/ml	Unchanged	+	Porgador (1993a)
	TS/A	3600 U/ml	↓	+	Cavallo (1993)
IL-6	3LL	3 U/ml	↓	+	Porgador (1992)
IFN-α	3Cl	256–512 U/ml	↓	+	Ferrantini (1993)
	TS/A	8–1024 U/ml	↓	+	Ferrantini (1994)
IFN-γ	TS/A	2–6000 U/ml	↑	NA	Lollini (1993)
	3LL	128–256 U/ml	↓	+	Porgador (1993b)
	3LL	2 U/ml	↑	+	Porgador (1993b)
TNF	ESB	130–200 pg/ml	↑	NA	Qin (1993)

IL, interleukin; IFN, interferon; TNF, tumor necrosis factor; 3LL, Lewis lung carcinoma; TS/A, mammary adenocarcinoma; 3Cl–8, Friend leukemia; ESb, T lymphoma; ↑, augmented metastasis; ↓, decreased metastasis; +, immunization with cytokine-producing tumor cells was more effective than immunization with nonproliferating parental cells; NA, not analyzed.

3 Local Cytokines Inhibit the Metastatic Potential of Transfected Tumor Cells

3.1 Interleukin-2

The IL-2 gene has been expressed in a variety of tumor cells, including the metastatic cell lines 3LL (subclone D122), B16 (subclone F10.9), and TS/A. IL-2 secretion by approximately ten different tumor cell lines reproducibly suppressed the growth of s.c. injected locally growing cells, and the cells often were completely rejected. The immunological mechanism involved T cells ($CD8^+$) and non-T cells (asialo $GM1^+$). By immunohistology of tumor tissue, macrophages or (depending on the tumor model) granulocytes have been detected. Their role in the antitumor response is not yet known. Secretion of up to 7 μ/ml IL-2 by transfected 3LL cells led to the reduced generation of spontaneous metastasis in the lung. Metastasis formation inversely correlated with the amount of secreted

IL-2. However, IL-2-secreting 3LL and B16 cells still formed metastasis when the cells were injected i.v. (PORGADOR et al. 1994). Similar results were obtained with TS/A cells producing relatively high amounts of IL-2 (3600 U/ml). Upon i.v. injection, IL-2 secretion by gene-modified tumor cells prolonged survival of tumor-bearing mice, but almost all mice died in the end (CAVALLO et al. 1993). The s.c. injected IL-2-producing cells were more efficiently eliminated than i.v. injected cells.

3.2 Interleukin-6

IL-6 is a multifunctional cytokine regulating immune response, hematopoiesis, and acute phase reaction. It stimulates growth and differentiation of B and T cells (VAN SNICK1990). Systemic application of recombinant IL-6 reduced the number of metastatic lesions in experiments with several methylcholanthrene (MCA)-induced sarcoma and a colon carcinoma cell line (MULE et al. 1990). 3LL cells transfected with the human IL-6 gene and secreting low amounts of IL-6 showed a phenotype with little or no metastasis either spontaneously or experimentally, whereas the mock-transfected cells were highly metastatic. However, expression of IL-6 by the cells also led to reduced growth in vitro, which might partially contribute to the diminished metastasis (PORGADOR et al. 1992).

3.3 Interferon-α

Interferons (INF) were discovered and named on the basis of their interference with viral infection. However, IFNs (several of which are known) have other activities, including regulation of immune response, inhibition of cell growth, and antitumor activity. INF-α has been widely used for treatment of certain malignancies (e.g., hairy cell leukemia; GRESSER 1989). FERRANTINI et al. (1993, 1994) transfected Friend leukemia cells 3Cl-8 and TS/A cells with murine IFN-α cDNA and showed that production of IFN-α by 3Cl-8 cells strongly diminished tumor growth in mice, regardless of whether the cells were injected s.c., i.p., or i.v. The experimental metastasis of TS/A–IFN-α was also significantly inhibited in comparison to that of the parental TS/A cells.

4 Vaccination Against Tumor Metastasis with Cytokine-Producing Cells

The next question addressed was whether mice immunized with cytokine-producing tumor cells as vaccine could prevent the growth of subsequently s.c. injected parental cells of the metastatic type. DRANOFF et al. (1993) transduced murine melanoma B16-F10 cells, which metastasize preferentially to the lung, with several cytokine genes. They found that irradiated GM-CSF-producing tumor cells stimulated tumor immunity against a s.c. challenge with the parental cells. Whether the tumor cells still metastasized in the immunized mice or whether they were directly eradicated at the injection site was not analyzed. PORGADOR et al. (1992) transfected 3LL cells with the IL-6 gene and observed that immunization with IL-6-expressing cells dose-dependently protected mice from metastatic challenge by the i.v. injected parental cells. FERRANTINI et al. (1993, 1994) found that injection of INF-α-secreting 3Cl-8 cells s.c. into DBA/2 mice induced a tumor-specific protective response against the s.c. challenge of nonmodified 3Cl-8 cells, while control cell-injected mice developed rapidly growing tumors and died with liver and spleen metastasis. In mice immunized with IL-2-secreting TS/A cells, protection was relatively efficient against a subsequent s.c. challenge, but to a lesser extent against i.v. challenge with the parental tumor cells (CAVALLO et al. 1993). However, at least for s.c. challenge repeated immunization with irradiated parental tumor cells had nearly the same effect as immunization with IL-2-secreting cells.

5 Cure of Established Micro-metastasis with Irradiated Cytokine-Producing Cells

Because metastasis is the major cause of death in cancer patients, experiments have been performed to demonstrate the curative potential of cytokine gene-modified tumor cells for treatment of metastasis-bearing mice. PORGADOR et al. (1994) performed experiments with tumor cell lines 3LL and B16; mice were injected in the footpad with the parental tumor cells and 11 days later with cytokine (IL-2, IL-6, IFN-γ)-producing cells, and lung metastasis was analyzed. The primary tumor had already been removed after it had reached a certain size. They observed reduced increase in lung weight in mice immunized with irradiated cytokine-secreting cells in comparison to those injected with parental tumor cells. How that correlated with increased survival was not analyzed.

In the TS/A tumor model, immunization with IL-2-producing cells starting 1 day after contralateral challenge with the parental tumor cells had only a moderate effect on the percentage of surviving mice, which further dropped if immunization was started later (CAVALLO et al. 1993). Similarily, FERRANTINI et al. (1993) observed that treatment with IFN-α-producing 3Cl-8 cells of mice challenged 1–3 days previously with parental tumor cells significantly prolonged survival, but did not cure the vast majority of the mice on a long-term basis.

Taken together, the above-mentioned results are encouraging; however, they also illustrate one of the major obstacles in cancer immunotherapy. Beyond a critical time point of the tumor growing in the mice (which never exceeded approximately 2 weeks), immunotherapy was virtually ineffective, regardless of which approach was investigated. Factors such as tumor load or tumor-induced immune suppression may be responsible for this. Another important point is that even though the above experiments clearly demonstrate an improved therapeutic effect of the cytokine-secreting tumor cells in comparison to the use of nonproliferating parental tumor cells as vaccines, they do not take into account the extensive clinical and experimental experience with vaccines consisting of tumor cells plus adjuvants, e.g., bacille Calmette-Guérin (BCG) or *Corynebacterium parvum*. In 1981 Dye et al. demonstrated in the P815 tumor model, conceivably an inherently immunogenic tumor, that injection of a tumor cell–*C. parvum* mixture not only led to rejection of these tumor cells, but also to those injected at a distant site at the same time. Moreover, similar to the experiments with cytokine gene-modified cells, a complete decay of systemic immunity was observed when the treatment with tumor cells and adjuvant was started 6 days after injection of P815 cells (Dye et al. 1981). Therefore, we compared the vaccine effect of several cytokine gene-transfected tumor cells (plasmacytoma J558L) with that of a tumor cell–*C. parvum* mixture. Similar to observations made by others, immunization with cytokine-producing cells, but not with parental irradiated cells, conferred some protection against subsequent challenge with wild-type tumor cells. This effect, however, was less obvious with higher challenge doses and, importantly, no qualitative difference between the cytokines and the adjuvants was found (Hock et al. 1993). Similar results have been obtained in the TS/A tumor model (Allione et al. 1994). Because tumor cell–adjuvant vaccines have been tested in numerous clinical trials with moderate and hardly reproducible success, we argue that gene-modified tumor cell vaccines require further improvement. By no means, however, should this notion be interpreted as a pessimistic view on the part of the authors regarding cancer immunotherapy. Profound new knowledge of immune regulation, mode of antigen presentation, T cell activation pathways, the existence of tumor antigens (at least in certain tumors), the accessibility of approximately 50 cytokines, and molecular insight into tumor development has paved the way for completely new strategies.

6 Production of Cytokine by Tumor Cells Leads to Augmented Metastasis

An adverse effect of cytokines expressed by gene-transfected tumor cells, namely tumor promotion instead of suppression, has been observed with some cytokines.

6.1 Tumor Necrosis Factor

TNF was identified by its direct cytotoxic effects on some tumor cells; however, its antitumor activity in vivo seems to be mainly mediated by immunological mechanisms and/or hemorrhagic necrosis (Havell 1988). TNF is now known as a multifunctional cytokine which plays an important role in inflammation, immunity, and hematopoiesis (BEUTLER and CERAMI 1989). Several groups have transfected different apparently nonmetastatic tumor cell lines with human or mouse TNF genes and demonstrated that local TNF effectively suppressed tumor growth in most cases (reviewed in BLANKENSTEIN 1994b). In order to analyze the influence of local TNF expression on metastasis, we chose ESB (L5178Y-M) cells as a tumor model for TNF gene transfection. This cell line is a highly metastasizing variant of EB, an MCA-induced DBA/2 T lymphoma.

TNF gene expression was achieved by infection of ESB cells with a TNF gene containing retrovirus vector, which contains murine TNF cDNA under the control of β-actin promotor (QIN et al. 1993). Bulk culture and several clones of TNF-secreting ESB cells were analyzed for their spontaneous metastasis after injection of the cells s.c. into syngeneic mice. As shown in Fig. 1, injection of 2.5×10^6

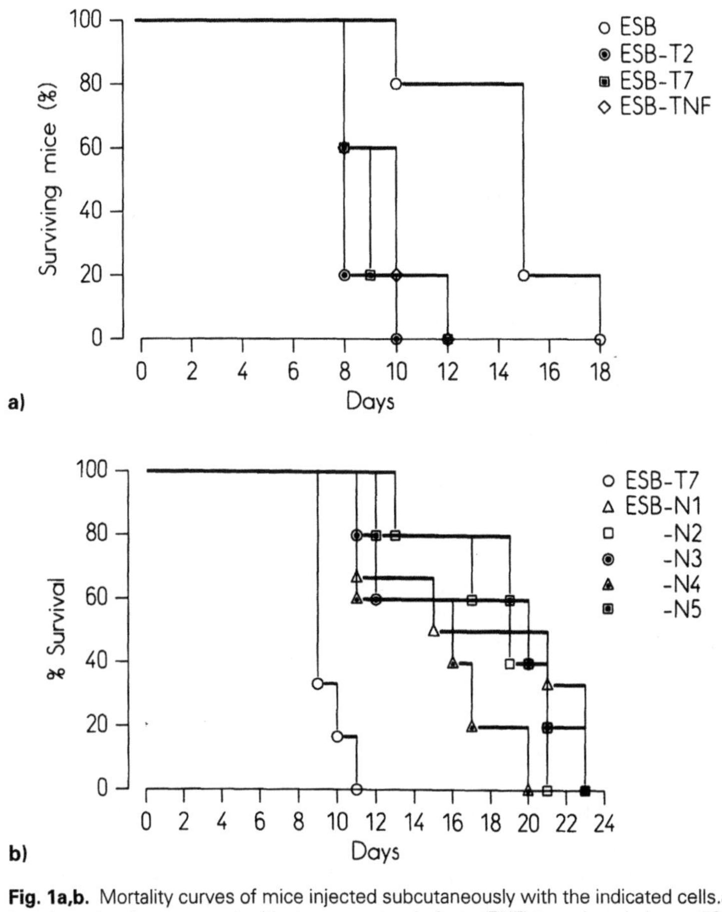

Fig. 1a,b. Mortality curves of mice injected subcutaneously with the indicated cells.**a** Parental *ESB* (T lymphoma) cells compared with tumor necrosis factor(TNF)-transduced clones ESB-T2 and ESB-T7 and bulk culture ESB-TNF. **b** TNF-transduced ESB-T7 cells compared with five clones of mock-infected ESB cells. Each group consisted of five or six mice. (With permission from QIN et al. 1993)

parental or mock-transfected ESB cells killed mice in an average of 18 days. However, secretion of 100–200 pg TNF/ml by ESB cells led to an accelerated death of mice. All mice bearing TNF-producing cells died within 8–12 days. Analysis of mice injected with TNF-producing cells revealed increased numbers of liver metastasis in comparison to mice injected with control cells. The enhancement of spontaneous metastasis of ESB tumor cells was due to the production of TNF, since an anti-TNF monoclonal antibody (mAb) specifically reverted the augmented liver metastasis of ESB-TNF cells. Of note, expression of TNF in similar amounts in the parental, nonmetastazing EB cells led to effective tumor suppression, and most of the mice completely rejected EB-TNF cells (QIN et al. 1993). The mechanism by which TNF increased hepatic metastases may involve any of its multiple effects on endothelial cells. TNF upregulates vascular cell adhesion molecule (VCAM), intercellular adhesion molecule (ICAM), and endothelial leukocyte adhesion molecule (ELAM)-1 on endothelial cells and directly causes transendothelial cell migration of polymorphonuclear neutrophils (PMN). ESB cells express several ligands (lymphocyte function-associated antigen-1, LFA-1; very late activation antigen-4, VLA-4; Mel-14) for adhesion molecules on endothelial cells, which could facilitate early arrest in the circulatory system, and ESB-derived TNF could enable stabilization of this interaction by upregulation of adhesion molecules on endothelial cells. Furthermore, local TNF may stimulate transendothelial migration of ESB cells, as has been observed for neutrophils in in vitro systems.

Interestingly, MIELE et al. (1994) observed that treatment of human malignant melanoma cells with TNF or INF-γ upregulated ICAM-1 (a counter-receptor for LFA-1) expression, which correlated with an increase of lung metastasis formation in nude mice.

6.2 Interferon-γ

Similar to TNF in gene transfer experiments, IFN-γ can strongly suppress tumor growth and induce tumor immunity (see above) or augment metastasis. By exogenous application of IFN-γ to B16 cells, which resulted in increased MHC class I expression, it was reported that INF-γ enhanced experimental metastasis of the tumor cells (TANIGUCHI et al. 1987). TS/A cells transfected with the murine INF-γ gene produced more experimental metastasis than the mock-transfected cells (LOLLINI et al. 1993; FERRANTINI 1994). When spontaneous metastasis was analyzed, cells with low but not high INF-γ secretion showed significantly more metastases. The enhancement of tumor metastasis by INF-γ may be related to upregulation of MHC class I expression and resistance of tumor cells to NK cell-mediated lysis (TANIGUCHI et al. 1987). Similarly, using 3LL cells as the tumor model, PORGADOR et al. (1993a,b) observed that cells with high INF-γ production manifested significantly reduced metastatic growth in comparison to the parental 3LL-D122 cells when tumor cells were i.v. injected into syngeneic mice. However, they observed that the INF-γ gene-transfected cells, which did not secret detectable amounts of IFN-γ but nevertheless showed elevated MHC class I expression, did metastasize equally well or even more rapidly than parental and mock-transfected cells, both in experimental and spontaneous metastasis experiments. The INF-γ-producing cells were equally sensitive to NK cell lysis as the parental cells in vitro in this experiment.

7 Conclusion

As we have shown, cytokine gene-transfected tumor cells are an interesting tool to analyze the function of a cytokine in vivo and to evaluate the immunotherapeutic potential of such cells. A variety of cytokines increase the immunogenicity of tumor cells following gene transfer. Cytokine gene-modified tumor cells can be effective as a vaccine against a subsequent tumor challenge and, at least to some extent, against small, preexisting tumors. However, the strength of these vaccines, e.g., in comparison to classical adjuvants as vaccines or with regard to the question of whether they are able to revert tumor-induced immune suppression (Tcell anergy), needs to be further investigated. Quite different effects have been observed in different tumor models, even when investigating the same cytokine. This becomes most obvious when TNF or IFN-γ are investigated. When transfected into different tumor cell lines, both cytokines can variously suppress tumor growth or augment metastasis of the cells. Whereas metastasis promotion by IFN-γ seems to involve upregulated MHC -class I expression and reduced lysis by NK cells, TNF seems to facilitate tumor cell migration.

Acknowledgments. This work was supported by the Deutsche Krebshilfe, Mildred Scheel Stifung, e.V.

References

Allione A, Consalvo M, P, Lollini PL, Cavallo F, Giovarelli M, Forni M, Gulino A, Colombo MP, Dellabona P, Hock H, Blankenstein T, Rosenthal FM, Gansbacher B, Musso T, Forni G (1994) Immunizing and curative potential of replicating and nonreplicating murine mammary adenocarcinoma cells engineered with interleukin (IL)-2, IL-4, IL-6, IL-7, IL-10, tumor necrosis factor alpha, granulocyte-macrophage colony-stimulating factor, and gamma-interferon gene or admixed with conventional adjuvants. Cancer Res 54: 6022–6026

Beutler B, Cerami A (1989) The biology of cachectin/TNF-a primary mediator of the host response. Annu Rev Immunol 7: 625–655

Blankenstein T (1994a) Increasing tumour immunogenicity by genetic modification. Eur J Cancer 30A: 1182–1187

Blankenstein T (1994b) Observations with tumor necrosis factor gene-transfected tumours. Folia Biol Praha 40: 19–28

Cavallo F, Di Pierro F, Giovarelli M, Gulino A, Vacca A, Stoppacciaro A, Forni M, Modesti A, Forni G (1993) Protective and curative potential of vaccination with interleukin-2-gene-transfected cells from a spontaneous mouse mammary adenocarcinoma. Cancer Res 53: 5067–5070

Colombo MP, Forni G (1994) Cytokine gene transfer in tumor inhibition and tumor therapy: where are we now? Immunol Today 15: 48

Dranoff G, Jaffee E, Lazenby A, Golumbek P, Levitsky H, Brose K, Jackson V, Hamada H, Pardoll D, Mulligan RC (1993) Vaccination with irradiated tumor cells engineered to secrete murine granulocyte-macrophage colony-stimulating factor stimulates potent, specific, and long-lasting antitumor immunity. Proc Natl Acad Sci USA 90: 3539–3543

Dye ES, North RJ, Mills CD (1981) Mechanisms of anti-tumor action of *Corynebacterium parvum*. I. Potentiated tumor-specific immunity and its therapeutic limitations. J Exp Med 154: 609–620

Ferrantini M, Proietti E, Santodonato L, Gabriele L, Peretti M, Plavec I, Meyer F, Kaido T, Gresser I, Belardelli F (1993) α1-Interferon gene transfer into metastatic Friend leukemia cells abrogated tumorgenictity in immunocompetent mice: antitumor therapy by means of interferon-producing cells. Cancer Res 53: 1107–1112

Ferrantini M, Giovarelli M, Modesti A, Musiani P, Modica A, Venditti M, Peretti E, Lollini P-L, Nanni P, Forni G, Belardelli F (1994) IFN-α1 gene expression into a metastatic murine adenocarcinoma (TS/A) results in $CD8^+$ T cell-mediated tumor rejection and development of antitumor immunity. J Immunol 153: 4604–4615

Gresser I (1989) Antitumor effects of interferon. Acta Oncol 28: 347–353

Havell EA, Fiers W, North RJ (1988) The antitumor function of tumor necrosis factor (TNF). 1. Therapeutic action of TNF against an established murine sarcoma is indirect, immunologically dependent, and limited by severe toxicity. J Exp Med 167: 1067–1085

Hock H, Dorsch M, Kunzendorf U, Überla K, Qin Z, Diamantstein T, Blankenstein T (1993) Vaccinations with tumor cells genetically engineered to produce different cytokines: effectivity not superior to a classical adjuvant. Cancer Res 53: 714–716

Lollini P-L, Bosco MC, Cavallo F, De Govanni C, Govarelli M, Landuzzi L, Musiani P, Modesti A, Nicaletti G, Palmieri G, Sautoni A, Young HA, Forni G, Nanni P (1993) Inhibition of tumor growth and enhancement of metastasis after transfection of the gamma-interferon gene. Int J Cancer 55: 320–339

Miele ME, Bennett CF, Miller BE, Welch DR (1994) Enhanced metastatic ability of TNF-α-treated malignant melanoma cells is reduced by intercellular adhesion molecule-1 (ICAM-1, CD54) antisense oligonucleotides. Exp Cell Res 214: 231–241

Mule JJ, McIntosh JK, Jablons DM, Rosenberg SA (1990) Antitumor acitivity of recombinant interleukin 6 in mice. J Exp Med 171: 629–636

Pardoll DM (1993) New strategies for enhancing the immunogenicity of tumors. Curr Opin Immunol 5: 719

Porgador A, Tzehoval E, Katz A, Vadai E, Revel M, Feldman M, Eisenbach L (1992) Interleukin 6 gene transfection into Lewis lung carcinoma tumor cells suppresses the malignant phenotype and confers immunotherapeutic competence against parental metastatic cells. Cancer Res 52: 3679–3686

Porgador A, Gansbacher B, Bannerji R, Tzehoval E, Gilboa E, Feldman M, Eisenbach L (1993a) Antimetastatic vaccination of tumor-bearing mice with IL-2-gene-inserted tumor cells. Int J Cancer 53: 471–477

Porgador A, Bannerji R, Watanabe Y, Feldman M, Gilboa E, Eisenbach L (1993b) Anti-metastatic vaccination of tumor-bearing mice with two types of IFN-γ gene-inserted tumor cells. J Immunol 150: 1458–1470

Porgador A, Feldman M, Eisenbach L (1994) Immunotherapy of tumor metastasis via gene therapy. Nat Immun 13: 113–130

Qin Z, Krüger-Krasagakes S, Kunzendorf U, Hock H, Diamantstein T, Blankenstein T (1993) Expression of tumor necrosis factor by different tumor cell lines results either in tumor suppression or augmented metastasis. J Exp Med 178: 355–360

Rosenberg SA, Longo DL, Lotze MT (1989) Principles and applications of biologic therapy. In: De Vita VT, Hellman S, Rosenberg SA (eds) Cancer: principles and practice of oncology, 3rd edn. Lippincott, Philadelphia, pp 301–347

Taniguchi K, Petersson M, Höglund, P, Kiessling R, Klein G, Kärre K (1987) Interferon-γ induces lung colonization by intravenously inoculated B16 melanoma cells in parallel with enhanced expression of class I major histocompatibility complex antigens. Proc Natl Acad Sci USA 84: 3405–3409

Van Snick J (1990) Interleukin-6: an overview. Annu Rev Immunol 8: 253–278

Antitumor and Antimetastatic Activity of Interleukin-12

J.A. HENDRZAK and M.J. BRUNDA

1 Introduction

Interleukin-12 (IL-12) was identified independently by two groups. The first group, at Hoffmann-La Roche, was studying a cytokine, named cytotoxic lymphocyte maturation factor (CLMF), that activated cytolytic lymphocyte activities (STERN et al. 1990). The second group, at the Wistar Institute, was evaluating a cytokine, originally named natural killer (NK) cell stimulatory factor (NKSF), that could stimulate NK cell functions (KOBAYASHI et al. 1989). Once CLMF and NKSF were characterized and cloned, it became clear that this was in fact one molecule, which is now called IL-12. IL-12 has profound effects on both NK and T cells, including the ability to stimulate proliferation, cytolytic activity, and cytokine induction. Furthermore, it plays a major role in regulating the induction of T cell subsets and thus is an important component of the immune response to foreign antigens. Based on these properties, IL-12 was evaluated in animal models of malignancies and shown to have potent antitumor and antimetastatic activities. In this chapter, we will summarize some of the biological effects of IL-12 and show how these relate to the in vivo effects of this cytokine on malignancies. Additional

Department of Oncology, Hoffmann-La Roche Inc., Nutley, NJ 07110, USA

information on IL-12 can be found in a number of review articles (Brunda 1994; Gately et al. 1994a; Brunda and Gately 1994;Trinchieri 1994; Gately and Brunda 1995; Brunda and Gately 1995; Hendrzak and Brunda 1995).

2 Structure, Molecular Biology, and Production of Interleukin-12

IL-12 is unique among known cytokines since it is a disulfide-linked heterodimeric protein; the molecular masses of the two subunits of IL-12 are 40 kDa (p40) and 35 kDa (p35), respectively (Kobayashi et al. 1989; Stern et al. 1990; Podlaski et al. 1992). The human p40 subunit of IL-12 is 306 amino acids long and contains ten cysteine residues and four potential N-linked glycosylation sites, while the human p35 chain is 197 amino acids long with seven cysteine residues and three potential N-linked glycosylation sites (Guber et al. 1991; Wolf et al. 1991; Podlaski et al. 1992). The genes for human p40 (chromosome 5, 5q31–q33) and p35 (chromosome 3, 3p12-3q 13.2) are found on separate chromosomes (Sieburth et al. 1992). There is no sequence homology between the p35 and p40 subunits (Gubler et al. 1991; Wolf et al. 1991), but the p35 subunit is distantly related to several other cytokines, while p40 has sequence homology with a few receptors (Gearing and Cosman 1991; Merberg et al. 1992; Schoenhaut et al. 1992). It has been hypothesized that IL-12 may have arisen from a disulfide-linked complex of a ligand with its receptor (Gearing and Cosman 1991). Heterodimeric disulfide-linked IL-12 is necessary for the optimal biological activity of IL-12 (Gubler et al. 1991; Wolf et al. 1991; Podlaski et al. 1992; Schoenhaut et al. 1992). The p40 and p35 subunits of murine IL-12 share 70% and 60% amino acid sequence identity to their human counterparts, respectively (Schoenhaut et al. 1992).

IL-12 was originally purified and cloned from Epstein-Barr virus (EBV)-transformed human B cell lines (Kobayashi et al. 1989; Stern et al. 1990; Gubler et al. 1991; Wolf et al. 1991). Following appropriate stimulation, additional human EBV-transformed B cells (D'Andrea et al. 1992; Valiante et al. 1992), myeloid (Kubin et al. 1994a) and keratinocyte (Muller et al. 1994; Aragone et al. 1994) cell lines, and murine B cell (Mengel et al. 1992), macrophage (Yoshida et al. 1994), and thymic stromal (Godfrey et al. 1994) cell lines can produce IL-12 mRNA or protein. In normal human or murine cell populations, IL-12 has been shown to be produced mainly by monocytes/macrophages (D'Andrea et al. 1992; Tripp et al. 1993; Hsieh et al. 1993; Smith et al. 1994; Gazzinelli et al. 1993; Vieira et al. 1994; Reiner et al. 1994; Ma et al. 1994). Other cell types that have been reported to produce IL-12 include B cells (D'Andrea et al. 1992), granulocytes (Cassatella et al. 1995), Langerhans cells (Trinchieri 1994), dendritic cells (Trinchieri 1994; Kanangat et al. 1995), mast cells (Smith et al. 1994), and keratinocytes (Muller et al. 1994; Aragone et al. 1994). A variety of stimuli, primarily infectious agents or their products, can induce IL-12 production both in vitro and in vivo (D'ANDREA et al. 1992; TRIPP et al. 1993; GAZZINELLI et al. 1993; KUBIN et al. 1994a; HEINZEL et al. 1994). IL-12 production is both positively (KUBIN et al. 1994a; 1994 YOSHIDA et al. 1994) and negatively (KUBIN et al. 1994a; D'ANDREA et al. 1995; VAN DER POUW KRAAN et al. 1995) regulated by other cytokines and factors.

3 Structure of and Signaling by the Interleukin-12 Receptor

IL-12 receptors (IL-12R) have been detected on activated human T cells (DESAI et al. 1992; CHIZZONITE et al. 1992) and on resting or activated NK cells (DESAI et al. 1992; NAUME et al. 1993a), while no IL-12R have been found on resting T cells, resting or activated B cells, or monocytes (DESAI et al. 1992; GATELY and BRUNDA 1995). The IL-12R can be upregulated on T cells by mitogen or alloantigen stimulation (DESAI et al. 1992) and on NK cells activated by IL-2 (DESAI et al. 1992) or IL-12 (NAUME et al. 1993a). Multiple forms of the IL-12R exist with apparent affinities of 5–20 p*M*, 50–200 p*M* and 2–6 n*M* (CHUA et al. 1994). One component of the human IL-12R has been cloned and is a type I transmembrane protein with an extracellular domain of 516 amino acids and a cytoplasmic domain of 91 amino acids (CHUA et al. 1994). The cloned IL-12R component binds IL-12 with an equivalent apparent affinity of 2-5 n*M* (CHUA et al. 1994) and is comparable to the low-affinity IL-12R; antibodies directed against this protein specifically inhibit IL-12-mediated biological activities (CHUA et al. 1994; GATELY and BRUNDA 1995). Work is in progress to search for other components of the IL-12R that result in high-affinity binding of IL-12.

Little is known about the signaling events that occur following interaction of IL-12 with its receptor. Tyrosine kinases appear to be important in signal transduction by IL-12, since tyrosine kinase inhibitors block several IL-12-induced functional activities in human NK cells (GEROSA et al. 1993). Phosphorylation of a number of proteins, including a 44-kDa member of the mitogen-activated protein (MAP) kinase family (PIGNATA et al. 1994), TYK2 (BACON et al. 1995), JAK2 (BACON et al. 1995), and $p56^{lck}$ tyrosine kinase (PIGNATA et al. 1993), occurs rapidly after exposure of NK or T cells to IL-12. IL-12 has also been shown to induce or inhibit apoptosis under various experimental conditions (CLERICI et al. 1994; ORTALDO et al. 1995; RADRIZZANI et al. 1995).

4 Biological Activities of Interleukin-12

4.1 Cell Proliferation

IL-12 can stimulate the proliferation of both T and NK cells. IL-12 stimulated the proliferation of activated T cells, T cell clones, cytotoxic T cells lines, and tumor-infiltrating lymphocytes (TIL) (Gately et al. 1991; Perussia et al. 1992; Andrew et al. 1993; Mehrotra et al. 1993; Nabioullin et al. 1994; Yanagida et al. 1994; Alzona et al. 1995). Proliferation of both $CD4^+$ and $CD8^+$ T cell receptor (TCR)-$\alpha\beta^+$ T cells or T cell clones and TCR-$\gamma\delta^+$ lymphoblasts has been observed with IL-12 (Gately et al. 1991; Perussia et al. 1992; Mehrotra et al. 1993; Nabioullin et al. 1994). Maximal proliferation of T cells by IL-12 was only observed in the $CD30^+$ T cell population (Alzona et al. 1995), and IL-12 does not stimulate proliferation of resting T cells (Gately et al. 1991). IL-12 was active at lower concentrations than IL-2, but the maximal stimulation observed with IL-12 ranged from 30% to 70% of that obtained with IL-2 (Gately et al. 1991). The effect of IL-12 is independent of IL-2, since antibodies to either IL-2 or the IL-2R did not inhibit IL-12-induced proliferation (Gately et al. 1991; Perussia et al. 1992; Andrew et al. 1993; Mehrotra et al. 1993). Costimulation by IL-12 and the B7–CD28 complex results in optimal proliferation of human T cells (Kubin et al. 1994b) and murine T cell clones (Murphy et al. 1994). Synergistic proliferation of various T cell populations has been observed with IL-12 and suboptimal concentrations of IL-2, (Gately et al. 1991; Perussia et al. 1992; Andrew et al. 1993; Nabioullin et al. 1994) but IL-12 inhibits proliferation if combined with high concentrations of IL-2 (Perussia et al. 1992; Nabioullin et al. 1994).

IL-12 is a relatively poor stimulator of NK cell proliferation compared to cytokines such as IL-2 or IL-7 (Gately et al. 1991; Perussia et al. 1992; Naume et al. 1992; Robertson et al. 1992; Naume et al. 1993a). Combining IL-12 with IL-2 resulted in either additive (Perussia et al. 1992; Naume et al. 1992) or inhibitory (Naume et al. 1992; Robertson et al. 1992) effects; as with T cells, these differences are explained in part by the dose of IL-2 used, since IL-12 enhanced proliferation at low doses of IL-2 but inhibited proliferation at high doses of IL-2 (Perussia et al. 1992). In contrast to its inhibitory effects on IL-2- or IL-7-induced proliferation, IL-12 synergizes with IL-4 to stimulate enhanced proliferation of NK cells (Robertson et al. 1992; Naume et al. 1993a). NK cell proliferation induced by IL-12 may differ from that induced by IL-2 or IL-7.

4.2 Cytolytic Activity

The ability of IL-12 to augment/induce lytic activity of cytotoxic T lymphocytes (CTL) was one of the first properties described for this cytokine (Wong et al. 1988; Gately et al. 1992). In some experiments, the effect of IL-12 on CTL was inhibited by antibodies to IL-2 (Gately et al. 1992), but in other studies no inhibition of CTL was found with antibodies to the IL-2R (MEHROTRA et al. 1993). Under some conditions, synergy of IL-12 with IL-2 was observed, while under other conditions inhibition of CTL was found (MEHROTRA et al. 1993). Murine IL-12 can augment alloantigen-stimulated CTL by a mechanism which is interferon (IFN)-γ, IL-2, and NK cell independent (BLOOM and HORVATH 1994). IL-12 can enhance the generation of alloantigen-specific CTL in vivo (GATELY et al. 1994b).

IL-12 can augment NK cell cytolytic activity both in vitro (KOBAYASHI et al. 1989; GATELY et al. 1992; SCHOENHAUT et al. 1992; ROBERTSON et al. 1992; CHEHIMI et al. 1993) and in vivo (GATELY et al. 1994b). IL-12-mediated enhancement of NK activity is independent of IL-2, IFN-α, IFN-β, and IFN-γ (ROBERTSON et al. 1992; CHEHIMI et al. 1993); TNF-α is also not involved in the activation of mature human NK cells by IL-12 (CHEHIMI et al. 1993), but it is involved in IL-12-mediated activation of immature NK cells (JEWETT and BONAVIDA 1994). Under various conditions, treatment of NK cells with a combination of IL-12 and IL-2 resulted in synergistic (ROSSI et al. 1994), additive (CESANO et al. 1993), or less than additive enhancement of cytotoxicity (CHEHIMI et al. 1992). IL-12 can also enhance activity of NK cells from patients with various malignancies (SOIFFER et al. 1993).

Lymphokine-activated killer (LAK) cell activation has also been observed following incubation of cells for several days with IL-12 in vitro (GATELY et al. 1992; NAUME et al. 1993a; NABIOULLIN et al. 1994). Activation of both NK cells (GATELY et al. 1992; NAUME et al. 1992, 1993a) and $CD8^+$ T cells (NABIOULLIN et al. 1994) by IL-12 has been reported. LAK cell induction from either peripheral blood leukocytes (PBL) or purified NK cells is inhibited by antibodies to TNF-α, while antibodies to IFN-γ exhibit some inhibitory effect and antibodies to IL-2 do not inhibit LAK cell induction (GATELY et al. 1992; NAUME et al. 1992). IL-4 does not inhibit the generation of IL-12-induced LAK activity (NAUME et al. 1993a). IL-12 is important for induction of LAK cell activity by IL-2, since IL-2-induced LAK activity is blocked by neutralizing anti-IL-12 antibody (GATELY et al. 1992).

The mechanism or mechanisms mediating the augmentation of cytolytic activity induced by IL-12 on NK and T cells are currently under investigation. Increased expression of perforin and granzymes (CESANO et al. 1993; SALCEDO et al. 1993; ASTE-AMEZAGA et al. 1994) and an increased number of cytoplasmic granules (CHEHIMI et al. 1993; CESANO et al. 1993) has been found in a number of cell types exposed to IL-12. In addition, IL-12-induced enhancement of stimulus-dependent granule exocytosis from NK/LAK cells has been reported (BONNEMA et al. 1994). Increased target cell binding was found in one study (RABINOWICH et al. 1993), but not in another (CESANO et al. 1993). IL-12 upregulates a number of cell surface molecules, including CD2, CD11a, CD54, CD56, CD69, CD71, HLA-DR, the p75-kDa tumor necrosis factor (TNF) receptor, and receptors for IL-2 (α,β, and γ-chains), IL-4, and IL-12 (NAUME et al. 1992, 1993; ROBERTSON et al. 1992; GEROSA et al. 1993; JEWETT and BONAVIDA 1994; YANAGIDA et al. 1994; ALLAVENA et al. 1994; NAKARAI et al. 1994). Antibodies to CD18 inhibit cytotoxicity of IL-12-stimulated NK cells, suggesting the importance of this molecule in recognition of target cells (ROSSI et al. 1994). IL-12 also induces directional migration of NK cells and activated T cells (ALLAVENA et al. 1994).

4.3 Cytokine Induction

IL-12 can induce IFN-γ production from both human/murine NK and T cells in vitro (Kobayashi et al. 1989; Chan et al. 1991, 1992; Schoenhaut et al. 1992; Perussia et al. 1992; Naume et al. 1993b; Hsieh et al. 1993; Tripp et al. 1993; Gazzinelli et al. 1993; Carson et al. 1994; Aste-Amezaga et al. 1994; Alzona et al. 1995). IL-12 is a more potent inducer of IFN-γ than either IL-2 or IL-7 (Naume et al. 1993b), and the ability of IL-12 to induce IFN-γ production is independent of IL-2 (Chan et al. 1991). Synergistic induction of IFN-γ is obtained from cells stimulated with IL-12 and a variety of other stimuli (Chan et al. 1991; Perussia et al. 1992; Hsieh et al. 1993; Tripp et al. 1993; Gazzinelli et al. 1993; Carson et al. 1994), while a number of cytokines downregulate IL-12-induced IFN-γ production (Chan et al. 1991; D'Andrea et al. 1993; Tripp et al. 1993; Gazzinelli et al. 1993). The molecular mechanism by which IL-12 mediates its effect on IFN-γ induction is not fully understood, but upregulation of mRNA expression by transcriptional and post-transcriptional effects have been observed (Chan et al. 1992; Nagy et al. 1994). Substantial circulating serum levels of IFN-γ have also been found in mice injected with IL-12 (Gately et al. 1994b; Brunda et al. 1995a).

In addition to IFN-γ, IL-12 has been shown to induce a number of other cytokines under some conditions. Following in vitro exposure of IL-12, T or NK cells increased production of IL-2 (Perussia et al. 1992), TNF (Perussia et al. 1992; Naume et al. 1992; Aste-Amezaga et al. 1994), granulocyte-macrophage colony-stimulating factor (GM-CSF; Naume et al. 1993b; Carson et al. 1994; Aste-Amezaga et al. 1994), IL-8 (Naume et al. 1993b; Valiante and Trinchieri 1993), macrophage CSF (Aste-Amezaga et al. 1994), and IL-3 (Naume et al. 1993b); in contrast to the high levels of IFN-γ induced by IL-12, the levels of these cytokines are generally low. In several studies, increased levels of cytokines have been observed by treatment with IL-12 and other stimuli (Naume et al. 1993b; Valiante and Trinchieri 1993; Carson et al. 1994; Aste-Amezaga et al. 1994). Mice injected with IL-12 have circulating serum levels of TNF-α (Orange et al. 1994) and mRNA for TNF-α (Morris et al. 1994; Orange et al. 1995) as well as a large increase in splenic mRNA and spleen cells secreting IL-10 (Morris et al. 1994). The in vivo induction of IL-10 following injection of IL-12 may be a regulatory response in the animal, since IL-10 can inhibit the production and biological effects of IL-12 (Tripp et al. 1993; Hsieh et al. 1993; D'Andrea et al. 1993).

4.4 Induction of T-Helper-1 and -2 Cells

T-helper (Th)1 and Th2 cells differ in their ability to secrete cytokines (Mosmann and Coffman 1989; Del Prete et al. 1994), with Th1 cells producing IFN-γ, IL-2, and TNF-β and Th2 cells producing IL-4, IL-5, IL-6, and IL-10; these differences in cytokine secretion patterns lead to distinct effector functions (Mosmann and Coffman 1989; Del Prete et al. 1994); Th1 cells are usually associated with cell-mediated immunity, while Th2 cells provide help for immunoglobulin production

(MOSMANN and COFFMAN 1993; DEL PRETE et al. 1994). IL-12 has the unique cytokine property of regulating the development of both murine (HSIEH et al. 1993; SEDER et al. 1993; SCHMITT et al. 1994a, 1994b; CROFT et al. 1994; MURPHY et al. 1994) and human (MANETTI et al. 1993, 1994) Th1/Th2 $CD4^+$ T cells. In contrast to IL-12, IL-4 induces the development of Th2 cells (DEL PRETE et al. 1994). Naive murine $CD4^+$ T cells primed with IL-12 during antigenic exposure in vitro produce the Th1 cytokine IFN-γ, while cells primed with IL-4 produce the Th2 cytokine IL-4 (HSIEH et al. 1993; SEDER et al. 1993; SCHMITT et al. 1994a, 1994b; CROFT et al. 1994); the effect of IL-4 dominates the effect produced by IL-12 (HSIEH et al. 1993; SEDER et al. 1993; SCHMITT et al. 1994a). The effects of IL-12 are mediated directly on T cells in these models (SEDER et al. 1993; SCHMITT et al. 1994a) and can be IFN-γ dependent (SCHMITT et al. 1994b) or independent (SEDER et al. 1993) in various model systems. IL-12 also has a number of direct effects on murine Th1, but not Th0 or Th2 clones (MURPHY et al. 1994; YANAGIDA et al. 1994; GERMANN et al. 1993; KENNEDY et al. 1994). Evidence in various in vivo situations has confirmed the importance of IL-12 in regulating Th1/Th2 responses (CLERICI et al. 1994; HEINZEL 1994; OSWALD et al. 1994; ROMANI et al. 1994).

IL-12 also induces the generation of human Th1 cells from antigen-stimulated cells from atopic individuals and anti-IL-12 shifts cells toward a Th2 phenotype (MANETTI et al. 1993). If IL-12 is added during the cloning of T cells, severalfold higher levels of IFN-γ are induced in both $CD4^+$ and $CD8^+$ T cells, but there is no effect on IL-4 production (MANETTI et al. 1994). In contrast to its lack of effect on murine Th2 clones, IL-12 can stimulate the transient expression of IFN-γ in Th2 clones (MANETTI et al 1994; YSSEL et al. 1994). The ability of IL-12 to induce the generation of Th1 cells may be important in the induction of cell-mediated immune responses to tumors.

5 Evaluation of Interleukin-12 in Murine Tumor Models

5.1 Antitumor and Antimetastatic Activities

Based on the immunomodulatory properties of IL-12, its antitumor activity was assessed in mice with established subcutaneous tumors. When administered systemically or intratumorally, IL-12 has been shown to exert potent antitumor effects against a number of tumors, including B16F10 melanoma, Renca renal cell carcinoma, M5076 reticulum cell sarcoma, MCA-105 and MCA-207 sarcomas, MC-38 colon adenocarcinoma, Lewis lung carcinoma, and several other tumor types (BRUNDA and GATELY 1994; HENDRZAK and BRUNDA 1995). Therapeutic efficacy of IL-12 was demonstrated even against well-established tumors that were approximately 1 cm in diameter (BRUNDA et al. 1993; NASTALA et al. 1994). In some tumor models, such as the B16F10 melanoma model, the administration of IL-12 resulted in dose-dependent inhibition of tumor growth and an increased length of survival, but complete regressions were not observed (Brunda et al. 1993). In other models, such as the Renca renal carcinoma, the MCA-105 and MCA-207 sarcoma, and the MC-38 colon adenocarcinoma models, complete regressions did occur after either peritumoral or systemic administration of IL-12, with mice being immune to rechallenge (Brunda et al. 1993; Nastala et al. 1994; Gately and Brunda 1995). IL-12 has also recently been shown to have antitumor effects not only in mice with large, established tumors, but also in mice with minimal, residual lymphoma after treatment with high-dose therapy and bone marrow transplantation (Verbik et al. 1995).

The antimetastatic effects of IL-12 were initially demonstrated against experimental pulmonary metastases of the B16F10 melanoma (Brunda et al. 1993). In this model, there was a dose-dependent inhibition of metastases when IL-12 was administered 1 day after i.v. injection of tumor cells and, importantly, the number of metastases was also significantly reduced when IL-12 was administered 7 days after tumor cell injection, at which time metastases were already established. Recently, it has been shown that lipid arabinomannan extracted from *Mycobacterium tuberculosis* (Z-100) has inhibitory activity on pulmonary metastases in mice bearing the B16F10 melanoma, and this antitumor activity may be mediated by the induction of IL-12 (Kobayashi et al. 1995). In addition, systemic IL-12 administration was shown to reduce experimental pulmonary metastases of the MC-38 adenocarcinoma when treatment was initiated 10 days after inoculation of tumor cells (Nastala et al. 1994). Treatment of mice with IL-12 also reduced the number of experimental hepatic metastases of the M5076 reticulum cell sarcoma, resulting in increased survival (Brunda et al. 1993). Finally, IL-12 has been shown to inhibit spontaneous pulmonary metastases of the Lewis lung carcinoma (Stern et al. 1993) and spontaneous hepatic metastases of the M5076 reticulum cell sarcoma (Gately et al. 1994a). In this hepatic metastasis model, IL-12 was effective when treatment was initiated 4 weeks after s.c. implantation of tumor cells.

Gene therapy models have recently been developed to study the effects of IL-12 secreted locally at the tumor site. Fibroblasts genetically engineered to secrete IL-12 were shown to suppress the growth of BL-6 melanoma cells simultaneously injected into the same site, with efficacy related to the amount of IL-12 expressed by the transfected fibroblasts (Tahara et al. 1994). In another report, mice were inoculated s.c. with IL-12-transduced C-26 murine colon carcinoma cells, and tumor onset was delayed as compared with control mice (Martinotti et al. 1995). Furthermore, the i.v. injection of these C-26/IL-12 cells resulted in a similar delay in the formation of lung and liver metastases and death of the mice. Thus, IL-12 has pronounced antitumor activity in a number of s.c. tumor and metastasis models, following administration by several routes.

5.2 Mechanism of Antitumor Activity

The mechanism by which IL-12 mediates its antitumor effects is not fully understood and may vary with the tumor type. Several studies have shown that IL-12 does not have any direct antiproliferative effects on murine tumor cell lines in vitro (O'TOOLE et al. 1993; TAHARA et al. 1994; BRUNDA et al. 1995a). These data are consistent with the possibility that IL-12 mediates its activity primarily through the immune system. Analysis of the antitumor mechanism of IL-12 in murine tumor models has confirmed that its biological activities on immune effector cells are clearly involved.

Evidence to date indicates that T cells may be the critical immune effector cells involved in mediating the antitumor effects of IL-12. In both the B16F10 and Renca models, the antitumor efficacy of IL-12 was substantially reduced in T cell-deficient nude mice (BRUNDA et al. 1993). In addition, as mentioned previously, mice cured of tumors by treatment with IL-12 were specifically immune to subsequent rechallenge with the same tumor, but not with other tumors (BRUNDA et al. 1993; NASTALA et al. 1994; GATELY and BRUNDA 1995) Furthermore, in gene therapy immunization models, locally secreted IL-12 at the tumor site was found to induce protective immunity against a subsequent tumor challenge. For example, when irradiated BL-6 tumor cells mixed with IL-12-secreting fibroblasts were injected as a vaccine and mice were challenged with nonirradiated BL-6 cells, tumor establishment was delayed (TAHARA et al. 1994). In addition, it has recently been reported that mice immunized with transfectants expressing B7-1 and secreting IL-12 were protected against a second tumor challenge, and the authors hypothesized that the induction of both cytotoxic T cells and a memory Th1-type response were involved (CHEN et al. 1995). However, it should be noted that IL-12 was an effective antitumor agent in T and B cell-deficient severe combined immunodeficient (SCID) mice bearing the X5563 B cell lymphoma, suggesting that T cells are not always essential for the antitumor activity of IL-12 (O'TOOLE et al. 1993).

The roles of $CD4^+$ and $CD8^+$ cell subsets have also been examined. Using monoclonal antibodies to deplete individual T cell subsets in tumor-bearing mice treated with IL-12, it was shown that $CD8^+$ T cells were required to inhibit the growth of the Renca tumor (BRUNDA et al. 1993), but either $CD4^+$ or $CD8^+$ T cells were sufficient to inhibit the growth of the MCA-207 tumor (NASTALA et al. 1994). Interestingly, $CD8^+$ T cell depletion in mice injected with C-26 colon carcinoma cells transduced with IL-12 genes had no effect on the time of tumor onset or survival, whereas depletion of $CD4^+$ T cells led to delayed tumor appearance and tumor regression in 40% of the mice (MARTINOTTI et al. 1995). Examination of tumor cell infiltrates showed very few T cells in nondepleted mice, but abundant $CD8^+$ T cell infiltration in mice depleted of $CD4^+$ T cells. Further analysis revealed that $CD4^+$ T cells did not inhibit $CD8^+$ T cell priming in vivo, and these primed T cells inhibited the formation of C-26/IL-12 tumors. The authors concluded that in the non-$CD4^+$ T cell-depleted mice, C-26/IL-12 cells activate $CD8^+$ T cells, but $CD4^+$ T cells inhibit $CD8^+$ T cell infiltration into the tumor, which may be essential for tumor regression. It should be noted, however, that an increased number of NK cells also infiltrated the tumors in mice depleted of CD4⁺ T cells. Additional support for the role of tumor-infiltrating CD8⁺ T cells in the antitumor activity of IL-12 was demonstrated in the MCA-207 sarcoma model (NASTALA et al. 1994). Tumor-bearing mice were treated intratumorally with IL-12, and the regressing tumors were found to be characterized by a predominantly CD8⁺ T cell infiltrate. In addition, a decreased number of CD4⁺ T cells and accumulation of macrophages was observed within the BL-6 tumors of mice previously vaccinated with irradiated tumor cells and IL-12-secreting fibroblasts to induce antitumor immunity (TAHARA et al. 1994). The exact role of the T cell populations involved in mediating the antitumor effects of IL-12 requires further investigation.

NK cells appear to play a role in the antitumor activity of IL-12 against some tumors but not others. Depletion of NK cells with anti-asialo GM1 antibody resulted in a significant reduction in time of tumor onset in mice injected s.c. with C-26 colon carcinoma cells transduced with IL-12 genes and reduced survival in mice with C-26/IL-12 tumor metastases (MARTINOTTI et al. 1995), Moreover, IL-12 displayed antitumor activity against the X5563 B cell lymphoma in SCID mice, suggesting that NK cells may be involved in the activity of IL-12 (O'TOOLE et al. 1993). In contrast, the efficacy of IL-12 was maintained in NK-deficient beige mice bearing the B16F10 melanoma and in Renca or MCA-207 tumor-bearing mice depleted of NK cells by treatment with the anti-asialo GM1 antibody (BRUNDA et al. 1993; NASTALA et al. 1994). Thus the importance of NK cells in the antitumor mechanism of IL-12 remains controversial and most probably varies with the tumor model studied.

A more recent area of investigation is the role of macrophages in mediating the antitumor effects of IL-12. Macrophages were found to infiltrate Renca (BRUNDA and GATELY 1994; TANNENBAUM et al. 1995) and BL-6 melanoma (TAHARA et al. 1994) tumors in IL-12-treated mice and were possibly recruited by expression of the chemokine inflammatory protein (IP)-10 within the tumor (TANNENBAUM et al. 1995). In addition, IL-12 did not directly activate the tumoricidal function of mouse peritoneal macrophages in vitro, but cytolytic peritoneal macrophages were obtained from mice injected with IL-12 (J.A. HENDRZAK et al. submitted). This macrophage activity is mediated in part by the induction of IFN-γ, but other cytokines may also be involved. One potential mediator of macrophage tumoricidal activity is nitric oxide (STUEHR and NATHAN 1989); IL-12 treatment induced nitric oxide production in vivo (NASTALA et al. 1994), and IFN-γ in combination with other cytokines can promote nitric oxide production by macrophages (COX et al. 1992).

In addition to augmenting the cytolytic activity of T cells, NK cells, and possibly macrophages, IL-12 induces high levels of IFN-γ, which can mediate a variety of antitumor effects. The levels of IFN-γ induced by IL-12 in tumor-bearing mice were similar to those in normal mice (BRUNDA et al. 1995a). IFN-γ may be produced by T cells, NK cells, or both cell types in IL-12-treated tumor-bearing mice, and, although IFN-γ production suggests that a Th1 response may be involved, the induction of other Th1 cytokines has not been addressed. Treatment of mice with a neutralizing anti-IFN-γ antibody significantly reduced the antitumor activity of IL-12 in the Renca (Brunda et al. 1995a), MCA-38 (Nastala et al. 1994), and B16F10 (M. Brunda, unpublished observations) tumor models. However, the administration of IL-12 to normal or tumor-bearing nude mice resulted in much higher serum IFN-γ levels than in euthymic controls, yet IL-12 displayed little antitumor activity in nude mice (Brunda et al. 1993, 1995a). Since it was also shown that the administration of IFN-γ to Renca tumor-bearing mice did not result in antitumor effects comparable to those achieved with IL-12, it is apparent that, while necessary for the antitumor action of IL-12, IFN-γ alone is not sufficient (Brunda et al. 1995a). IFN-γ may exert its effects by upregulating major histocompatibility complex (MHC) class I and II molecules (Wong et al. 1983), activating macrophages (Varesio et al. 1984), and generating $CD8^+$ cytolytic T cells (Giovarelli et al. 1988). It is therefore possible that T cells or NK cells infiltrating tumors in IL-12-treated mice secrete IFN-γ, which in turn enhances the functions of macrophages or T cells present at the tumor site. Recently, it has been reported that IFN-γ can also mediate the nonimmunogenic, antiangiogenic effect of IL-12, which was involved in the inhibition of basic fibroblast growth factor-induced corneal neovascularization in mice (Voest et al. 1995).

A variety of other cytokines have been reported to be upregulated following injection of IL-12 into normal mice (Sect. 3.3; Hendrzak and Brunda 1995). IL-12 induced low levels of serum TNF-α in mice, but administration of the neutralizing anti-TNF-α antibody to mice bearing the MC-38 colon carcinoma did not decrease the antitumor efficacy of IL-12 (Nastala et al. 1994). However, in this report, it was also shown that the depletion of TNF-α and IFN-γ together partially decreased the antitumor effect of IL-12 on MC-38 pulmonary metastases. Additional work is necessary to examine the role of other cytokines in IL-12-mediated antitumor effects.

Further investigation using murine models in which IL-12 is delivered locally to the tumor site may provide a clearer understanding of the mechanism by which IL-12 inhibits tumor growth. As discussed above, the local delivery of IL-12 by peritumoral administration or by gene therapy can lead to the development of antitumor immunity. In addition, this form of therapy often results in tumor infiltration by leukocytes, especially $CD8^+$ T cells, and recent evidence suggests that this may enhance the antitumor activity of IL-12 (Martinotti et al. 1995). Finally, the production of IFN-γ and other cytokines induced by IL-12 could be measured at the tumor site, and the antitumor effects of any cytokines detected could then be examined.

5.3 Additional Clinical Issues Addressed

Several issues associated with the clinical use of IL-12, such as treatment-related toxicities, have been addressed using murine tumor models. Fortunately, few toxicities have been associated with IL-12 treatment. When tumor-bearing mice were treated with a regimen associated with antitumor efficacy (1 μg/day, 5 days per week), mice developed a mild leukopenia, a mild anemia, and moderately elevated levels of hepatic transaminases, which correlated with mononuclear cell infiltrates in the liver (M. Brunda and T. Anderson, unpublished results). Minimal apparent toxicity has similarly been demonstrated at effective doses of IL-12 in murine adenocarcinoma and sarcoma models (Nastala et al. 1994). More extensive studies performed in normal mice and monkeys have revealed some additional toxicities at high doses (Gately et al. 1994a; Brunda and Gately 1995; Hendrzak and Brunda 1995) but there were no major toxicities to prevent the use of IL-12 for the treatment of human malignancies.

Several studies have also been performed to compare the antitumor efficacy of IL-12 with that of other cytokines currently in use in cancer therapy. At maximum tolerated doses, IL-12 was more effective than IL-2 against the B16F10 melanoma and the Lewis lung carcinoma and comparable to IL-2 against the colon 38 and M5076 reticulum cell sarcoma (Stern et al. 1993; Brunda et al. 1995b; M. Brunda and L. Luistro, unpublished results). IL-12 was active over a broader range of doses than IL-2 in the Renca model (Brunda et al. 1995b), while high doses of IL-12 or IL-2 were equally effective against MC-38 metastases, but IL-2 treatment resulted in systemic toxicities (Nastala et al. 1994). In addition, IL-12 exhibited greater antitumor activity than IFN-α against both the B16F10 melanoma and the Renca renal tumors (Brunda et al. 1995b).

Another issue relevant to the clinical use of IL-12 that has been addressed is the efficacy of IL-12 in combination with cytokines or other anticancer agents. Suboptimal doses of IL-12 and either IL-2 or IFN-α resulted in additive or synergistic effects in tumor-bearing mice, but these effects were not better than those achieved by IL-12 alone (Brunda et al. 1995b), and higher doses of these cytokines resulted in enhanced toxicities. Furthermore, the combination of low doses of IL-12 and IL-2 was ineffective in reducing MC-38 pulmonary metastases, and no synergy was observed at higher doses of these cytokines (Nastala et al. 1994). IL-12 may be used more effectively in combination with other cancer treatments, such as cellular therapies. Human killer T cells transferred in combination with IL-12 increased the survival of SCID mice bearing acute myelogenous leukemia (Cesano et al. 1994), and IL-12 enhanced the antitumor activity of human NK cells in combination with IL-2 in human melanoma-bearing SCID mice (Hill et al. 1994). Recently, it has also been shown that IL-12 has the ability to enhance the antitumor activity of the immunomodulator bacille Calmette-Guerin (BCG; Hunter-Mayer et al. 1995) as well as various chemotherapeutic agents and radiation therapy (Emi et al. 1995). Further testing must be performed, however, to determine whether such combinations will prove to be too toxic.

In addition to the data obtained from murine tumor models, antitumor effects of human IL-12 have been observed after in vitro incubation with immune effector cells obtained from cancer patients. IL-12 was found to restore defective NK activity of peripheral blood mononuclear cells (PBMC) from patients with metastatic soild tumors (Soiffer et al. 1993) as well as stimulate NK cell activity of hairy cell leukemia patients (Bigda et al. 1993). In addition, IL-12 was recently shown to reverse aberrant cytokine production by PBMC from patients with Sezary syndrome, an advanced form of cutaneous T cell lymphoma (ROOK et al. 1995). PBMC from cancer patients produced significantly less IL-12 and IFN-γ when stimulated in vitro than PBMC from normal controls, but treatment of patients' PBMC with IL-12 resulted in increased production of IFN-γ, decreased production of IL-4, and increased NK activity (ROCK et al. 1995). Finally, a possible use for IL-12 in the treatment of cancer patients with minimum residual disease has been suggested by its ability to enhance the cytolytic function of PBMC from patients with hematologic malignancies after allogeneic bone marrow transplant (SOIFFER et al. 1993) and by its ability, in combination with IL-2, to enhance the NK activity of cord blood cells, which could be used for transplantation (CHANG et al. 1995).

In summary, IL-12 is a cytokine with unique immunomodulatory properties which mediate potent antitumor and antimetastatic effects in a number of murine tumor models. Both T cells and NK cells have been shown to be involved in the seemingly complex antitumor mechanism of IL-12; therefore, it is likely that IL-12 stimulates the proliferation and cytolytic activity of these cells in tumor-bearing mice. In addition, IL-12 may enhance the cytolytic activity of macrophages in vivo via an indirect mechanism. IL-12 also stimulates the induction of cytokines, especially IFN-γ, in tumor-bearing mice, and IFN-γ has been shown to be necessary but not sufficient for the antitumor response. The activation of T cells, together with the induction of IFN-γ, suggests that IL-12 may induce a Th1-type response in at least some tumor models. Furthermore, some of these biological effects of IL-12 have also been observed when IL-12 was incubated in vitro with immune cells from cancer patients. Since IL-12 therapy is associated with few toxicities, may be more effective than cytokines currently in use, and may be used in combination with other anticancer agents, it is hoped that the efficacy of IL-12 observed in murine tumor models will some day translate to success in the clinical setting.

References

Allavena P, Paganin C, Zhou D, Bianchi G, Sozzani S, Mantovani A (1994) Interleukin-12 is chemotactic for natural killer cells and stimulates their interaction with vascular endothelium. Blood 84: 2261–2268

Alzona M, Jack H-M, Fisher RI, Ellis TM (1995) IL-12 activates IFN-γ production through the preferential activation of $CD30^+$ T cells. J Immunol 154: 9–16

Andrew JVR, Schoof DD, Bertagnolli MM, Peoples GE, Goedegebuure PS, Eberlein TJ (1993) Immunomodulatory effects of interleukin–12 on human tumor-infiltrating lymphocytes. J Immunother 14: 1–10

Aragone Y, Riemann H, Bhardwaj RS, Schwarz A, Sawada Y, Yamada H, Luger TA, Kubin M, Trinchieri G, Schwarz T (1994) IL-12 is expressed and released by human keratinocytes and epidermoid carcinoma cell lines. J Immunol 153: 5366–5372

Aste-Amezaga M, D'Andrea A, Kubin M, Trinchieri G (1994) Cooperation of natural killer cell stimulatory factor/interleukin-12 with other stimuli in the induction of cytokines and cytotoxic cell-associated molecules in human T and NK cells. Cell Immunol 156: 480–492

Bacon CM, McVicar DW, Ortaldo JR, Rees RC, O'Shea JJ, Johnston JA (1995) Interleukin 12 (IL-12) induces tyrosine phosphorylation of JAK2 and TYK2: differential use of Janus family tyrosine kinases by IL-2 and IL-12. J Exp Med 181: 399–404

Bigda J, Mysliwska J, Dziadziuszko R, Bigda J, Mysliwski A, Hellmann A (1993) Interleukin 12 augments natural killer-cell mediated cytotoxicity in hairy cell leukemia. Leuk Lymphoma 10: 121–125

Bloom ET, Horvath JA (1994) Cellular and molecular mechanisms of the IL-12- induced increase in allospecific murine cytolytic T cell activity. Implications for the age-related decline in CTL. J Immunol 152: 4242–4254

Bonnema JD, Rivlin KA, Ting AT, Schoon RA, Abraham RT, Leibson RJ (1994) Cytokine-enhanced NK cell-mediated cytotoxicity. Positive modulatory effects of IL-2 and IL-12 on stimulus-dependent granule exocytosis. J Immunol 152: 1883–1887

Brunda MJ (1994) Interleukin-12. J Leukoc Biol 55: 280–288

Brunda MJ, Gately MK (1994) Anti-tumor activity of interleukin-12. Clin Immunol Immunopathol 71: 253–255

Brunda MJ, Gately MK (1995) Interleukin-12: potential role in cancer therapy. In: De Vita VT, Hellman S, Rosenberg SA (eds) Important advances in oncology. Lippincott, Philadelphia, pp 3–18

Brunda MJ, Luistro L, Warrier RR, Wright RB, Hubbard BR, Murphy M, Wolf SF, Gately MK (1993) Antitumor and antimetastatic activity or interleukin 12 against murine tumors. J Exp Med 178: 1223–1230

Brunda MJ, Luistro L, Hendrzak JA, Fountoulakis M, Garotta G, Gately MK (1995a) Role of interferon-γ in mediating the antitumor efficacy of interleukin-12. J Immunother 17: 71–77

Brunda MJ, Luistro L, Warrier RR, Gately MK (1995b) Antitumor activity of interleukin-12 compared to and combined with other cytokines. Ninth International Congress of Immunology, July 23–29, 1995, San Francisco, CA, USA

Carson WE, Giri JG, Lindermann MJ, Linett ML, Ahdieh M, Paxton R, Anderson D, Eisenmann J, Grabstein K, Caligiuri MA (1994) Interleukin (IL) 15 is a novel cytokine that activates human natural killer cells via components of the IL-2 receptor. J Exp Med 180: 1395–1403

Cassatella MA, Meda L, Gasperini S, D'Andrea A, Ma X, Trinchieri G (1995) Interleukin-12 production by human polymorphonuclear leukocytes. Eur J Immunol 25: 1–5

Cesano A, Visonneau S, Clark SC, Santoli D (1993) Cellular and molecular mechanisms of activation of MHC nonrestricted cytotoxic cells by IL-12. J Immunol 151: 2943–2957

Cesano A, Visonneau S, Cioe L, Clark SC, Rovera G, Santoli D (1994) Reversal of acute myelogenous leukemia in humanized SCID mice using a novel adoptive transfer approach. J Clin Invest 94: 1076–1084

Chan SH, Perussia B, Gupta JW, Kobayashi M, Young D, Pospisil M, Young HA, Wolf SF, Yound D, Clark SC, Trinchieri G (1991) Induction of interferon γ production by natural killer cell stimulatory factor: characterization of the responding cells and synergy with other inducers. J Exp Med 173: 869–879

Chan SH, Kobayashi M, Santoli D, Perussia B, Trinchieri G (1992) Mechanisms of IFN-γ induction by natural killer cell stimulatory factor (NKSF/IL-12). Role of transcription and mRNA stability in the synergistic interaction between NKSF and IL-2. J Immunol 148: 92–98

Chang L, Bruner V, van de Ven C, Cairo MS (1995) Synergistic enhancement of natural killer (NK) cell activity with cord blood cells by IL-2⁺IL-12: a future immunotherapy approach post cord BMT. Proc Am Assoc Cancer Res 36: 466

Chehimi J, Starr SE, Frank I, Rengaraju M, Jackson SJ, Llanes C, Kobayashi M, Perussia B, Young D, Nickbarg E, Wolf SF, Trinchieri G (1992) Natural killer (NK) cell stimulatory factor increases the cytotoxic activity of NK cells from both healthy donors and human immunodeficiency virus-infected patients. J Exp Med 175: 789–796

Chehimi J, Valiante NM, D'Andrea A, Rengaraju M, Rosado Z, Kobayashi M, Perussia B, Wolf SF, Starr SE, Trinchieri G (1993) Enhancing effect of natural killer cell stimulatory factor (NKSF/interleukin-12) on cell-mediated cytotoxicity against tumor-derived and virus-infected cells. Eur J Immunol 23: 1826–1830

Chen PW, Geer DC, Podack ER, Ksander BK (1995) Tumor cells transfected with B7–1 + IL-12 cDNa induce protective immunity. Proc Am Assoc Cancer Res 36: 247

Chizzonite R, Truitt T, Desai BB, Nunes P, Podlaski FJ, Stern AS, Gately MK (1992) IL-12 receptor. I. Characterization of the receptor on phytohemagglutinin-activated human lymphoblasts. J Immunol 148: 3117–3124

Chua AO, Chizzonite R, Desai BB, Truitt TP, Nunes P, Minetti LJ, Warrier RR, Presky DH, Levine JF, Gately MK, Gubler U (1994) Expression cloning of a human IL-12 receptor component. A new member of the cytokine receptor superfamily with strong homology to gp130. J Immunol 153: 128–136

Clerici M, Sarin A, Coffman RL, Wynn TA, Blatt SP, Hendrix CW, Wolf SF, Shearer GM, Henkart PA (1994) Type 1/type 2 cytokine modulation of T-cell programmed cell death as a model for human immunodeficiency virus pathogenesis. Proc Natl Acad Sci USA 91: 11811–11815

Cox GW, Melillo G, Chattopadhyay U, Mullet D, Fertel RHL Varesio L (1992) Tumor necrosis factor-α-dependent production of reactive nitrogen intermediates mediates IFN-γ plus IL-2-induced murine macrophage tumoricidal activity. J Immunol 149: 3290–3296

Croft M, Carter L, Swain SL, Dutton RW (1994) Generation of polarized antigen-specific populations: reciprocal action of interleukin (IL)-4 and IL-12 in promoting type 2 versus type 1 cytokine profiles. J Exp Med 180: 1715–1728

Del Prete G, Maggi E, Romagnani S (1994) Human Th1 and Th2 cells: functional properties, mechanisms of regulation, and role in disease. Lab Invest 70: 299–306

Desai BB, Quinn PM, Wolitzky AG, Mongini PKA, Chizzonite R, Gately MK (1992) IL-12 receptor. II. Distribution and regulation of receptor expression. J Immunol 148: 3125–3132

D'Andrea A, Rengaraju M, Valiante NM, Chehimi J, Kubin M, Aste M, Chan SH, Kobayashi M, Young D, Nickbarg E, Chizzonite R, Wolf SF, Trinchieri G (1992) Production of natural killer cell stimulatory factor (interleukin 12) by peripheral blood mononuclear cells. J Exp Med 176: 1387–1398

D'Andrea A, Aste-Amezaga M, Valiante NM, Ma X, Kubin M, Trinchieri G (1993) Interleukin 10 (IL-10) inhibits human lymphocyte interferon γ-production by suppressing natural killer cell stimulatory factory/IL-12 synthesis in accessory cells. J Exp Med 178: 1041–1048

D'Andrea A, Ma X, Aste-Amezaga M, Paganin C, Trinchieri G (1995) Stimulatory and inhibitory effects of interleukin (IL)-4 and IL-13 on the production of cytokines by human peripheral blood mononuclear cells: priming for IL-12 and tumor necrosis α production. J Exp Med 181: 537–546

Emi Y, Ara G, Holden SA, Clinton S, Teicher BA (1995) Interleukin-12 (IL-12) enhances antitumor activity of chemotherapy and radiotherapy in murine models. Proc Am Assoc Cancer Res 36: 480

Gately MK, Brunda MJ (1995) Interleukin-12: a pivotal regulator of cell-mediated immunity. In: Kurzrock R, Talpaz M (eds) Cytokines: interleukins and their receptors. Kluwer, Norwell (in press)

Gately MK, Desai BB, Wolitzky AG, Quinn PM, Dwyer CM, Podlaski FJ, Familletti PC, Sinigaglia F, Chizzonite R, Gubler U, Stern AS (1991) Regulation of human lymphocyte proliferation by a heterodimeric cytokine, IL-12 (cytotoxic lymphocyte maturation factor). J Immunol 147: 874–882

Gately MK, Wolitzky AG, Quinn PM, Chizzonite R (1992) Regulation of human cytolytic lymphocyte responses by interleukin-12. Cell Immunol 143: 127–142

Gately MK, Gubler U, Brunda MJ, Nadeau RR, Anderson TD, Lipman JM, Sarmiento U (1994a) Interleukin-12: a cytokine with therapeutic potential in oncology and infectious diseases. Ther Immunol 1: 187–196

Gately MK, Warrier RR, Honasoge S, Carvajal DM, Faherty DA, Connaughton SE, Anderson TD, Sarmiento U, Hubbard BR, Murphy M (1994b) Administration of recombinant IL-12 to normal mice enhances cytolytic lymphocyte activity and induces production of IFN-γ in vivo. Int Immunol 6: 157–167

Gazzinelli RT, Hieny S, Wynn TA, Wolf S, Sher A (1993) Interleukin 12 is required for the T-lymphocyte-independent induction of interferonγ by an intracellular parasite and induces resistance in T-cell-deficient hosts. Proc Natl Acad Sci USA 90: 6115–6119

Gearing DP, Cosman D (1991) Homology of the p40 subunit of natural killer cell stimulatory factor (NKSF) with the extracellular domain of the interleukin-6 receptor. Cell 66: 9–10

Germann T, Gately MK, Schoenhaut DS, Lohoff M, Mattner F, Fischer S, Jin S-C, Schmitt E, Rude E (1993) Interleukin-12/T cell stimulating factor, a cytokine with multiple effects on T helper type (TH1) but not Th2 cells. Eur J Immunol 23: 1762–1770

Gerosa F, Tommasi M, Benati C, Gandini G, Libonati M, Tridente G, Carra G, Trinchieri G (1993) Differential effects of tyrosine kinase inhibition in CD69 antigen expression and lytic activity induced by rIL-2, rIL-12 and rIIFN-α in human NK cells. Cell Immunol 150: 382–390

Giovarelli M, Santoni A, Jemma C, Musso T, Giuffrida AM, Cavallo G, Landolfo S, Forni G (1988) Obligatory role of IFN-γ in induction of lymphokine-activated and T lymphocyte killer activity, but not in boosting of natural cytotoxicity. J Immunol 141: 2831–2836

Godfrey DI, Kennedy J, Gately MK, Hakimi J, Hubbard BR, Zlotnik A (1994) IL-12 influences intrathymic T cell development. J Immunol 152: 2729–2735

Gubler U, Chua AO, Schoenhaut DS, Dwyer CM, McComas W, Motyka R, Nabavi N, Wolitzky AG, Quinn PM, Familletti PC, Gately MK (1991) Coexpression of two distinct genes is required to generate secreted, bioactive cytotoxic lymphocyte maturation factor. Proc Natl Acad Sci USA 88: 4143–4417

Heinzel FP (1994) Interleukin 12 and the regulation of $CD4^+$ T-cell subset responses during murine leishmaniasis. Parasitol Today 10: 190–192

Heinzel FP, Berko RM, Ling P, Hakimi J, Schoenhaut DS (1994) Interleukin 12 is produced in vivo during endotoxemia and stimulates synthesis of gamma interferon. Infect Immun 62: 4244–4249

Hendrzak JA, Brunda MJ (1995) Interleukin-12: biological activity, therapeutic utility and role in disease. Lab Invest 72: 619–637

Hill LL, Perussia B, McCue PA, Korngold R (1994) Effect of human natural killer cells on the metastatic growth of human melanoma Xenografts in mice with severe combined immunodeficiency. Cancer Res 54: 763–770

Hsieh C-S, Macatonia SE, Tripp CS, Wolf SF, O'Garra A, Murphy KM (1993) Development of Th1 CD4 T cells through IL-12 produced by Listeria-induced macrophages. Science 260: 547–549

Hunter-Mayer SE, O'Donnell M, Szilvasi A, Leonard J, Clinton SK (1995) Combination interleukin-12 and bacille Calmette-Guerin immune therapy of bladder carcinoma in mice. Proc Am Assoc Cancer Res 36: 480

Jewett A, Bonavida B (1994) Activation of the human immature natural killer cell subset by IL-12 and its regulation by endogenous TNF-α and IFN-γ secretion. Cell Immunol 154: 273–286

Kanangat S, Nair S, Babu JS, Rouse BT (1995) Expression of cytokine mRNA in murine splenic dendritic cells and better induction of T-cell-derived cytokines by dendritic cells than by macrophages during in vitro costimulation assay using specific antigens. J Leukoc Biol 57: 310–316

Kennedy MK, Picha KS, Shanebeck KD, Anderson DM, Grabstein KH (1994) Interleukin-12 regulates the proliferation of Th1, but not Th2 or Th0, clones. Eur J Immunol 24: 2271–2278

Kobayashi M, Fitz L, Ryan M, Hewick RM, Clark SC, Chan S, Loudon R, Sherman F, Perussia B, Trinchieri G (1989) Identification and purification of natural killer cell stimulatory factor (NKSF), a cytokine with multiple biological effects on human lymphocytes. J Exp Med 170: 827–845

Kobayashi M, Nakajima N, Pollard RB, Suzuki F (1995) Tumor immunotherapeutic lipid-arabinomannan extracted from *Mycobacterium tuberculosis* (Z-100) induces the production of IL-12. Proc Am Assoc Cancer Res 36: 467

Kubin M, Chow JM, Trinchieri G (1994a) Differential regulation of interleukin-12 (IL-12), tumor necrosis factor α, and IL-1β production in human myeloid leukemia cell lines and peripheral blood mononuclear cells. Blood 83: 1847–1855

Kubin M, Kamoun M, Trinchieri G (1994b) Interleukin 12 synergizes with B7/CD28 interaction in inducing efficient proliferation and cytokine production of human T cells. J Exp Med 180: 211–222

Ma Y, Seiler KP, Tai KF, Yang L, Woods M, Weis JJ (1994) Outer surface lipoproteins of *Borrelia burgdorferi* stimulate nitric oxide production by the cytokine-inducible pathway. Infect Immun 62: 3663–3671

Manetti R, Parronchi P, Giudizi MG, Piccinni M-P, Maggi E, Trinchieri G, Romagnani S (1993) Natural killer cell stimulatory (interleukin 12[IL-12]) induces T helper type 1 (Th1)-specific immune responses and inhibits the development of IL-4-producing Th cells. J Exp Med 177: 1199–1204

Manetti R, Gerosa F, Giudizi MG, Biagiotti R, Parronchi P, Piccinni M-P, Sampognaro S, Maggi E, Romagnani S, Trinchieri G (1994) Interleukin 12 induces stable priming for interferon γ (IFN-γ) production during differentiation of human T helper (Th) cells and transient IFN-γ production in established Th2 cell clones. J Exp Med 179: 1273–1283

Martinotti A, Stoppacciaro A, Vagliani M, Melani C, Spreafico F, Wysocka M, Parmiani G, Trinchieri G, Colombo MP (1995) CD4 T cells inhibit in vivo the CD8-mediated immune response against murine colon carcinoma cells transduced with interleukin-12 genes. Eur J Immunol 25: 137–146

Mehrotra PT, Wu D, Crim JA, Mostowski HS, Siegel JP (1993) Effects of IL-12 on the generation of cytotoxic activity in human $CD8^+$ T lymphocytes. J Immunol 151: 2444–2452

Mengel J, Dare L, Dare GM, Delgado M, Nomizo A, Silva JS, Campos-Neto A (1992) An activated murine B cell lymphoma line (A-20) produces a factor-like activity which is functionally related to human natural killer cell stimulatory factor. Eur J Immunol 22: 3173–3178

Merberg DM, Wolf SF, Clark SC (1992) Sequence similarity between NKSF and the IL-6/G-CSF family. Immunol Today 13: 77–78

Morris SC, Madden KB, Adamovicz JJ, Gause WC, Hubbard BR, Gately MK, Finkelman FD (1994) Effects of IL-12 on in vivo cytokine gene expression and Ig isotype selection. J Immunol 152: 1047–1056

Mosmann TR, Coffman RL (1989) Heterogeneity of cytokine secretion patterns and functions of helper T cells. Adv Immunol 46: 111–147

Muller G, Saloga J, Germann T, Bellinghausen I, Mohamadzadeh M, Knop J, Enk AH (1994) Identification and induction of human keratinocyte-derived IL-12. J Clin Invest 94: 1799–1805

Murphy EE, Terres G, Macatonia SE, Hsieh C-S, Mattson J, Lanier L, Wysocka M, Trinchieri G, Murphy K, O'Garra A (1994) B7 and interleukin 12 cooperate for proliferation and interferon γ production by mouse T helper clones that are unresponsive to B7 costimulation. J Exp Med 180: 223–231

Nabioullin R, Sone S, Nii A, Haku T, Ogura T (1994) Induction mechanism of human blood CD8⁺ T cell proliferation and cytotoxicity by natural killer cell stimulatory factor (interleukin-12). Jpn J Cancer Res 85: 853–861

Nagy E, Buhlmann JE, Henies T, Waugh M, Rigby WFC (1994) Selective modulation of IFN-γ mRNa stability by IL-12/NKSF. Cell Immunol 159: 140–151

Nakarai T, Robertson MJ, Streuli M, Wu Z, Ciardelli TL, Smith KA, Ritz J (1994) Interleukin 2 receptor gamma chain expression on resting and activated lymphoid cells. J Exp Med 180: 241–251

Nastala CL, Edington HD, McKinney TG, Tahara H, Nalesnik MA, Brunda MJ, Gately MK, Wolf SF, Schreiber RD, Storkus WJ, Lotze MT (1994) Recombinant IL-12 administration induces tumor regression in association with IFN-γ production. J Immunol 153: 1697–1706

Naume B, Gately MK, Espevik T (1992) A comparative study of IL-12 (cytotoxic lymphocyte maturation factor): IL-2, and IL-7-induced effects on immunomagnetically purified CD56⁺ NK cells. J Immunol 148: 2429–2436

Naume B, Gately MK, Desai BB, Sundan A, Espevik T (1993a) Synergistic effects of interleukin 4 and interleukin 12 on NK cell proliferation. Cytokine 5: 38–46

Naume B, Johnsen A-C, Espevik T, Sundan A (1993b) Gene expression and secretion of cytokines and cytokine receptors from highly purified CD56⁺ natural killer cells stimulated with interleukin-2, interleukin-7 and interleukin-12. Eur J Immunol 23: 1831–1838

Orange JS, Wolf SF, Biron CA (1994) Effects of IL-12 on the response and susceptibility to experimental viral infections. J Immunol 152: 1253–1264

Orange JS, Salazar-Mather TP, Opal SM, Spencer RL, Miller AH, McEwen BS, Biron CA (1995) Mechanism of interleukin 12-mediated toxicities during experimental viral infections: role of tumor necrosis factor and glucocorticoids. J Exp Med 181: 901–914

Ortaldo JR, Mason AT, O'Shea JJ (1995) Receptor-induced death in human natural killer cells: involvement of CD16. J Exp Med 181: 339–344

Oswald IP, Caspar P, Jankovic D, Wynn TA, Pearce EJ, Sher A (1994) IL-12 inhibits Th2 cytokine responses induced by eggs of *Schistosoma mansoni*. J Immunol 153: 1707–1713

O'Toole M, Wolf SF, O'Brien C, Hubbard B, Herrmann S (1993) Effect of in vivo IL-12 administration on murine tumor cell growth. J Immunol 150: 294A

Perussia B, Chan SH, D'Andrea A, Tsuji K, Santoli D, Pospisil M, Young D, Wolf SF, Trinchieri G (1992) Natural killer (NK) cell stimulatory factor or IL-12 has differential effects on proliferation of TCR-αβ⁺, TCR-γδ⁺ T lymphocytes, and NK cells. J Immunol 149: 3495–3502

Pignata C, Prasad KVS, Robertson MJ, Levine H, Rudd CE, Ritz J (1993) Fcγ IIIA-mediated signaling involves src-family lck in human natural killer cells. J Immunol 151: 6794–6800

Pignata C, Sanghera J, Cossette L, Pelech SL, Ritz J (1994) Interleukin-12 induces tyrosine phosphorylation and activation of 44-kD mitogen-activated protein kinase in human T cells. Blood 83: 184–190

Podlaski FJ, Nanduri V, Hulmes JD, Pan Y-CE, Levin W, Danho W, Chizzonite R, Gately MK, Stern AS (1992) Molecular characterization of interleukin 12. Arch Biochem Biophys 294: 230–237

Rabinowich H, Herberman RB, Whiteside TL (1993) Differential effects of IL12 and IL2 on expression and function of cellular adhesion molecules on purified human natural killer cells. Cell Immunol 152: 481–498

Radrizzani M, Accornero P, Amidei A, Aiello A, Delia D, Kurrle R, Colombo MP (1995) IL-12 inhibits apoptosis induced in human Th1 clone by gp 120/CD4 cross-linking and CD3/TCR activation or by IL-2 deprivation. Cell Immunol 161: 14–21

Reiner SL, Zheng S, Wang Z-E, Stowring L, Locksley RM (1994) Leishmania promastigotes evade interleukin 12 (IL-12) induction by macrophages and stimulate a broad range of cytokines from CD4 T cells during initiation of infection. J Exp Med 179: 447–456

Robertson MJ, Soiffer RJ, Wolf SF, Manley TJ, Donahue C, Young D, Herrmann SH, Ritz J (1992) Responses of human natural killer (NK) cell stimulatory factor (NKSF):cytolytic activity and proliferation of NK cells are differentially regulated by NKSF. J Exp Med 175: 779–788

Romani L, Mencacci A, Tonnetti L, Spaccapelo R, Cenci E, Puccetti P, Wolf SF, Bistoni F (1994) IL-12 is both required and prognostic in vivo for T helper type 1 differentiation in murine candidiasis. J Immunol 153: 5167–5175

Rook AH, Kubin M, Cassin M, Vonderheid EC, Vowels BR, Wolfe JT, Wolf SF, Singh A, Trinchieri G, Lessin SR (1995) IL-12 reverses cytokine and immune abnormalities in Sezary syndrome. J Immunol 154: 1491–1498

Rossi AR, Pericle F, Rashleigh S, Janiec J, Djeu JY (1994) Lysis of neuroblastoma cell lines by human natural killer cells activated by interleukin-2 and interleukin-12. Blood 83: 1323–1328

Salcedo TW, Azzoni L, Wolf SF, Perussia B (1993) Modulation of perforin and granzyme messenger RNA expression in human natural killer cells. J Immunol 151: 2511–2520

Schmitt E, Hoehn P, Germann T, Rude E (1994a) Differential effects of interleukin-12 on the development of naive mouse CD4[+] T cells. Eur J Immunol 24: 343–347

Schmitt E, Hoehn P, Huels C, Goedert S, Palm N, Rude E, Germann T (1994b) T helper type 1 development of naive CD4[+] T cells requires the coordinate action of interleukin-12 and interferon-γ and is inhibited by transforming growth factor-β. Eur J Immunol 24: 793–798

Schoenhaut DS, Chua AO, Wolitzky AG, Quinn PM, Dwyer CM, McComas W, Familletti PC, Gately MK, Gubler U (1992) Cloning and expression of murine IL-12. J Immunol 148: 3433–3440

Seder RA, Gazzinelli R, Sher A, Paul WE (1993) Interleukin 12 acts directly on CD4[+] T cells to enhance priming for interferon γ production and diminishes interleukin 4 inhibition of such priming. Proc Natl Acad Sci USA 90: 10188–10192

Sieburth D, Jabs EW, Warrington JA, Li X, Lasota J, LaForgia S, Kelleher K, Huebner K, Wasmuth JJ, Wolf SF (1992) Assignment of NKSF/IL 12, a unique cytokine composed of two unrelated subunits, to chromosomes 3 and 5. Genomics 14: 59–62

Smith TJ, Ducharme LA, Weis JH (1994) Preferential expression of interleukin-12 or interleukin-4 by murine bone marrow mast cells derived in mast cell growth factor or interleukin-3. Eur J Immunol 24: 822–826

Soiffer RJ, Robertson MJ, Murray C, Cochran K, Ritz J (1993) Interleukin-12 augments cytolytic activity of peripheral blood lymphocytes from patients with hematologic and solid malignancies. Blood 82: 2790–2796

Stern AS, Podlaski FJ, Hulmes JD, Pan Y-CE, Quinn PM, Wolitzky AG, Familletti PC, Stremlo DL, Truitt T, Chizzonite R, Gately MK (1990) Purification to homogeneity and partial characterization of cytotoxic lymphocyte maturation factor from human B-lymphoblastoid cells. Proc Natl Acad Sci USA 87: 6808–6812

Stern LL, Tarby CM, Tamborini B, Truitt GA (1993) Preclinical development of IL-12 as an anticancer drug: comparison to IL-2. Proc Am Assoc Cancer Res 34: 520

Stuehr DJ, Nathan CF (1989) Nitric oxide: a macrophage product responsible for cytostasis and respiration inhibition in tumor target cells. J Exp Med 169: 1543–1555

Tahara H, Zeh HJ, Storkus WJ, Pappo I, Watkins SC, Gubler U, Wolf SF, Robbins PD, Lotze MT (1994) Fibroblasts genetically engineered to secrete interleukin 12 can suppress tumor growth and induce antitumor immunity to a murine melanoma in vivo. Cancer Res 54: 182–189

Tannenbaum CS, Wicker N, Tubbs R, Olenki T, Finke J, Bukowski R, Hamilton TA (1995) Patterns of chemokine expression and leukocyte infiltration in murine models of cytokine immunotherapy. Proc Am Assoc Cancer Res 36: 495

Trinchieri G (1994) Interleukin-12: a cytokine produced by antigen-presenting cells with immunoregulatory functions in the generation of T-helper cells type 1 and cytotoxic lymphocytes. Blood 84: 4008–4027

Tripp CS, Wolf SF, Unanue ER (1993) Interleukin 12 and tumor necrosis factor α are costimulators of interferon γ production by natural killer cells in severe combined immunodeficiency mice with listeriosis, and interleukin 10 is a physiologic antagonist. Proc Natl Acad Sci USA 90: 3725–3729

Valiante NM, Trinchieri G (1993) Identification of a novel signal transduction surface molecule on human cytotoxic lymphocytes. J Exp Med 178: 1397–1406

Valiante NM, Rengaraju M, Trinchieri G (1992) Role of the production of natural killer cell stimulatory factor (NKSF/IL-12) in the ability of B cell lines to stimulate T and NK cell proliferation. Cell Immunol 145: 187–198

van der Pouw Kraan TCTM, Boeije LCM, Smeenk RJT, Wijdenes J, Aarden LA (1995) Prostaglandin-E2 is a potent inhibitor of human interleukin 12 production. J Exp Med 181: 775–779

Varesio L, Blasi E, Thurman GB, Talmadge JE, Wiltrout RH, Herberman RB (1984) Potent activation of mouse macrophages by recombinant interferon-gamma. Cancer Res 44: 4465–4470

Verbik DJ, Stinson WW, Brunda MJ, Kessinger A, Joshi SS (1995) Therapeutic effects of in vivo administration of interleukin-12 against residual lymphoma. Proc Am Assoc Cancer Res 36: 480

Vieira LQ, Hondowicz BD, Adonso LC, Wysocka M, Trinchieri G, Scott P (1994) Infection with Leishmania major induces interleukin-12 production in vivo. Immunol Lett 40: 157–161

Voest EE, Kenyon BM, Truitt GA, O'Reilly MS, D'Amato RJ, Folkman J (1995) Interleukin 12 inhibits angiogenesis is vivo. J Natl Cancer Inst 87: 581–586

Wolf SF, Temple PA, Kobayashi M, Young D, Dicig M, Lowe L, Dzialo R, Fitz L, Ferenz C, Hewick RM, Kelleher K, Herrmann SH, Clark SC, Azzoni L, Chan SH, Trinchieri G, Perussia B (1991) Cloning of cDNA for natural killer cell stimulatory factory, a heterodimeric cytokine with multiple biologic effects on T and natural killer cells. J Immunol 146: 3074–3081

Wong GHW, Clark-Lewis I, McKimm-Breschkin JL, Harris AW, Schrader JW (1983) Interferon gamma induces enhanced expression of Ia and H-2 antigens on B-lymphoid, macrophage and myeloid cell lines. J Immunol 131: 788–793

Wong HL, Wilson DE, Jenson JC, Familletti PC, Stremlo DL, Gately MK (1988) Characterization of a factor(s) which synergizes with recombinant interleukin 2 in promoting allogeneic human cytolytic T-lymphocyte responses in vitro. Cell Immunol 111: 39–54

Yanagida T, Kato T, Igarashi O, Inoue T, Nariuchi H (1994) Second signal activity of IL-12 on the proliferation and IL-2R expression of T helper cell-1 clone. J Immunol 152: 4919–4928

Yoshida A, Koide Y, Uchijima M, Yoshida TO (1994) IFN-γ induces IL-12 mRNA by a murine macrophage cell line, J774. Biochem Biophys Res Commun 198: 857–861

Yssel H, Fasler S, de Vries JE, Malefyt RdW (1994) IL-12 transiently induces IFN-γ transcription and protein synthesis in human $CD4^+$ allergen-specific Th2 T cell clones. Int Immunol 6: 1091–1096

Curing Metastases? Gene and Peptide Therapy

LEA EISENBACH

1 Background

Major histocompatibility complex (MHC) molecules are peptide receptors. Their function is to collect peptides inside the cell and transport them to the cell surface, where the complex of peptide and MHC molecule may be recognized by T lymphocytes (RAMMENSEE et al. 1993). In normal cells MHC-associated peptides are derived from normal, self proteins. During their differentiation, T cells become tolerant – by elimination or inactivation (anergy) – to complexes of self MHC and self peptides (Von Boehmer 1992). Thus when new peptides derived from infectious agents or changed self proteins occur later, these can be recognized by T cells (ZINKERNAGEL and DOHENTY 1974). However, overexpressed, aberrantly expressed, and even normal self proteins can induce T cell responses under conditions which revert anergy (Nanda and Sercarz 1995). Since the specific immune response is regulated by T cells, the trimolecular complex of T cell receptor (TCR), MHC molecule, and peptides serves as a control switch for the immune system. Control of tumor growth and tumor dissemination by the immune system, particularly by the cellular immune system, is the basis for development of antitumor vaccines (FOLEY 1953; PREHN and MAIN 1957; KLEIN et al. 1960; OLD et al. 1962). The role MHC molecules and their associated peptides play in tumor contol is important for development of targeted antigen-specific vaccines for cancer therapy.

Department of Immunology, Weizmann Institute of Science, Rehovot 76100, Israel

2 Major Histocompatibility Complex Structure and Function

There are two distinct, classes of MHC molecules, class I and class II. Class I consists of a 45-kDa heavy chain and a 12-kDa, noncovalently attached β_2-microglobulin light chain (KLEIN 1986). X-ray crystallography has shown that the assembled cell surface molecule forms a cup-like structure with a β-pleated sheet structure at the bottom and two α-helix structures as walls, forming a closed groove that binds peptides (BJORKMAN et al. 1987a,b). Human leukocyte antigen (HLA)-A,B,C are the classical human class I molecules, and H-2K, H-2D, and H-2L are the mouse equivalents. Class II MHC are 30-kDa heterodimers, and both α- and β-chains are transmembranal; HLA-DR, DQ, and DP are human class II MHC, and H-2A and H-2E are their murine equivalents. Crystallography showed that, in general, the three-dimensional structure of class II molecule is similar to class I; however the groove formed in class II is open on both sides, accomodating longer peptides than class I molecules (BROWN et al. 1993; STERN et al. 1994; STERN and WILEY 1994).

MHC molecules are very polymorphic, and tens to hundreds of alleles exist in a species. Human have up to six classical class I, additional nonclassical class I genes and six to ten expressed class II molecules, while homozygous mice have two to three classical class I genes and one to two expressed class II heterodimers (KLEIN 1986; BODMER et al. 1994). Most polymorphic residues in class I and class II molecules are clustered at the bottom and on the α-helices of the peptide-binding groove. Thus each MHC molecule will bind a relatively limited set of peptides, which will undergo appropriate interactions with amino acid residues of the groove (BJORKMAN et al. 1987a). Rammensee and colleagues were the first to isolate MHC-bound peptides by isolating first the MHC-peptide complexes on anti-MHC antibody columns and then using acid to release the associated peptides. Pool sequencing of these peptides showed that a high frequency of a given amino acid appeared in certain positions of the peptides; these were defined as anchor positions. For class I, one anchor position usually exists in the middle of the peptide and a second at the C terminus of the peptide. Occasionally, a third amino acid is involved in binding and is called an auxiliary anchor (reviewed in RAMMENSEE et al. 1995).

MHC molecules function as peptide receptors. Class I molecules are synthesized in the endoplasmic reticulum (ER) and newly synthesized heavy and light chains are assembled with preprocessed peptides in the ER. The peptides loaded on class I are mainly derived from cytosolic proteins, most often from endogenous, cellular proteins which are degraded by the proteosomes (JACKSON and PETERSON 1993; ROCK et al 1994; DICK et al. 1994). The high specificity of the proteosome may be influenced by two proteosome subunits encoded in the MHC region, LMP2 and LMP7, which are regulated by γ-interferon (IFN-γ) and which replace two resident proteosome subunits, x and y, after IFN-γ treatment (HOWARD and SEELING 1993). Processed peptides, according to one hypothesis, are protected by a chaperone of the heat-shock family in the cytoplasm until they are transported into the lumen of the ER. Entry into the ER is mediated by an ER transmembranal dimer of transporters in antigen processing, TAP1-TAP2, which transport peptides in an adenosine triphosphate (ATP)-dependent process (NEEFJES and MOMBURG 1993; SRIVATSAVA et al. 1994; MOMBURG et al. 1994). The discharge of peptides from the TAP transporter was suggested to occur directly to class I molecules bound to TAP or alternatively to involve a second chaperone of the heat-shock family (ORTMANN et al. 1994; SUH et al. 1994; SRIVASTAVA et al. 1994). Stabilized, assembled class I MHC–peptides complexes migrate through the Golgi apparatus to the cell surface. Commonly, class I molecules present endogenous peptides to CD8-expressing T cells but in macrophages, at least, exogenous antigens can be presented by class I MHC by a direct phagosome to cytosal pathway (PFIEFER et al. 1993; ROCK et al. 1993; KOVACSOVICS-BANKOWSKI and ROCK 1995).

Class II molecules mostly present exogenous protein peptides to CD4- expressing T cells. Processing of endocytosed proteins in endosomes or lysosomes require acidic pH. The class II α- and β-chains assemble in the ER with a nonpolymorphic invariant (Ii) chain which stabilizes the structure and prevents early loading. In the class II loading compartment, the Ii chain is removed and specific peptides are bound before the complex is expressed on the cell surface (CRESSWELL 1994).

3 Defective Antigen Presentation in Tumors

Loss of MHC class I expression is a common finding on tumors of many origins and may represent a way for tumor cells to escape T cell immune surveillance. Reduced expression of HLA-A, -B, and -C was found in a variety of colorectal tumors, in breast carcinomas, in lung carcinomas, especially small cell lung carcinomas, in melanomas, and in other tumors (reviewed in ELLIOT et al. 1989; EISENBACH and FELDMAN 1991; FELDMAN and EISENBACH 1991, 1993). Comparison of HLA expression between primary and metastatic lesions from the same patients, including breast, colon, bladder , and renal carcinomas, showed that metastatic lesions generally express low class I MHC compared to primary tumors (CORDON-CARDO et al. 1991). Most studies concerning human tumors evaluated HLA expression by immunohistochemsitry using antibodies that recognize HLA-A,-B, and -C. However, allele-specific losses were observed in colorectal carcinomas, in gastric carcinomas, and in laryngeal tumors by the use of allele-specific antibodies (REES et al. 1988; SMITH et al. 1988, 1989; MOMBURG et al. 1986; NATALI et al. 1989; LOPEZ-NEVOT et al. 1989; RODRIGUEZ et al. 1994). Multiple mechanisms were described for down-regulation of MHC class I expression in tumors. Reduced transcription of the class I heavy chain and/or β_2-microglobulin often occurs (PLAKSIN et al. 1988; KUSHTAI et al. 1988; SCHRIER et al. 1983; BERNARDS et al. 1986;

BLANCHET et al. 1991; HENSELING et al. 1990; LASSAM and JAY 1989; SHEMESH et al. 1991; VAN'T VEER et al. 1993). Mutations in the β_2-microglobulin gene were recently described for a number of colorectal tumor lines and fresh tumors (BROWNING et al. 1993; BICKNELL et al. 1994). In small cell lung carcinomas and in prostrate cell lines, down-regulation of TAP1 and TAP2 expression was demonstrated to be an important mechanism responsible for low cell surface expression of class I MHC (SOONG et al. 1991; SOONG and HUI 1992; RESTIFO et al. 1993a,b; SANDA et al. 1995; ROTEMYEHUDAR et al. 1994a).

In murine tumors we and others have shown a correlation between low class I MHC expression and high tumorgenicity or high malignancy. Low expression is principally controlled transcriptionally. The crucial promoter regulatory sequences are NFκB, AP1/CREB, and IFN response elements (ISRE), which are either not occupied in vivo (YAMIT-HEZI et al. 1994) or are occupied by negative regulators (PLAKSIN et al. 1993). However, downregulation of TAP1 and TAP2 has been documented in adenovirus 12-transformed cells which express low class I MHC (SHEMESH et al. 1991). Post-transcriptional mechanisms involving retention of class I MHC heavy chain have also been shown in adenovirus 12-transformed cells (ROTEM-YEHUDAR et al. 1994b).

4 Improvement of Tumor Immunogenicity by Gene Transfer

Low antigenicity of tumors defective in class I MHC presentation can be reconstituted by transfection with MHC class I or TAP genes (HUI et al. 1984; TANAKA et al. 1985; WALLICH et al. 1985; PLAKSIN et al. 1988; PORGADOR et al. 1989; MANDELBOIM et al. 1992; SPIES and DEMARS (1991). In contrast, increasing tumor immunogenicity, i.e., the ability to induce tumor-specific T cell responses, often necessitates more complex manipulation. Efficient T cell activation requires two signals. The first is received through T cell receptors after engaging antigenic peptides on MHC class I molecules. The second is a costimulatory signal originally propsoed for B cells (BRETSCHER and COHN 1970) or actual antigen-presenting cells (APC) such as macrophages, dendritic cells, or Langerhans cells (JANEWAY 1989; NOSSAL 1989; LAFFERTY and WOOLNOUGH 1977). APC proper are MHC class II-expressing cells. Antigens endocytosed or phagocytosed into these cells are processed in an endosomal/lysosomal compartment and peptides are presented on MHC class II. The interaction between the MHC class II-peptide complex and TCR of $CD4^+$ helper T cells, in conjunction with other costimulatory molecules such as B7-1 on the APC and CD28 on the T cell, activates helper cells to produce necessary cytokines (CHEN et al. 1992; TOWNSEND and ALLISON 1993; BASKAR et al. 1993). These cytokines serve as the second signal for activation of class I-restricted cytotoxic cells.

Conflicting data exist with regard to the question of which cells can present class I-peptide complexes for induction of cytotoxic T lymphocyte (CTL). A number of studies have shown that tumor cells or virally infected cells are first killed by nonspecific effectors such as natural killer (NK) cells, macrophages, or granulocytes, and their antigens are taken up by APC (BEVAN 1976; MATZINGER and BEVAN 1977; CARBONE and BEVAN 1990; HUANG et al. 1994). Although the exact mechanism by which such antigens enter the class I MHC presentation pathway is not clear in most APC, a direct phagosome to cytosol pathway for exogenous antigens presented on MHC class I was recently demonstrated in macrophages (KOVASOVICS-BANKOWSKI and ROCK 1995). This mechanism postulates that potentially all tumors can be immunogenic even if they do not express class I MHC or process their own peptides (HUANG et al. 1994). On the other hand, a large body of evidence shows that transduction of tumor cells with MHC class I genes, with MHC class II genes or with B7-1 genes dramatically increases their immunogenicity (PLAKSIN et al. 1988; PORGADOR et al. 1989; OSTRAND-ROSENBERG et al. 1990; CHEN and ANAN THASWAMY 1993; BASKAR et al. 1988; CHEN et al. 1994; LI et al. 1994).

Had antigen presentation occurred exclusively by host APC the expression of such molecules by the tumor should have been ignored by the immune system or should lead to anergy as suggested before (LANZAVECCHIA 1988; MUELLER et al. 1989).

It was recently demonstrated that tumor cells of various origins-epithelial, fibroblastic, and others – can present antigen directly for induction of CTL (COHEN and KIM 1994; TOWNSEND and ALLISON 1993; KUNDIG et al. 1995). While denovo expression of MHC class I, class II, or B7-1 on tumor cells may serve to increase their intrinsic immunogenicity, transduction by a variety of cytokine genes was shown to increase tumor cell functional immunogenicity. Murine tumor cells modified by transduction of interleukin (IL)-1, IL-2, IL-4, IL-6, IL-7, IL-10, IL-12, IFN-α, IFN-γ, tumor necrosis factor TNF-α, granulocyte colony-stimulating factor (G-CSF), macrophage CSF(M-CSF), granulocyte-macrophage CSF (GM-CSF), or 10-kDa inflammatory protein (IP-10) were shown to be rejected in vivo or manifested reduced growth rates (reviewed in PARDOLL 1993; BLANKENSTEIN 1994; PORGADOR et al. 1994). Some of the cytokines induce nonspecific immunocyte infiltrates consisting of macrophages, eosinphils, neutrophils, or NK cells at the site of rejection, while in other cases $CD4^+$ and $CD8^+$ bearing lymphocytes were also identified (TEPPER et al. 1989; BANNERJI et al. 1994; COLOMBO and TORNI 1994). With the exception of IL-5, TNF, and G-CSF, all cytokine-transduced cells were shown to induce systemic immunity in mice that rejected the challenge with cytokine-modified tumor cells (BLANKENSTEIN 1994).Systemic immunity was always dependent on $CD8^+$cells and in a number of systems also on $CD4^+$ expressing cells (BLANKENSTEIN 1994).

However, when a more stringent comparison between immunogenicity of irradiated cytokine-transduced cells and irradiated parental cells was performed, cytokines such as IL-1, IL-2, IL-6, IFN-γ, GM-CSF, and IL-12 were mainly shown to actually increase the immunogenicity of weakly immunogenic tumors (DOUVDEVANI et al. 1992; PORGADOR et al. 1992, 1993a,b; DRANOFF et al. 1993). A more crucial question is whether increased immunity of cytokine-transduced cells can be translated to the design of antitumor vaccines. Vaccines based on IL-2, IFN-γ,

GM-CSF, and IL-12 gene transduced tumor cells were shown to protect mice against a subsequent challenge with parental tumor cells or to reject small established tumors (DRANOFF et al. 1993; CAVALLO et al. 1992; CONNOR et al. 1993; TAHARA et al. 1994). In metastatic tumors such as the murine Lewis lung carcinoma (3LL-D122) carcinoma, the murine B16 melanoma, or the rat R3327-Mat LyLu prostate carcinoma, treatment of animals bearing small established tumors with vaccines secreting IL-2, IL-6, or IFN-γ prevents metastasis (PORGADOR et al. 1992, 1993a,b; VIEWEG et al. 1994). However, established metastases treated after the removal of a large primary tumor can be rejected in the 3LL-D122 system only by IFN-γ transduced 3LL-D122 cells and in the B16-F10.9 system only by combined vaccination with MHC class I-transfected and IL-2 transduced cells (PORGADOR et al. 1993b, 1995).

5 Tumor Antigens and Tumor Peptides

The ability of gene-modified tumor cells to induce effective antitumor responses in animal models indicated that tumor-associated antigens which can be recognized by CTL are probably expressed in the majority of animal tumors. Tumor-specific CTL have also been detected in several human cancers (YASUMURA et al. 1993; VIRET et al. 1993). A small group of TAA peptides have been identified to date, mostly of human mealnomas. The most successful approach, developed by T. Boon and colleagues, uses cDNA libraries of human tumor cell lines cloned in a pcDNAI/AMP expression vector. Pools of plasmids from this library are transiently transfected into COS-7 cells together with relevant HLA class I genes. These transfectants are then reacted with CTL lines (mostly autologous to the tumor) and identified by secretion of cytokines such as TNF. Positive plasmid pools are subdivided until a single clone that transfers sensitivity to CTL lysis is identified. Using this or similar methods, the melanoma antigens melanoma antigen 1 (MAGE1), MAGE3, BAGE, GAGE, tyrosinase, melan A/MART-1, glycoprotein (gp) 100/Pmel17, and gp75 were identified (GAUGLER et al. 1994; VAN DER BRUGGEN et al. 1991, 1994a, b BOËL et al. 1995; VAN DEN EYDENE et al. 1995; BRICHARD et al. 1993; WOLFEL et al. 1994; COULIE et al. 1994;KAWAKAMI et al. 1994 a,b; BAKKER et al. 1994; WANG et al. 1995). These TAAs can be subdivided into two groups; one group consists of normal melanoma proteins, including tyrosinase, MelanA/MART1, gp100/pmel 17, and gp75, while the other TAAs are embryonal proteins of unknown functions which are not expressed in adult tissues, except for the testis. Recently, a new TAA gene, MUM-1, was cloned from a human melanoma; this gene, which is expressed in many tissues, serves as a TAA only in a mutated form (COULIE et al. 1995). In murine tumors, mutagenesis of P815 mastocytoma cells was shown to give rise to two mutated TAAs, while the nonmutagenized P815 cells express a nonmutated TAA protein (LURQUIN et al. 1989; SIBILLE et al. 1990; SZIKORA et al. 1990).

In recent years, the novel approaches developed for isolation of MHC-bound peptides by chemical methods have yielded many new peptides. The first three natural MHC class I ligands were identified in 1990 and were all of viral origin (ROTZSCHKE et al. 1990a; VAN BLEEK and NATHENSON 1990). In these instances, knowledge of the viral protein sequence was mandatory for peptide identification. Since then, the number of MHC ligands that have been characterized has been growing exponentially (ROTZSCHKE et al. 1991; UDAKA et al. 1992; VAN BLEEK and NATHENSON 1991; SHIBATA et al. 1992; JARDETZKY et al. 1991; HUNT et al. 1992; HENDERSON et al. 1992; WEI and CRESSWELL 1992; RAMMENSEE et al. 1995).

Most ligands known to date are self peptides, i.e., peptides derived from normal cellular proteins, such as histones, heat-shock proteins, enzyme, and others. Relatively few natural class I MHC ligands derived from foreign proteins, i.e., from pathogens have been isolated (RAMMENSEE et al. 1995).

Although over 100 different MHC class I haplotypes have been identified in humans (BODMER et al. 1992), only three to six of these structural isoforms are expressed in any one individual. Two to three MHC class alleles are expressed in inbred homozygous mice. Thus each isoform must bind a wide variety of peptides. Most mouse and human cells express $1–10\times10^5$ class I molecules per cell. From the level of class I expression present and knowledge of the occupancy and the diversity of the peptide repertoire a single peptide–class I complex is calculated to be expressed in the range 100–4000 copies per cell, and every cell can present about 2000 distinct peptides (ENGELHARD 1994; HUCZKO et al. 1993). As a result, there are two basic problems that prevent the routine isolation of tumor-associated antigen peptides. The first problem is the amount: only a small fraction (0.1%) of MHC bound peptides are considered to be TAA peptides, embedded in a large amount of normal cellular peptides. The second problem concerning the isolation of TAA peptides is heterogenicity: the TAA peptides are embedded in an ocean of variable self peptides; hence the Edman degradation sequence of Class I MHC associated peptides would require 5–10, nmol (approximaely 5–10 mg) of starting material. It seems that successful identification of TAA peptides necessitates both large quantities of tumor cells and a sensitive bioassay to follow active peptides. Therefore, a variety of techniques have been employed to isolate and charcterize peptides constituently bound to class I molecules.

Peptides can be dissociated from class I molecules by treatment with trifluoracetic acid (TFA) or acetic acid (VAN BLEEK and NATHENSON 1990, 1991; ROTZSCHKE et al. 1990a, b, 1991; UDAKA et al. 1992; SHIBATA et al. 1992; FALK et al. 1990, 1991, reviewed in RAMMENSEE et al. 1995). Three alternative approaches for isolation of MHC class I peptides were employed based on acid wash. Treatment of whole cell lysate with TFA brings the peptides into solution, and they can then be detected by the respective T cells (ROTZSCHKE et al. 1990a). This was the approach used for isolation of minor histocompatibility antigens and virus-derived peptides from cells (ROTZSCHKE et al. 1990a; FALK et al. 1990). Acid extraction of whole cells has the advantage of yielding not only MHC-bound peptides but also other peptides, e.g., potential intermediates of the processing pathway (ROTZSCHKE et al. 1990b, 1991; UDAKA et al. 1992; FALK et al. 1990). In the second approach,

MHC molecules with their associated peptides are first purified by anti-MHC antibody columns and then acid extraction of the peptides is performed. The resulting peptide mixture is much purer than in the first approach, and the peptides lend themselves more readily to further isolation and sequence analysis (Jardetyzky et al. 1991; Hunt et al. 1992; Henderson et al. 1992; (Wei and Cresswell 1992). In the third approach, brief pH 3.3 acid treatment of live cells brings peptides into solution without the need of MHC class I purification, and the neutralized cell culture can be left for further peptide extractions (Zeh et al. 1994). The resulting peptide fraction is much purer, since it does not contain the MHC class I molecules themselves; however, the peptide quantitiy is very limited, and no TAA peptides have been isolated so far using this approach.

Conventionally, the TAA peptides are isolated from the peptide mixture by high-performance liquid chromatography (HPLC) fractionation. However, the sensitivity of HPLC separation is limited and in many cases the final fraction contained several peptides that cannot be separated and hence cannot be sequenced. To overcome the limitation of sensitivity, Engelhard's and Hunt's groups joined forces by sharing their facilities and expertise to characterize naturally processed peptides associated with class I molecules using a combination of microcapillary reversed phase (RP)-HPLC separation coupled with electrospary ionization tandem mass spectrometry (Hunt et al. 1992; Huczko et al. 1993; Henderson et al. 1992, 1993). Their collaborative efforts have facilitated the identification and characterization of peptides isolated from as few as 10^8 cells. By using these sophisticated instruments, they have been able to determine amino-sequences, beginning with only a few hundred fentamoles of peptides (Hunt et al. 1992; Huczko et al. 1993; Henderson et al. 1993). The potential of this system has been best demonstrated by the determination of the amino acid sequence of endogenous peptides recognized by a xenoreactive CTL (Henderson et al. 1993) and by the successful identification of a tumor peptide antigen restricted by class I MHC molecules.

Cox et al. (1994) have used peptide fractionation to identify a shared HLA-A2-restricted melanoma peptide recognized by CTL lines derived from T cells of patients' lymph nodes. Because each of the several bioactive fractions contained over 50 peptides on mass spectroscopic (MS) analysis, a second dimension HPLC fractionation using different elution conditions was required. Tandem MS was employed to separate and sequence a number of peptides in one bioreactive fraction. One of the peptides sequenced was derived from the melanocyte-specific gp100 that was previously identified by genetic means. This peptide reconstituted recognition by melanoma-specific T cell lines from four separate patients (Cox et al. 1994). The main disadvantage of this method is the need for sophisticated equipment which is not widely available. In our studies, the quantitative problems were overcome through the isolation of a sufficient amount of peptide using acid extraction of tumor cells growing in vivo in the mouse. Hence 200 g tumor tissue, which is equivalent to about 2×10^{12} tumor cells growing in tissue culture was grown in nude mice (mature T cell-deficient mice), thus avoiding other unrelated cells in the growing tumors. The amount of TAA peptides was further increased by preparing whole cell peptide extract rather than using MHC affinity columns, followed by several rounds of HPLC purification. Purified peptide fractions were tested for their ability to lyse unrelated MHC-compatible tumor cells following incubation with bulk CTL, which enabled us to overcome the problem of generating CTL clones manipulated in vitro. Such manipulation might lead to the development of the non-immunodominant clones instead of the immunodominant ones. Using our approach – the in vivo tumor propagation/bulk CTL – we were the first to isolate TAA peptides from a spontaneous mouse carcinoma (MANDELBOIM et al. 1994). The resulting peptides, MUT1, (Phe-Glu-Gln-Asn-Thr-Ala-Gln-Pro)and MUT2, (Phe-Glu-Gln-Asn-Thr-Ala-Gln-Ala), were shown to be derived from the lung gap junction protein connexin 37. The normal homogous peptide, Con37 (Phe-Glu-Cys-Asn-Thr-Ala-Gln-Pro), is different in position 3 of the peptide from MUT1 and in positions 3 and 8 from MUT2. We have shown that MUT1 and MUT2 bind to H-2K^b class I, while Con37 does not. These new TAA peptides define new binding motifs to the H-2K^b molecule. While the classical K^b motifs contain a Tyr in position 5, a Leu, Ile, Val, or Met in position 8, and a Tyr in position 3 as an auxillary anchor, in MUT1 or MUT2 Glu in position 2 and Thr in position 5 substitute for Tyr in position 5. Pro or Ala in position 8 substitutes for hydrophobic residues in classical peptides, and Gln but not cys in position 3 substitutes for the K^b auxillary anchor motif and determines binding (MANDELBOIM et al. 1995b). Synthetic MUT1 and MUT2, when loaded on the antigen processing-deficient cell line RMA-S (of C57BL/6 origin), stabilize high expression of H-2K^b. Immunization of syngeneic mice with RMA-S-loaded peptides or with peptides in incomplete Freund's adjuvant induces high levels of antitumor CTL (MANDELBOIM et al. 1994). Recently, we have shown that such immunized mice do not develop lung metastases when challenged with the highly metastatic 3LL clone D122. Moreover, mice were injected in their footpad with D122 cells subsequently underwent surgery of the primary tumor 6–6.5 mm), and were treated after surgery with peptides; they developed reduced levels of metastases when treated with peptides in adjuvant or completely rejected established metastases when treated with peptides loaded on RMA-S cells (MANDELBOIM et al. 1995a). Thus peptide vaccines seem to be an effective vaccine modality to treat even established micrometastases.

6 Conclusion

The ultimate goal of tumor vaccine design is the generation of antigen-specific vaccines. Our growing knowledge on specific TAAs and their peptides will enable the design of peptide or recombinant protein vaccines for cancer treatment. It is not as yet clear by what modes such vaccines will be administered and what the role of APC in vaccination will be. Cytokines might be needed to guide the immune system in the development of an effective antitumor CTL response. Vaccines based on autologous tumor cells modified by MHC class I, MHC class II, B7–1, or cytokine genes will be useful before specific TAA are identified for many cancers. Moreover, vaccines based on allogenic tumor cell lines, modified by cytokine genes or DNA vaccines designed to modify tumors in vivo, are being developed. These will hopefully lead to improved cancer treatment.

References

Bakker AB, Schreurs MW, de Boer AJ, Kawakami Y, Rosenberg SÀ, Adema GJ, Figdor CG (1994) Melanocyte lineage-specific antigen gp100 is recognized by melanoma-derived tumor-infiltrating lymphocytes. J Exp Med 179: 1005–1009

Bannerji R, Arroyo CD, Cordon-Cardo C, Gilboa E (1994) The role of IL-2 secreted from genetically modified tumor cells in the establishment of antitumor immunity. J Immunol 152: 2324–2332

Baskar S, Ostrand-Rosenberg S, Nabavi N, Nadler LM, Freeman GJ, Glimcher LH (1993) Constitutive expression of B7 restores immunogenicity of tumor cells expressing truncated major histocompatibility complex class II molecules. Proc Natl Acad Sci USA 90: 5687–5690

Bernards R, Dessain S, Weinberg R (1986) N-myc amplification causes down-modulation of MHC class I antigen expression in neuroblastoma. Cell 47: 667–674

Bevan MJ (1976) Cross-priming for a secondary cytotoxic response to minor H antigens with H-2 congenic cells which do not cross-react in the cytotoxic assay. J Exp Med 143: 1283–1288

Bicknell DC, Rowan A, Bodmer WF (1994) Beta 2-microglobulin gene mutations: a study of established colorectal cell lines and fresh tumors. Proc Natl Acad Sci USA 91: 4751–4756

Bjorkman PJ, Saper MA, Samraoui B, Bennett WS, Strominger JL, Wiley DC (1987) Structure of the human class I histocompatibility antigen, HLA-A2. Nature 329: 506–512

Bjorkman PJ, Saper MA, Samraoui B, Bennett WS, Strominger JL, Wiley DC (1987) The foreign antigen binding site and T cell recognition regions of class I histocompatibility antigens. Nature 329: 512–518

Blanchet O, Bourge JF, Zinszner H, Tatari Z, Degos L, Paul P (1991) DNA binding of regulatory factors interacting with MHC-class-I gene enhancer correlates with MHC-class-I transcriptional level in class-I-defective cell lines. Int J Cancer 6: 138–145

Blankstein T (1994) Increasing tumor immunogenicity by genetic modification. Eur J Cancer 30: 1182–1187

Bodmer JG, Marsh SG, Albert ED, Bodmer WF, Dupont B, Erlich HA, Mach B, Mayr WR, Parham P, Sasazuki T (1992) Nomenclature for factors of the HLA system 1991. Hum Immunol 34: 4–18

Bodmer JG, Marsh SG, Albert ED, Bodmer WF, Dupont B, Erlich HA, Mach B, Mayr WR, Parham P, Sasazuki T, Schreuder GMT, Strominger JL, Svejgaard A, Terasaki PI (1994) Nomenclature for factors of the HLA system 1994. Tissue Antigens 44: 1–18

Boel P, Wildmann C, Sensi ML, Brasseur R, Renauld JC, Coulie P, Boon T, van der Bruggen P (1995) BAGE: a new gene encoding an antigen recognized on human melanomas by cytolytic T lymphocytes. Immunity 2: 167–175

Bretscher P, Cohn M (1970) A theory of self-nonself discrimination. Paralysis and induction involve the recognition of one and two determinants on an antigen, respectively. Science 169: 1042–1045

Brichard V, Van Pel A, Wolfel T, Wolfel C, De Plaen E, Lethe B, Coulie P, Boon T (1993) The tyrosinase gene codes for an antigen recognized by autologous cytolytic T lymphocytes on HLA-A2 melanomas. J Exp Med 178: 489–495

Brown JH, Jardetzky TS, Gorga JC, Stern LJ, Urban RG, Strominger JL, Wiley DC (1993) Three-dimensional structure of the human class II histocompatibility antigen HLA-DRI. Nature 364: 33–39

Browning MJ, Kraussa P, Rowan A, Hill AB, Bicknell DC, Bodmer JG, Bodmer WF (1993) Loss of human leukocyte antigen expression on colorectal tumor cell lines: implications for anti-tumor immunity and immunotherapy. J Immunother 14: 163–168

Carbone FR, Bevan MJ (1990) Class I-restricted processing and presentation of exogenous cell-associated antigen in vivo. J Exp Med 171: 377–387

Cavallo F, Giovarelli M, Gulino A, Vacca A, Stoppacciaro A, Modesti A, Forni G (1992) Role of neutrophils and CD4+T lymphocytes in the primary and memory response to nonimmunogenic murine mammary adenocarcinoma made immunogenic by IL-2 gene. J Immunol 149: 3627–3635

Chen L, Ashe S, Brady W, Hellstrom I, Hellstrom K, Ledbetter J, McGowan P, Linsley P (1992) Costimulation of antitumor immunity by the B7 counterreceptor for the T lymphocyte molecules CD28 and CTLA- 4. Cell 71: 1093–1102

Chen L, McGowan P, Ashe S, Johnston J, Li Y, Hellstrom I, Hellstrom K (1994) Tumor immunogenicity determines the effect of B7 costimulation on T cell-mediated tumor immunity. J Exp Med 179: 523–532

Chen PW, Ananthaswamy H (1993) Rejection of K1735 murine melanoma in syngeneic hosts requires expression of MHC class I antigens and either class II antigens or IL-2. J Immunol 151: 244–255

Cohen EP, Kim TS (1994) Neoplastic cells that express low levels of MHC class I determinants escape host immunity. Semin Cancer Biol 5: 419–428

Colombo MP, Forni G (1994) Cytokine gene transfer in tumor inhibition and tumor therapy: where are we now? Immunol Today 15: 48–51

Connor J, Bannerji R, Saito S, Heston W, Fair W, Gilboa E (1993) Regression of bladder tumors in mice treated with interleukin 2 gene-modified tumor cells. J Exp Med 177: 1127–1134

Cordon-Cardo C, Fuks Z, Drobnjak M, Moreno C, Eisenbach L, Feldman M (1991) Expression of HLA-A, B, C antigens on primary and metastatic tumor cell populaitons of human carcinomas. Cancer Res 51: 6372–6380

Coulie PG, Brichard V, Van Pel A, Wolfel T, Schneider J, Traversari Mattei SC, De Plaen E, Lurquin C, Szikora JP, Ranauld JC, Boon T (1994) A new gene coding for a differentiation antigen recognized by autologous cytolytic T lymphocytes on HLA-A2 melanomas. J Exp Med 180: 35–42

Coulie PG, Lehmann F, Lethe B, Herman J, Lurquin C, Andrawiss M, Boon T (1995) A mutated intron sequence codes for an antigenic peptide recognized by cytolytic T lymphocytes on a human melanoma. Proc Natl Acad Sci USA 92: 7976–7980

Cox AL, Skipper J, Chen Y, Henderson RA, Darrow TL, Shabanowitz J, Engelhard VH, Hunt DF, Slingluff CL, Seigler HF (1994) Identification of a peptide recognized by five melanoma-specific human cytotoxic T cell lines. Science 264: 716–719

Cresswell P (1994) Assembly, transport and function of MHC class II molecules. Annu Rev Immunol 12: 259–293

Dick LR, Aldrich C, Jameson SC, Moomaw CR, Pramanik BC, Doyle CK, Demartino GN, Bevan MJ, Forman JM, Slaughter CA (1994) Proteolytic processing of ovalbumin and beta-galactosidase by the proteasome to yield antigenic peptides. J Immunol 152: 3884–3894

Douvdevani A, Huleihel M, Zoller M, Segal S, Apte RN (1992) Reduced tumorigenicity of fibrosarcomas which constitutively generate IL-1 alpha either spontaneously or following IL-1 alpha gene transfer. Int J Cancer 51: 822–830

Dranoff G, Jaffee E, Lazenby A, Golumbek P, Levitsky H, Brose K, Jackson V, Hamada H, Pardoll DM, Mulligan R (1993) Vaccination with irradiated tumor cells engineered to secrete murine ganulocytemacrophate colony-stimulating factor stimulates potent, specific and long-lasting anti-tumor immunity. Proc Natl Acad Sci USA 90: 3539–3543

Eisenbach L, Feldman M (1991) Tumor cell surface determinaitons in host immunity against metastastes. Semin Cancer Biol 2: 179–188

Elliot BE, Carlow DA, Rodricks AM, Wade A (1989) Perspectives on the role of MHC antigens in normal and malignant cell development. Adv Cancer Res 53: 181–245

Englehard VH (1994) Structure of peptides associated with MHC class I molecules. Curr Opin Immunol 6: 13–23

Falk K, Rotzschke O, Rammensee HG (1990) Cellular peptide composition governed by major histocompatibility complex class I molecules. Nature 348: 248–251

Falk K, Rotzschke O, Stevanovic S, Jung G, Rammensee HG (1991) Allele-specific motifs revealed by sequencing of self-peptides eluted from MHC molecules. Nature 351: 290–296

Feldman M, Eisenbach L (1991) MHC class I genes controlling the metastatic phenotype of tumor cells. Semin Cancer Biol 2: 337–346

Feldman M, Eisenbach L (1993) MHC control of growth and metastic properties of tumor cells. In: Solheim BG, Ferrone S, Moller E (eds) The HLA system in clinical transplantation. Basic concepts and importance. Springer, Berlin Heidelberg New York, pp 185–198

Foley EJ (1953) Antigenic properties of methylcholanthrene induced tumors in mice strain of origin. Cancer Res 13: 835–837

Gaugler B, Van den Eynde B, Van der Bruggen P, Romero P, Gaforio JJ, De Plaen E, Lethe B, Brasseur F, Boon T (1994) Human gene MAGE-3 codes for an antigen recognized on a melanoma by autologous cytolytic T lymphocytes. J Exp Med 179: 921–930

Henderson RA, Michel H, Sakaguchi K, Shabanowitz J, Appella E, Hunt DF, Engelhard VH (1992) HLA-A2.1-associated peptides from a mutant cell line: a second pathway of antigen presentation. Science 255: 1264–1266

Henderson RA, Cox AL, Sakaguchi K, Appella E, Shabanowitz J, Hunt DF, Engelhard VA (1993) Direct identification of an endogenous peptide recognized by multiple HLA-A2.1-specific cytotoxic T cells. Proc Natl Acad Sci USA 90: 10275–10279

Henseling U, Schmidt W, Scholer HR, Gruss P, Hatzopoulos AK (1990) A transcription factor interacting with the class I gene ehancer is inactive in tumorigenic cell lines which suppress major histocompatibility complex class I genes. Mol Cell Biol 10: 4100–4109

Howard JC, Seelig A (1993) Antigen processing. Peptides and the proteasome. Nature 365: 211–212

Huang A, Golumbek P, Ahmadzadeh M, Jaffee E, Pardoll D, Levitsky H (1994) Role of bone marrow-derived cells in presenting MHC class I-restricted tumor antigens. Science 264: 961–965

Huczko EL, Bodnar WM, Benjamin D, Sakaguchi K, Zhu NZ, Shabanowitz J, Henderson RA, Appella E, Hunt DF, Englehard VH (1993) Characteristics of endogenous peptides eluted from the class I MHC molecule HLA-B7 determined by mass spectrometry and computer modeling. Immunol 151: 2572–2587

Hui KM, Grosveld F, Festenstein H (1984) Rejection of transplantable AKr leukemia cells following MHC DNA mediated cell transformation. Nature 311: 750–752

Hunt DF, Henderson RA, Shabanowitz J, Sakaguchi K, Michel H, Sevilir N, Cox AL, Appella E, Engelhard VH (1992) Characterization of peptides bound to the class I MHC molecule HLA-A2.1 by mass spectrometry. Science 255: 1261–1263

Jackson MR, Peterson PA (1993) Assembly and intracellular transport of MHC class I molecules. Annu Rev Cell Biol 9: 207–235

Janeway CA (1989) The role of CD4 in T-cell activation: accessory molecule or co-receptor? Immunol Today 10: 234–238

Jardetzky TS, Lane WS, Robinson RA, Madden DR, Wiley DC (1991) Identification of self peptides bound to purified HLA-B27. Nature 353: 326–329

Kawakami Y, Eliyahu S, Sakaguchi K, Robbins PF, Rivoltini L, Yanelli JR, Appella E, Rosenberg SA, (1994) Identification of the immunodominant peptides of the MART-1 human melanoma antigen recognized by the majority of HLA-A2-restricted tumor infiltrating lymphocytes. J Exp Med 180: 347-352

Kawakami Y, Eliyahu S, Delgado CH, Robbins PF, Rivoltini L, Topalian SL, Miki T, Rosenberg SA (1994) Cloning of the gene coding for a shared human melanoma antigen recognized by autologous T cells infiltrating into tumor. Proc Natl Acad Sci USA 91: 3515–3519

Klein J (1986) Natural history of the major histocompatibility complex. Wiley, New York

Klein G, Sjogren HO, Klein E, Hellstrom KE (1960) Demonstration of resistance against methycholantherene induced sarcoma in primary autochthonous host. Cancer Res 20: 1561–1572

Kovacsovics-Bankowski M, Rock KL (1995) A phagosome-to-cytosol pathway for exogenous antigens presented on MHC class I molecules. Science 267: 243–246

Kundig TM, Bachmann MF, Dipaolo C, Simard JJ, Battegay M, Lother H, Gessner A, Kuhlcke K, Ohashi PS, Hengartner H, Zinkernagel RM (1995) Fibroblasts as efficient antigen-presenting cells in lymphoid organs. Science 268: 1343–1347

Kushtai G, Barzilay JH, Feldman M, Eisenbach L (1988) The c-fos proto-oncogene in murine 3LL carcinoma clones controls the expression of MHC genes. Oncogene 2: 119–127

Lafferty K, Woolnough J (1977) The origin and mechanism of the allograft reaction. Immunol Rev 35: 231–262

Lanzavecchia A (1988) Exploiting the immune system's own strategies for immunotherapy. Immunol Today 9: 167–168

Lassam N, Jay G (1989) Suppression of MHC class I RNA in highly oncogenic cells occurs at the level of transcription initiation. J Immunol 143: 3792–3797

Li Y, McGowan P, Hellstrom I, Hellstrom K, Chen L (1994) Costimulation of tumor-reactive CD4+ and CD8+T lymphocytes by B7, a natural ligand for CD28, can be used to treat established mouse melanoma. J Immunol 153: 421–428

Lopez-Nevot MA, Esteban F, Ferron A, Gutierrez J, Oliva MR, Romero C, Huelin C, Ruiz-Cabella F, Garrido F (1989) HLA class I gene expression on human primary tumours and autologous metases: demonstration of selective losses of HLA antigens on colorectal, gastric and laryngeal carcinomas. Br J Cancer 59: 221–226

Lurquin C, Van Pel A, Mariame B, De Plaen E, Szikora JP, Janssens C, Reddehase M, Lejeune J, Boon T (1989) Structure of the gene of tum-transplantation antigen P91A. the mutated exon encodes a peptide recognized with Ld by cytolytic T cells. Cell 58: 293–303

Mandelboim O, Feldman M, Eisenbach L (1992) H-2K double transfectants of tumor cells as antimetastatic cellular vaccines in heterozygous recipients. Implications for the T cell repertoire. J Immunol 148: 3666–3673

Mandelboim O, Berke, G, Fridkin M, Feldman, M, Eisenstein M, Eisenbach L (1994) CTL induction by a tumour-associated antigen octapeptide derived from a murine lung carcinoma. Nature 369: 67–71

Mandelboim O, Vadai E, Fridkin M, Katz-Hillel A, Feldman M, Berke G, Eisenbach L (1995) Regression of establish murine carcinoma metastases following vaccination with tumour associated antigen peptides. Nature Med (in press)

Mandelboim O, Eisenstein M, Kopko H, Fridkin M, Berke G, Eisenbach L (1995) Alternative H-2k^b binding motifs of antigenic peptides defined by a chemically modified peptide with functional antitumor activity (submitted)

Matzinger P, Bevan MJ (1977) Induction of H-2-restricted cytotoxic T cells: in vivo induction has the appearance of being unrestricted. Cell Immunol 33: 92–100

Momburg F, Moller P, Moldenhauer G, Hammerling GJ (1986) Loss of HLA-A,B,C in colorectal carcinoma is related to the degree of de-differentiation. J Immunogenet 13: 195–199

Momberg F, Neefjes JJ, Hammerling GJ (1994) Peptide selection by MHC-encoded TAP transporters. Curr Opin Immunol 6: 32–37

Mueller DL, Jenkins MK, Schwartz RH (1989) Clonal expansion versus functional clonal inactivation: a costimulatory signalling pathway determines the outcome of T cell antigen receptor occupancy. Annu Rev Immunol 7: 445–480

Nanda NK, Sercarz EE (1995) Induction of anti-self-immunity to cure cancer. Cell 82: 13–17

Natali PG, Nicotra MR, Bigotti A, Venturo I, Marcenaro L, Giacomini P, Russo C (1989) Selective changes in expression of HLA class I polymorphic determinants in human solid tumors. Proc Natl Acad Sci USA 86: 6719–6723

Neefjes JJ, Momburg F (1993) Cell biology of antigen presentation. Curr. Opin. Immunol 5: 27–34

Nossal GJ (1989) Immunologic tolerance: collaboration between antigen and lymphokines. Science 245: 147–153

Old LJ, Boyse EA, Clark DA, Carswell EA (1962) Antigenic properties of chemically induced tumors. Ann NY Acad Sci 101: 80–85

Ortmann B, Androlewicz MJ, Cresswell P (1994) MHC class I/beta 2-microglobulin complexes associate with TAP transporters before peptide binding. Nature 368: 864–867

Ostrand-Rosenberg S, Thakur A, Clements V (1990) Rejection of mouse sarcoma cells after transfection of NHC class II genes. J Immunol 144: 4068–4071

Pardoll DM (1993) Cancer vaccines. Immunol Today 14: 310–316

Pfeifer JD, Wick MJ, Roberts RL, Findlay K, Normark SJ, Harding CV (1993) Phagocytic processing of bacterial antigents for class I MHC presentation to T cells. Nature 361: 359–362

Plaksin D, Gelber C, Feldman M, Eisenbach L (1988) Reversal of the metastatic phenotype in Lewis lung carcinoma cells after transfection with syngeneic H-2Kb gene. Proc Natl Acad Sci USA 85: 4463–4467

Plaksin D, Baeuerle P, Eisenbach L (1993) KBF1 (p50 NF-kappa B homodimer) acts as a repressor of H-2Kb gene expression in metastatic tumor cells. J Exp Med 177: 1651–1662

Porgador A, Feldman M, Eisenbach L (1989) H-2Kb transfection of B16 melanoma cells results in reduced tumourigenicity and metastatic competence. J Immunogen 16: 291–303

Porgador A, Tzehoval E, Katz A, Vadai E, Revel M, Feldman M, Eisenbach L (1992) Interleukin 6 gene transfection into Lewis lung carcinoma tumor cells suppresses the malignant phenotype and confers immunotherapeutic competence against parental metastatic cells. Cancer Res 52: 3679–3686

Porgador A, Gansbacher B, Bannerji R, Tzehoval E, Gilboa E, Feldman M, Eisenbach L (1993a) Antimetastatic vaccination of tumor-bearing mice with Il-2-gene-inserted tumor cells. Int J Cancer 53: 471–477

Porgador A, Bannerji R, Watanabe Y, Feldman M, Gilboa E, Eisenbach L (1993b) Antimetastatic vaccination of tumor-bearing mice with two types of IFN-gamma gene-inserted tumor cells. J Immunol 150: 1458–1470

Porgador A, Feldman M, Eisenbach L (1994) Immunotherapy of tumor metastasis via gene therapy. Nat Immun 13: 113–130

Porgador A, Tzehoval E, Vadai E, Feldman M, Eisenbach L (1995) Combined vaccination by MHC class I and IL-2 transduced melanoma cells synergizes the cure of post-surgical lung metastases. Cancer Res (in press)

Prehn RT, Main JM (1957) Immunity to methylcholantherene induced sarcomas. J Nat Cancer Inst 18: 769–778

Rammensee HG, Falk K, Rotzschke O (1993) Peptides naturally presented by MHC class I molecules. Annu Rev Immunol 11: 213–244
Rammensee HG, Friede T, Stevanovic S (1995) MHC ligands and peptide motifs: first listing. Immunogenetics 41: 178–228
Rees RC, Buckle AM, Gelsthorpe K, James V, Potter CW, Rogers K, Jacob G (1988) Loss of polymorphic A and B locus HLA antigens in colon carcinoma. Br J Cancer 57: 374–377
Restifo NP, Esquivel F, Kawakami Y, Yewdell JW, Mule JJ, Rosenberg SA, Bennink JR (1993) Identification of human cancers deficient in antigen processing. J Exp Med 177: 265–272
Restifo NP Kawakami Y, Marincola F, Shamamian P, Taggarse A, Esquivel F, Rosenberg SA (1993) Molecular mechanisms used by tumors to escape immune recognition: immunogenetherapy and the cell biology of major histocompatibility complex class I. J Immunother 14: 182–190
Rock KL, Rothstein L, Gamble S, Fleischacker C (1993) Characterization of antigen-presenting cells that present exogenous antigens in association with class I MHC molecules. J Immunol 150: 438–446
Rock KL, Gramm C, Rothstein L, Clark K, Stein R, Dick L, Hwang D, Goldberg AL (1994) Inhibitors of the proteasome block the degradation of most cell protein and the generation of peptides presented of MHC class I molecules. Cell 78: 761–771
Rodriguez F, Peran F, Garrido F, Ruiz-Cabello F (1994) Upmodulation by estrogen of HLA class I expression in breast tumor cell lines. Immunogenetics 39: 161–167
Rotem-Yehudar R, Schechter H, Ehrlich R (1994) Transcriptional regulation of class-I major histocompatibility complex genes transformed in murine cells is mediated by positive and negative regulatory elements. Gene 144: 265–270
Rotem-Yehudar R, Winograd S, Sela S, Coligan JE, Ehrlich R (1994) Downregulation of peptide transporter genes in cell lines transformed with the highly oncogenic adenovirus 12. J Exp Med 180: 477–488
Rotzschke O, Falk K, Deres K, Schild H, Norda M, Metzger M, Jung G, Rammensee HG (1990) Isolation and analysis of naturally processed viral peptides as recognized by cytotoxic T cells. Nature 348: 252–254
Rotzschke O, Falk K, Wallny HJ, Faath S, Rammensee, HG (1990) Characterization of naturally occurring minor histocompatibility peptides including H-4 and H-Y. Science 249: 283–287
Rotzschke O, Falk K, Faath S, Rammensee HG (1991) On the nature of peptides involved in T cell alloreactivity. J Exp Med 174: 1059–1071
Sanda MG, Restifo NP, Walsh JC, Kawakami Y, Nelson WG, Pardoll DM, Simons JW (1995) Molecular characterization of defective antigen processing in human prostate cancer. J Natl Cancer Inst 87: 280–285
Schrier PL, Bernards R, Vaessen RT, Houweling A, Van der Eb, AJ (1983) Expression of class I major histocompatibility antigens switched off by highly onocogenic adenovirus 12 in transformed rat cells. Nature 305: 771–775
Schwartz RH (1989) Acquisition of immunologic self-tolerance. Cell 57: 1073–1081
Shemesh J, Rotem-Yehudar R, Ehrlich R (1991) Transcriptional and posttranscriptional regulation of class I major histocompatibility complex genes following transformation with human adenoviruses. J Virol 65: 5544–5548
Shibata K, Imarai M, Van Bleek GM, Joyce S, Nathenson SG (1992) Vesicular stomatitis virus antigenic octapaptide N52–59 is anchored into the groove of the H-2Kb molecule by the side chains of three amino acids and the main-chain atoms of the amino terminus. Proc Natl Acad Sci USA 89: 3135–3139
Sibille C, Chomez P, Wildmann C, Van Pel A, De Plaen E, Maryanski JL, de Bergeyck V, Boon T (1990) Structure of the gene of tum-transplantation antigen P198: a point mutation generates a new antigenic peptide. J Exp Med 172: 35–45
Smith ME, Bodmer WF, Bodmer JG (1988) Selective loss of HLA-A,B,C locus products in colorectal adenocarcinoma. Lancet ii: 824–825
Smith ME, Marsh SG, Bodmer JG, Gelsthorpe K, Bodmer WF (1989) Loss of HLA-A,B,C allele products and lymphocyte function-associated antigen 3 in colorectal neoplasia. Proc Natl Acad Sci USA 86: 5557–5561
Soong TW, Oei AA, Hui KM (1991) Regulation of HLA class I gene expression in human colorectal carcinoma. Semin Cancer Biol 2: 23–33
Soong TW, Hui KM (1992) Locus-specific transcriptional control of HLA genes. J Immunol 149: 2008–2020
Spies T, DeMars R (1991) Restored expression of major histocompatibility class I molecules by gene transfer of a putative peptide transporter. Nature 351: 323–324

Srivastava PK, Udono H, Blachere NE, Li Z (1994) Heat shock proteins transfer peptides during antigen processing and CTL priming. Immunogenetics 39: 93–98

Stern LJ, Wiley DC (1994) Antigenic peptide binding by class I and class II histocompatibility proteins. Structure 2: 245–251

Stern LJ Brown JH Jardetzky TS, Gorga JC, Urban RG, Strominger JL, Wiley DC (1994) Crystal structure of the human class II MHC protein HLA-DR1 complexed with an influenza virus peptide. Nature 368: 215–221

Suh WK, Cohen-Doyle MF, Fruh K, Wang K, Peterson PA, Williams DB (1994) Interaction of MHC class I molecules with the transporter associated with antigen processing. Science 264: 1322–1326

Szikora JP, Van Pel A, Brichard V, Andre M, Van Baren N, Henry P, De Plaen E, Boon T (1990) Structure of the gene of tum-transplantation antigen P35B: presence of a point mutation in the antigenic allele. EMBO J 9: 1041–1050

Tahara H, Zeb HJ, Storkus WJ, Pappo I, Watkins SC, Gubler U, Wolf SF, Robbins RD, Lotze MT (1994) Fibroblasts genetically engineered of secrete interleukin 12 can suppress tumor growth and induce antitumor immunity to a murine melanoma in vivo. Cancer Res 54: 182–189

Tanaka K, Isselbacher KJ, Khoury G, Jay G (1985) Reversal of oncogenesis by the expression of a major histocompatibility complex class I gene. Science 228: 26–30

Tepper RI, Pattengale PK, Leder P (1989) Murine interleukin-4 displays potent anti-tumor activity in vivo. Cell 57: 503–512

Townsend SE, Allison JP (1993) Tumor rejection after direct costimulation of CD8+T cells by B7-transfected melanoma cells. Science 259: 368–370

Udaka K, Tsomides TJ, Eisen HN (1992) A naturally occurring peptide recognized by alloreactive CD8+ cytotoxic T lymphocytes in association with a class I MHC protein. Cell 69: 989–998

Van Bleek GM, Nathenson SG (1990) Isolation of an endogenously processed immunodominant viral peptide from the class I H-2Kb molecule. Nature 348: 213–216

Van Bleek GM, Nathenson SG (1991) The structure of the antigen-binding groove of major histocompatibility complex class I molecules determines specific selection of self-peptides. Proc Natl Acad Sci USA 88: 11032–11036

Van den Eynde B, Peeters O, De Backer O, Gaugler B, Lucas S, Boon T (1995) A new family of genes coding for an antigen recognized by autologous cytolytic T lymphocytes on a human melanoma. J Exp Med 182: 689–698

Van der Bruggen P, Traversari C, Chomez P, Lurquin C, De Plaen E, Van den Eynde B, Knuth A, Boon T (1991) A gene encoding an antigen recognized by cytolytic T lymphocytes on a human melanoma. Science 254: 1643–1647

Van der Bruggen P, Bastin J, Gajewski T, Coulie PG, Boel P, De Smet C, Traversari C, Townsend A, Boon T (1994) A peptide encoded by human gene MAGE-3 and presented by HLA-A2 induces cytolytic T lymphocytes that recognize tumor cells expressing MAGE-3. Eur J Immunol 24: 3038–3043

van der Bruggen P, Szikora JP, Boel P, Wildmann C, Somville M, Sensi M, Boon T (1994) Autologous cytolytic T lymphocytes recognize a MAGE-1 nonapeptide on melanomas HLA-Cw* 1601. Eur J Immunol 24: 2134–2140

van't Veer LJ, Beijersbergen RL, Bernards R (1993) N-myc suppresses major histocompatibility complex class I gene expression through down-regulation of the p50 subunit of NF-kappa B. EMBO J 12: 195–200

Vieweg J, Rosenthal FM Bannerji R, Heston WD, Fair WR, Gansbacher B, Gilboa E (1994) Immunotherapy of prostate cancer in the Dunning rat model: use of cytokine gene modified tumor vaccines. Cancer Res 54: 1760–1765

Viret C, Davodeau F, Guilloux Y, Bignon JD, Semana G, Breathnach R, Jotereau F (1993) Recognition of shared melanoma antigen by HLA-A2-restricted cytolytic T cell clones derived from human tumor-infiltrating lymphocytes. Eur J Immunol 23: 141–146

von Boehmer H (1992) Thymic selection: a matter of life and death. Immunol Today 13: 454–458

Wallich R, Bulbuc N, Hammerling GJ, Katzav S, Segal S, Feldman M (1985) Abrogation of metastatic properties of tumour cells by de novo expression of H-2K antigens following H-2 gene transfection. Nature 315: 301–305

Wang RF, Robbins PF, Kawakami Y, Kang XQ, Rosenberg SA (1995) Identification of a gene encoding a melanoma tumor antigen recognized by HLA-A31-restricted tumor-infiltrating lymphocytes. J Exp Med 181: 799–804

Wei ML, Cresswell P (1992) HLA-A2 molecules in an antigen-processing mutant cell contain signal sequence-derived peptides. Nature 356: 443–446

Wolfel T, Van Pel A, Brichard V, Schneider J, Seliger B, Meyer zum Buschenfelde KH, Boon T (1994) Two tyrosinase nonapeptides recognized on HLA-A2 melanomas by autologous cytolytic T lymphocytes. Eur J Immunol 24: 759–764

Yamit-Hezi A, Plaksin D, Eisenbach I (1994) c-fos and c-jun overexpression in malignant cells reduces their tumorigenic potential and affects their MHC class I gene expression. Oncogene 9: 1065–1079

Yasumura S, Hirabayashi S, Schwartz DR, Toso JF, Johnson JT, Herbermann RB, Whiteside TL (1993) Human cytotoxic T-cell lines with restricted specificity for squamous cell carcinoma of the head and neck. Cancer Res 53: 1461–1468

Zeh HJ, Leder GH, Lotze MT, Salter RD, Tector M, Stuber G, Modrow S, Storkus WJ (1994) Flow-cytometric determination of peptide-class I complex formation. Identification of p53 peptides that bind to HLA-A2. Human Immunol 39: 76–86

Zinkernagel RM, Doherty PC (1974) Restriction of in vivo T cell mediated cytotoxicity in lymphocyte choriomeningitis within a syngeneic or semi allogeneic system. Nature 248: 701–703

Effect of Antisense Inhibition of Urokinase Receptor on Malignancy*

L. Ossowski

1 Role of Urokinase-Type Plasminogen Activator in Malignancy – General Introduction

Plasminogen activator of the urokinase type (uPA) has been shown to be overexpressed in most experimental malignant tumors and human cancers (Dano et al. 1985; DeBruin et al. 1987; Sappino et al. 1987; Markus 1988; Sim et al. 1988; Duffy et al. 1990; Janicke et al. 1990; Testa and Quigley 1990; Sappino et al. 1991; Mignatti and Rifkin 1993). The overproduction of uPA by tumor cells was detected in primary cultures, organ cultures, and cancer cell lines. However, examination by in situ analyses (both immunocytochemistry and in situ hybridization) of sections of human tumors consisting of multiple cell types identified in some instances cells other than cancer cells as the uPA producers. Depending on the type of cancer, and in some instances the laboratory and the reagents used to perform the analysis, tumor, stromal, and sometimes infiltrating macrophages were found to be the main source of uPA and/or its receptor (Grondahl-Hansen et al. 1991; Pyke et al. 1991, 1993; Bianchi et al. 1994; Carriero et al. 1994). Regardless of its source, however, it has been firmly established in several human cancers that high levels of uPA are predictive of more aggressive disease

Mount Sinai School of Medicine, Department of Medicine, Division of Neoplastic Diseases, Box 1178, New York, NY 10029, USA

*I dedicate this review to a very young colleague, Dr. Yu Wu, whose pain and premature death from breast cancer mobilized us all to resolve to do everything in our power to improve the outcome of cancer therapy.

with shorter relapse-free survival (Duffy et al. 1990; Janicke et al. 1990, 1991; Grondahl-Hansen et al. 1993; Kobayashi et al. 1994).

Two additional components of the plasminogen activation system, the uPA receptor (uPAR) and a specific inhibitor of plasminogen activators, PAI-1, display a distribution similar to that of uPA in human and animal cancers and have also been found to be reliable as independent predictors of the disease-free survival (Janicke et al. 1991; Grondahl-Hansen et al. 1993; Kobayashi et al. 1994).

The discovery of uPA overproduction by tumors led to a flurry of experiments in which its role as a determinant of the invasive phenotype of cancer cells has been shown in numerous in vitro and in vivo models (Ossowski and Reich 1983; Mignatti et al. 1986; Reich et al. 1988; Ossowski 1988a; Hearing et al. 1988; Axelrod et al. 1989; Yu and Schultz 1990). The existing evidence points to uPA as a crucial initiator of a proteolytic cascade leading to the generation of active plasmin and collagenases, which by degrading biological barriers remove the impediment to tumor cell spread. With the discovery of a specific cell surface receptor (Vassalli et al. 1985; Stoppelli et al. 1985), it was shown that receptor-bound uPA can remain active on the cell surface for several hours and that the bound form of uPA is more efficient in matrix degradation and invasion (Ossowski 1988b; Schlechte et al. 1989; Ossowski et al. 1991; Quax et al. 1991; Liu et al. 1995). Surface-bound uPA was shown to be protected from its physiological inhibitors in macrophages (Kirchheimer and Remold 1989), but this was not confirmed in any other cell types (Cubellis et al. 1989). The reason for this discrepancy remains unknown.

In addition to its matrix-degrading function, the binding of uPA or pro-uPA to its receptor has been shown to stimulate chemotaxis and/or chemokinesis in a variety of cell type such as neutrophils, tumor cells, and endothelium (Gudewicz and Gilboa 1987; Fibbi et al 1988; Odekon et al. 1992; Del Rosso et al. 1993; Busso et al. 1994; Stahl and Mueller 1994). The fascinating observation (Wei et al. 1994) that binding of uPA to its receptor confers vitronectin-binding specificity on uPAR suggests that tumor-secreted uPA, by binding to uPAR on endothelial cells and affecting their adhesion and migration, may lead to enhanced angiogenesis. In addition, not unlike the case for other glycophosphatidylinositol (GPI)-linked proteins, which are known to transduce signals through as yet unidentified pathways, the possibility of several signal transduction pathways through this receptor has also been considered. These involve tyrosine phosphyrylation, protein kinase C (PKC), de novo diacylglycerol (DAG) generation etc. (Del Rosso et al. 1993; Dumler et al. 1993, 1994; Anchini et al. 1994; Busso et al. 1994).

2 Experimental Approaches Aimed at Interfering with the Interaction of Urokinase-Type Plasminogen Activator with Its Receptor

The evidence cited above links two important functions of cancer cells, migration and invasion, with the cell surface-bound uPA activity. It is not surprising, therefore, that numerous attempts have been made to interfere with the interaction between uPA and its receptor as a means of specific targeting of cancer therapy . Targeting of uPA–uPAR interaction can be achieved either by competitive uPA antagonists or by specific reduction in the expression or in the binding ability of uPAR. If antagonists with sufficiently high affinity for uPAR can be found, it may be possible to achieve therapeutic efficacy without major side effects. This notion is based on the observation that mice in which the uPA gene has been knocked out by homologous recombination do not suffer severe general consequences (CARMELIET et al. 1994). Theoretically, the best antagonist would be an enzymatically inactive peptide with high uPAR-binding activity. Such peptide derived from the growth factor domain of mouse uPA and spanning the amino acids 17–34 has been tested in Lewis lung carcinoma (KOBAYASHI et al. 1994), and has been shown to significantly reduce lung metastasis from subcutaneously growing tumors. The peptide was ineffective in reducing lung metastasis when coinjected intravenously with cancer cells followed by daily injections. This observation suggests that surface uPA may be required for completion of specific, but not of all, steps of metastasis. It confirms our earlier finding in human carcinoma cells grown in the chick embryo (OSSOWSKI 1988a), in which we showed that uPA-blocking antibodies which drastically reduced the overall level of metastasis did not affect the ability of cells to extravasate, while both connective tissue invasion and, most likely, invasion of blood vessels (intravasation) were shown to be greatly reduced and thus dependent on active uPA (OSSOWSKI 1988a).

An overexpression by tumor cells of inactive or noncleavable forms of uPA via gene therapy may prove to be a sound alternative to the exogenous delivery of therapeutic doses of antagonists. This approach has been tested by transfecting human prostate carcinoma cells (PC3) with a construct expressing catalytically inactive human uPA in which one of the active site amino acids (serine 356) was substituted by alanine. Both "wild type-active" and "mutant-inactive" uPA expressing PC3 cells produced similar-sized tumors when injected subcutaneously into nude mice. However, while active uPA expressing PC3 uPA produced metastases in lymph nodes, bones, brain, and lungs, the metastatic potential of the mutant-expressing PC3 was drastically reduced for all organs tested (CROWLEY et al. 1993). Although this result underscores the immense potential of therapy with uPA antagonists, it does not resolve the question of tumor-specific, in vivo delivery of the genes.

Several other attempts at blocking uPAR have also been described. For example, a hybrid between amino terminal fragment (ATF) of uPA and human serum albumin has been shown to inhibit urokinase-dependent plasminogen activation and invasion through Matrigel (Becton Dickinson, Bedford, MA) (LU et al. 1994). In addition, recombinant soluble human uPAR was used to interfere with uPA binding to its receptor by scavenging the secreted uPA. This intervention was sufficiently potent to inhibit invasion by human ovarian carcinoma cells in vitro (WILHELM et al. 1994).

For all of these approaches to have therapeutic potential, methods for specific delivery of genes and their efficient expression in tumor cells or peptides which bind with high affinity to uPAR must be developed. In absence of such high-affinity antagonists, unrealistically high concentrations of these compounds would have to be delivered. An attempt to identify such peptides through random peptide bacteriophage display has recently been described (GOODSON et al. 1994). It is based on selecting from a bacteriophage library expressing 10^7 random 15-mere peptides those which interact with cells expressing surface human uPAR. At least one peptide with an IC_{50} of 10 n*M* was identified by this technology. It is interesting to note that, although none of the selected peptides contain cysteine residues, believed to be necessary for the proper folding and binding to uPAR, they nevertheless bind with relatively high affinity and specificity. This suggests that secondary structures mimicking those of the amino terminal fragment of uPA may be formed independently of disulfide bonding and offers an opportunity to search for better pharmaceutical agents.

3 Direct Targeting of Urokinase-Type Plasminogen Activator Receptor

It may be possible to directly target the uPAR expression in malignant cells. Since in some cases uPAR appears to be associated with the malignant phenotype, regardless of its uPA-binding potential (QUATTRONE et al. 1995), aiming the therapy toward receptor inactivation may prove to be more effective. Inactivation of uPAR can be achieved either through inhibition of its mRNA with antisense RNA, degradation by specific ribozymes, or proteolytic cleavage of the receptor uPA-binding domain. Switching off the uPAR gene expression by an antimessenger oligonucleotide has been shown to reduce invasion through Matrigel in human fibroblasts transformed with SV-40 virus (QATTRONE et al. 1995). Whether it will be possible to design oligonucleotides that are stable enough and clinically useful remains an open question. The ribozymes, although still highly experimental and not sufficiently specific or stable, may prove to be useful. A successful ribozyme synthesis directed at human uPAR RNA and its delivery to cultured human osteosarcoma cells using lipofectin has been described (KARIKO et al. 1994). While "free" ribozymes were degraded immediately upon delivery to tissue culture medium, in complex with lipofectin they both entered the cells rapidly and were protected from RNAses for up to 22 h, indicating that improvements in this approach are likely.

As mentioned, uPAR binds uPA through a specific sequence located in domain 1 of the receptor (RONNE et al. 1991). Under experimental and in vivo conditions, proteolytic enzymes have been shown to cleave this domain and render uPAR unable to bind its ligand (SOLBERG et al. 1994). uPAR was also shown to be a substrate for the streptococcal pyrogenic exotoxin B secreted by *Streptococcus pyogenes,* which can be inhibited by cysteine proteinase inhibitors (WOLF et al. 1994). Under conditions of streptococcal infection, this bacterial activity may block inflammatory response by monocytes and neutrophils, whose migration is dependent on surface uPA. It may, however, also provide a clue for new receptor-degrading family of enzymes which may have some future therapeutic potential.

Our own choice of intervention was to stably disable the uPAR gene expression by transfecting a highly malignant human epidermoid carcinoma cells, HEp3 (TOOLAN 1954), with a construct producing uPAR antimessenger RNA. We reasoned that since correlative evidence and some less direct experiments put uPAR at the center of invasive activity, blocking its synthesis may be a rate-limiting step in metastasis.

In spite of a vast literature on the use of antisense RNA-producing constructs, there is no clear consensus on the type of construct or the part of cDNA that should be used for this purpose. Good inhibitory effects are shown by some authors to be achieved with the 5'-fragments, while others find the 3'-ends more effective in inhibition. Whether it is gene, cell type, or construct dependent has not been entirely resolved. We prepared a construct by subcloning a 300-bp, polymerase chain reaction (PCR)-amplified 5'-fragment of uPAR cDNA, which included the ATG codon, into pCDM8 or PCDNA/I-NEO plasmids under cytomegalovirus (CMV) promoter in antisense orientation. (Transfection of cells with a full-length antisense construct did not produce any clones with reduced levels of uPAR). The pCDM8-transfected cells were cotransfected with a pSVneo plasmid which renders cells neomycin resistant. Cells transfected with the "empty" vectors (pCDM8 with pSVneo, or pCDNA/I-NEO) served as positive controls. Close to a 100 neo-resistant clones (controls and antisense transfected) were isolated and tested for surface uPAR expression. Among the 40 clones transfected with the vectors alone, the mean receptor site number per cell, determined by binding of ^{125}I-labeled pro-uPA and analyzed by a Scatchard analysis, was approximately 10^5, with less than 15% variation from the mean. Only five of the 60 tested antisense clones showed between 20% and 40% fewer uPAR sites, and only one clone had 74% fewer uPAR sites than the control. Many of the antisense-transfected clones grew very slowly and never reached sufficient density to be isolated and characterized. Since this was not observed in the vector-transfected cultures, we suspect that clones in which the uPAR suppression was maximal may be unable to grow even in vitro. The affinity (k_D) of pro-uPA binding to uPAR varied in different clones from 5×10^{-10} to 1×10^{-9} *M* and was unaffected by the type of the transfected construct. The tested clones did not differ with regard to the rate of growth in culture and the amount and type of protease content; all clones produced high levels (approximately 0.5 Ploug units/10^6 cells) of uPA and two gelatinases (72 and 92 kDA, respectively). Therefore, except for blocking of uPAR, the transfection or the clonal selection did not alter other specific phenotypic properties usually associated with malignancy. The reduction in pro-uPA binding paralleled reduced levels of uPAR protein and uPAR mRNA. These values did not, however, correlate with the levels of uPAR antisense RNA; a clone with only 30% reduction in uPAR had easily detectable levels of antisense RNA; while in the clone with the fewest receptors and very low level of uPAR mRNA, no antisense RNA could be detected (KOOK et al. 1994). Thus, a high copy number of integrated constructs or the presence of detectable levels of antisense RNA did not predetermine the reduction in uPAR mRNA. However, once a reduction in uPAR mRNA level was achieved, it correlated well with a reduced uPAR protein and uPA-binding activity. Similar observations were reported by others in epithelial growth factor receptor (EGFR) or *c-myb* antisense inhibition (Moroni et al. 1992; Raschella et al. 1992).

The metastatic properties of the clones we have isolated and characterized were tested in an in vivo model which we have developed and published (Ossowski 1988 a,b; Ossowski and Reich 1980, 1983). As described briefly below, this model can be used to simultaneously quantitate the overall potential of cells to form metastases and to quantitate individual steps in this process:

1. *Tumorigenicity*. Cells are inoculated onto the chorioallantoic membranes (CAM) of chick embryos and the eggs are incubated for 7 days. The size of nodules produced on the CAM is measured and recorded, and the presence of tumor cells in the nodules is assessed microscopically. The latter is very important, since even a drop of buffer on the CAM can invoke an inflammatory reaction which results, 7 days later, in a well-vascularized nodule.
2. *Metastasis*. Cells inoculated onto a wounded CAM of a chick embryo rapidly grow and disseminate through the vasculature to reach almost every organ of the embryo within 2–3 days (Ossowski 1988a; Ossowski and Reich 1980, 1983). Lungs are used for quantitation since they are the most predictable target for metastasis. Lungs of tumor-bearing chick embryos are removed, minced, and reinoculated on fresh CAM for an additional week of growth. Even if only a very small number of cells are present in the original lung "mince," this number is sufficiently enriched by the second cycle of growth and can be easily detected either microscopically or by measurement of human specific uPA levels (Ossowski and Reich 1983).
3. *Local invasion of connective tissue*. Wounded CAM that have been allowed to reseal in vivo are used for tumor cell inoculation. In this assay, cells are labeled with ^{125}Iodo deoxy Uridine (^{125}IUdR) for 24 h in culture. The labeled cells are inoculated onto CAM and after a 25 h incubation period the cells which have not invaded are removed by extensive wash and trypsinization and the excised CAM, containing cells which have penetrated into the tissue and are protected from trypsin action, are counted in a gamma counter. We showed previously that the resistance to tumor cell invasion was directly proportional to the recovery period ("resealing time") between CAM wounding and cell inoculation (Ossowski 1988b), making this model suitable for testing tumor cells with varied invasive potential.
4. *Extravasation and intravasation*. Extravasation is tested by injecting ^{125}IUdR–labeled cells intravenously and measuring their arrest in organs. However, since we showed previously that this step of metastasis was independent of active uPA (Ossowski 1988a), it will not be discussed further. From indirect evidence, intravasation, or the entry of tumor cells into the circulation, appears to be dependent on active uPA. Since we have only very recently developed a PCR-based test to detect intravasating human tumor cells in the chick embryo, we have not yet had the chance to test the effect of uPAR block on intravasation.

5. *Nude mouse model.* The HEp3 carcinoma grows rapidly in nude mice and produces lung metastases from a subcutaneous site in the majority of mice within a well-defined time period (4 weeks). The primary tumors show histologic and macroscopic signs of local invasion (Ossowski et al. 1987, 1991).

Thus, in the chick embryo model described above, it is possible to test the direct quantitative impact of inhibition of malignancy-linked genes such as uPAR on individual steps of metastasis. Using the same tumor cells, several of these effects can be confirmed in a mammalian, nude mouse model.

We have previously shown that HEp3 cells kept in culture reversibly lose their malignant potential, first losing their ability to form metastasis and then their ability to form tumors. Reexposure of these cells to in vivo conditions (chick embryos or nude mice) restores their malignant potential (Ossowski and Reich 1983). We expected that, because of the need to isolate and culture the transfected clones (controls and uPAR antisense) for prolonged periods in vitro, they would have to be reexposed to in vivo conditions to restore their malignant potential. We tested four antisense-inhibited and six control clones (Kook et al. 1994) and, more recently ten additional HEp3 clones transfected with a construct under β-actin promoter (Wen Yu and L. Ossowski, unpublished results) and found that while the majority of control clones recover turmorigenicity after different lengths of exposure to in vivo conditions, the antisense clones are unable to form tumors even after ten passages (10 weeks) on the CAM. This suggests that reduced surface uPAR may be responsible for the inability of these tumor cells to readapt to in vivo conditions and to form tumors. Alternatively, this can be viewed as a permanently latent state, a condition of obvious interest in cancer therapy. One of the reasons for the defect in establishing growth from a relatively small inoculum may be inferred from our finding that tumor cells in which uPAR sites have been reduced by antisense inhibition show a greatly diminished ability to invade connective tissue. We found that the clone in which the uPAR number was reduced to less than 30% of control showed a 75% reduction in invasiveness (Kook et al. 1994). The invasive ability of clones expressing intermediate number of receptors (between 50% and 80% of control) was appropriately less affected. Since none of the antisense-inhibited clones in the first group of transfectants regained tumorigenicity on the CAM, and therefore no primary tumors were generated, the impact of uPAR inhibition on metastasis could not be tested in the chick embryo. The short gestation period of chick embryos necessitates weekly passage of the inoculated cells even if no growth of the inoculum is observed. The possibility exists, therefore, that some cells may be lost during this procedure. This is not the case in the nude mice model, in which cells are left undisturbed following inoculation for as long as is needed. We therefore tested whether the clone with 70% reduced receptor numbers recovers tumorigenicty when injected subcutaneously into nude mice. Cells from a vector-transfected clone served as a positive control. Tumors initiated by the control cells were palpable within 6 days in all inoculated mice, while the cells of the antisense clone did not become palpable until 24 days after inoculation. Once palpable, however, the growth rates of all tumors was similar and rapid. Based on this outcome, two contrasting conclusions can be drawn with regard to uPAR participation in tumorigenicity: one would postulate that once a tumor is established (following a much extended latency period), uPAR is not required for further growth. A second postulate would predict that a selection process takes place during the latent period, enriching the population in cells which express sufficiently high levels of uPAR. By analyzing cells recovered from the antisense cell-initiated tumors and finding that both uPAR mRNA and uPA-binding activity rose to levels higher than those of in vitro-cultured positive control cells, we determined that the second conclusion was correct. We have not determined whether the loss of uPAR inhibition was due to inactivation of the CMV promoter driving the antisense transcription or other causes. Experiments are currently underway to examine cells transfected with a construct in which the CMV was replaced by a mammalian promoter. In agreement with the CAM invasion assay, the antisense-transfected cells produced nude mouse tumors that were determined by histologic analysis to be 40% less locally invasive. They also produced fewer and more limited lung metastases, but this observation will have to be confirmed with additional clones in a larger number of mice.

Overall these results show, therefore, that diminished expression of surface uPAR leads to a reduction in invasiveness of tumor cells and a substantial increase in tumor latency. This conclusion is reinforced by the observation that the invasive ability of the antisense-transfected cells is inversely proportional to the density of surface uPAR. A diminished malignant potential early in the establishment of tumors (or metastases) may shift the balance in favor of the host by blocking angiogenesis, preventing processing of prohormones or growth factors, or inhibiting formation of chemotactic proteolytic products. In addition, a reduction in their surface proteolytic activity may render tumor cells more susceptible to host defense mechanisms.

Within the limitation of these experimental models, the results shown allow us to conclude that uPAR is indispensable in the early stages of tumor growth and that, if applied to patients with small or clinically undetectable metastases, its blocking may induce a prolonged state of latency. Whether this conclusion will hold true when tested with additional clones, with other tumors and possibly using inducible promoters in which antisense level can be regulated, remains the object of our current investigation. Once this is established, the challenge will be to identify the mechanisms by which a relatively small inoculum of cancer cells enters the phase of invasive growth and to relate it through tissue invasion, angiogenesis, or other signalls to the uPAR function.

4 Conclusions

Experimental evidence from in vitro and in vivo models and the correlation between uPAR expression levels and cancer aggressiveness suggest that uPAR-bound urokinase is an important component of the invasive phenotype and that mere enhancement of soluble uPA production does not entirely substitute for surface-bound activity. Our results show that, at least in experimental models, it is possible to target specific genes for inactivation of expression by antisense RNA and that interference with one of the many steps in the proteolytic cascade results in a dramatic downmodulation of the malignant potential of tumor cells. Therefore, unless very efficient antagonists of uPA can be found, direct blocking of uPAR expression, possibly through antisense or ribozyme, would probably hold most promise for therapy. Our preliminary observation suggesting that a block greater than 70% in uPAR expression may be incompatible with in vivo growth should encourage the use of combined approaches to therapy. For example, uPAR expression can be lowered by antisense techniques, while uPA binding to the remaining receptors can be competed by the use of antagonists, leading to a more comprehensive block of the receptor function.

Acknowledgment. Most of the work from our laboratory discussed in this review was supported by USPHS research grant CA-40758 and The Samuel Waxman Cancer Research Foundation.

References

Anchini E, Fibbi G, Pucci M, Caldini R, Chevanne M, Del Rosso M (1994) Production of second messengers following chemotactic and mitogenic urokinase-receptor interaction in human fibroblasts transfected with human urokinase receptor. Exp Cell Res 213: 438–448

Axelrod JH, Reich R, Miskin R (1989) Expression of human recombinant plasminogen activators enhances invasion and experimental metastasis of H-ras-transformed NIH 3T3 cells. Mol Cell Biol 9: 2133–2141

Behrendt N, Ploug M, Patthy L, Houen G, Blasi F, Dano K (1991) The ligand binding domain of the cell surface receptor for urokinase-type plasminogen activator. J Biol Chem 266: 7842–7847

Bianchi E, Cohn RL, Thor AT, Todd RF 3rd, Mizukami IF, Lawrence DA, Ljung BM, Shuman MA, Smith HS (1994) The urokinase receptor is expressed in invasive breast cancer but not in normal breast tissue. Cancer Res 54: 861–866

Busso N, Masur SK, Lazega D, Waxman S, Ossowski L (1994) Induction of cell migration by pro-urokinase binding to its receptor: possible mechanism for signal transduction. J Cell Biol 126: 259–270

Carmeliet P, Schoonjans L, Kieckens L, Ream B, Degen J, Bronson R, Devos R, van den Ord J, Collen D, Mulligan R (1994) Physiological consequences of the loss of plasminogen activator gene formation in mice. Nature 368: 419–424

Carriero MV, Franco P, Del-Vecchio S, Massa O, Botti G, D'Aiuto G, Stoppelli MP, Salvatore M (1994) Tissue distribution of soluble and receptor bound urokinase in human breast cancer using a panel of monoclonal antibodies. Cancer Res 54: 5445–5454

Crowley CW, Cohen R, Lucas BK, Liu G, Shuman MA, Levinson AD (1993) Prevention of metastasis by inhibition of the urokinase receptor. Proc Natl Acad Sci USA 90: 5021–5025

Cubellis MV, Andreasen P, Ragno P, Mayer M, Dano K, Blasi F (1989) Accessibility of receptor-bound urokinase to type-1 palsminogen activator inhibitor. Proc Natl Acad Sci USA 86: 4828–4834

Dano K, Andreasen PA, Grondahl-Hansen J, Kristensen P, Nielsen L, Skriver L (1985) Plasminogen activators, tissue degradation and cancer. Adv Cancer Res 44: 139–266

DeBruin PA, Griffioen G, Verspaget HV, Verheijen JH (1987) Plasminogen activators and tumor development in the human colon. Activity levels in normal mucosa, adenomateous tissue, polyps and adenocarcinomas. Cancer Res 47: 4654–4657

Del Rosso M, Anichini E, Pedersen N, Blasi F, Fibbi G, Pucci M, Ruggiero M (1993) Urokinase-urokinase receptor interaction: non-mitogenic signal transduction in human epidermal cells. Biochem Biophys Res Commun 190: 347–352

Duffy MJ, Reilly D, O'Sullivan C, O'Higgins N, Fennely JJ (1990) Urokinase plasminogen activator and prognosis in breast cancer. Lancet 13: 108–111

Dumler I, Petri T, Schleuning W-D (1993) Interaction of urokinase-type plasminogen activator (uPA) with its cellular receptor (uPAR) induces phosphorylation on tyrosine of a 38 kDa protein. FEBS Lett 322: 37–40

Dumler I, Petri T, Schleuning W-D (1994) Induction of c-fos gene expression by urokinase-type plasminogen activator in human ovarian cancer cells. FEBS Lett 343: 103–106

Fibbi G, Ziche M, Morbidelli L, Magnelli L, Del Rosso M (1988) Interaction of Urokinase with specific receptors stimulates mobilization of bovine adrenal capillary endothelial cells. Exp Cell Res 179: 385–395

Goodson RJ, Doyle MV, Kaufman SE, Rosenberg S (1994) High affinity urokinase receptor antagonists identified with bacteriophage peptide display. Proc Natl Acad Sci USA 91: 7129–7133

Grondahl-Hansen J, Ralfkiaer E, Kirkeby L, Kristensen P, Lund LR, Dano K (1991) Localization of urokinase-type plasminogen activator in stromal cells in adenocarcinomas of the colon in humans. Am J Pathol 138: 111–117

Grondahl-Hansen J, Christensen IJ, Rosenquist C, Brunner N, Mouridsen HT, Dano K, Blichert-Toft M (1993) High levels of urokinase-type palsminogen activator and its inhibitor PAI-1 in cytosolic extracts of breast carcinomas are associated with poor prognosis. Cancer Res 53: 2513–2521

Gudewicz PW, Gilboa N (1987) Human urokinase-type plasminogen activator stimulates chemotaxis of human neutrophils. Biochem Biophys Res Commun 147: 1176–1181

Hearing VJ, Law LW, Corti A, Appella E, Blasi F (1988) Modulation of metastatic potential by cell surface urokinase of murine melanoma cells. Cancer Res 48: 1270–1278

Hollas W, Blasi F, Boyd D (1991) Role of urokinase receptor in facilitating extracellular matrix invasion by cultured colon cancer. Cancer Res 51: 3690–3695

Janicke F, Schmidt M, Hollrieder A, Babic R, Ulm K, Gossner W, Graeff H (1990) Urokinase-type plasminogen activator (uPA) is a predictor of early relapse in breast cancer. Fibrinolysis 4: 69–78

Janicke F, Graeff H, Schmitt M (1991) Clinical relevance of urokinase-type and tissue-type plasminogen activators and their type I inhibitor in breast cancer. Semin Thromb Hemost 17: 303–312

Kariko K, Megyeri K, Xiao Q, Barrnathan ES (1994) Lipofectin aided cell delivery of ribozyme targeted to human urokinase receptor mRNA. FEBS Lett 352: 41–44

Kirchheimer JC, Remold HG (1989) Functional characteristic of receptor-bound urokinase on human monocytes: catalytic efficiency and susceptibility to inactivation by plasminogen activator inhibitors. Blood 74: 1396–1402

Kobayashi H, Fujishiro S, Terao T (1994) Impact of urokinase-type plasminogen activator and its inhibitor type I on prognosis in cervical cancer. Cancer Res 54: 6539–6548

Kobayashi H, Gotoh J, Fujie M, Shinohara H, Moniwa N, Terao T (1994) Inhibition of metastasis of Lewis lung carcinoma by a synthetic peptide within growth factor domain of urokinase in the experimental and spontaneous metastasis model. Int J Cancer 57: 727–733

Kook Y-H, Adamsk J, Zelent A, Ossowski L (1994) The effect of antisense inhibition of urokinase receptor in human squamous cell carcinoma on malignancy. EMBO J 13: 3983–3991

Liu G, Shuman MA, Cohen RL (1995) Co-expression of urokinase, urokinase receptor and PAI-1 is necessary for optimum invasiveness of cultured lung cells. Int J Cancer 60: 501–506

Lu H, Yeh P, Guitton JD, Mabilat C, Desanlis F, Maury I, Legrand Y, Soria J, Soria C (1994) Blockage of the urokinase receptor on the cell surface: construction and characterization of hybrid protein consisting of the N-terminal fragment of human urokinase and human albumin. FEBS Lett 356: 56–59

Markus G (1988) The relevance of plasminogen activators to neopalstic growth. Enzyme 40: 158–172

Mignatti P, Rifkin DB (1993) Biology and biochemistry of proteinases in tumor invasion. Physiol Rev 73: 161–195

Mignatti P, Robbins E, Rifkin DB (1986) Tumor invasion through the human amniotic membrane: requirement of proteinase cascade. Cell 47: 487–498

Moroni MC, Willingham MC, Beguinot L (1992) EGF-R antisense blocks expression of the epidermal growth factor receptor and suppresses the transformed phenotype of a human carcinoma cell line. J Biol Chem 267: 2714–2722

Odekon LE, Sato Y, Rifkin DB (1992) Urokinase-type palsminogen activator mediates basic fibroblast growth factor-induced bovine endothelial cell migration independent of its proteolytic activity. J Cell Physiol 150: 258–263

Ossowski L (1988a) Plasminogen activator dependent pathways in the dissemination of human tumor cells in the chick embryo. Cell 52: 321–328

Ossowski L (1988b) In vivo invasion of modified membrane by tumor cells: the role of cell surface-bound urokinase. J Cell Biol 107: 2437–2445

Ossowski L (1992) Invasion of connective tissue by human carcinoma cell lines: requirement for urokinase, urokinase receptor, and interstitial collagenase. Cancer Res 52: 6754–6760
Ossowski L, Reich E (1980) Experimental model for quantitative study of metastasis. Cancer Res 40: 2300–2309
Ossowski L, Reich E (1983) Antibodies to plasminogen activator inhibit human tumor metastasis. Cell 35: 611–619
Ossowski L, Russo-Payne H, Gartner M, Wilson EL (1987) Inhibition of urokinase-type plasminogen activator by antibodies: the effect of dissemination of human tumor in the nude mouse. Cancer Res 51: 274–281
Osswski L, Clunie G, Masucci T, Blasi F (1991) In vivo paracrine interaction between urokinase and its receptor: effect on tumor cell invasion. J Cell Biol 115: 1107–1112
Pyke C, Graem N, Ralfkiaer E, Hoyer-Hansen G, Brunner N, Dano K (1993) Receptor for urokinase is present in tumor-associated macrophages in ductal breast carcinoma. Cancer Res 53: 1911–1915
Pyke C, Kristensen P, Ralkfiaer E, Ronne E, Grondahl-Hansen J, Eriksen J, Blasi F, Dano K (1991) Urokinase-type plasminogen activator is expressed in stromal cells and its receptor in cancer cells in invasive foci in human colon adenocarcinomas. Am J Pathol 138: 1059–1067
Quattrone A, Fibbi G, Anichini E, Pucci M, Zamperini A, Capaccioli S, Del Rosso M (1995) Reversion of the invasive phenotype of transformed human fibroblasts by anti-messenger oligonucleotide inhibition of urokinase receptor gene expression. Cancer Res 55: 90–95
Quax PHA, Padersen N, Masucci MT, Weeningverhoeff EJD, Dano K, Verheijnen JH, Blasi F (1991) Complementation between urokinase-producing and receptor-producing cells in extracellular matrix degradation. Cell Regul 2: 793–803
Raschella G, Negroni A, Skorski T, Pucci S, Nieborowska-Skorska M, Romeo A, Calabretta B (1992) Inhibition of proliferation by c-myb antisense RNA and Oligonucleotides in transformed neuro-ectodermal cell lines. Cancer Res 52: 4221–4226
Reich R, Thompson EW, Iwamoto Y, Martin GR, Deason JR, Fuller GC, Miskin R (1988) Effect of inhibitors of plasminogen activator, serine proteinases and collagenase IV on the invasion of basement membranes by metastatic cells. Cancer Res 48: 3307–3312
Roldan AL, Cubellis MV, Masucci MT, Behrendt N, Lund LR, Dano K, Appela E, Blasi F (1990) Cloning and expression of the receptor for human urokinase plasminogen activator, a central molecule in cell surface plasmin-dependent proteolysis. EMBO J 9: 467–474
Ronne E, Behrendt N, Ellis V, Ploug M, Dano K, Hoyer-Hansen G (1991) Cell-induce potentiation of the plasminogen activation system is abolished by a monoclonal antibody that recognizes the NH_2 amino terminal domain of urokinase receptor. FEBS Lett 288: 233–236
Sappino AP, Busso N, Belin D, Vassalli J-D (1987) Increase of urokinase-type plasminogen activator gene expression in human lung and breast carcinomas. Cancer Res 47: 4043–4046.
Sappino A-P, Belin D, Huarte J, Hirschel-Scholtz S, Saurat J-H, Vassalli J-D (1991) Differential protease expression by cutaneous squamous and basal cell carcinomas. J Clin Invest 88: 1073–1079
Schlechte W, Murano G, Boyd D (1989) Examination of the role of urokinase receptors in human colon cancer mediated laminin degradation. Cancer Res 49: 6064–6069
Sim P-S, Stephens RW, Fayle DRH, Doe WF (1988) Urokinase-type plasminogen activator in colorectal carcinomas and adenomatous polyps: quantitative expression of active and proenzyme. Int J Cancer 42: 483–488
Solberg H, Romer J, Brunner N, Sidenius N, Dano K, Hoyer-Hansen G (1994) A cleaved form of the receptor for urokinase-type plasminogen activator in invasive translated human and murine tumors. Int J Cancer 58: 877–881
Stahl A, Mueller BM (1994) Binding of urokinase to its receptor promotes migration and invasion of human melanoma cells in vitro. Cancer Res 54: 3066–3071
Stoppelli MP, Corti A, Sofientini A, Cassani G, Blasi F, Assoian RK (1985) Differentiation-enhanced binding of the amino terminal fragment of human urokinase plasminogen activator to a specific receptor on U937 monocytes. Proc Natl Acad Sci USA 82: 4939–4943
Testa JE, Quigley JP (1990) The role of plasminogen activator in aggressive tumor behavior. Cancer Metastasis Rev 9: 353–367
Toolan HW (1954) Transplantable human neoplasms maintained in cortisone-treated laboratory animals: HS#1, H.E.p#1, H.Ep.#2, H.Ep.#3 and H.Emb.#1. Cancer Res 14: 660–666
Vassalli J-D, Baccino D, Belin (1985) A cellular binding site for the Mr 55,000 form of the human plasminogen activator, urokinase. J Cell Biol 100: 86–92
Wei Y, Waltz DA, Drummond RJ, Rosenberg S, Chapman HA (1994) Identification of the urokinase receptor as an adhesion receptor for vitronectin. J Biol Chem 269: 32380–32388

Vilhelm O, Weidle U, Hohl S, Rottenberger P, Schmitt M, Graeff H (1994) Recombinant soluble urokinase receptor as a scavenger for urokinase-type plasminogen activator. Inhibition of proliferation and invasion of human ovarian cancer cells. FEBS Lett 337: 131–134

Volf BB, Gibson CA, Kapur V, Hussaini IM, Musser JM, Gonias SL (1994) Proteolytically active streptococcal pyrogenic exotoxin B cleaves monocytic cell urokinase receptor and releases an active fragment of the receptor from the cell surface. J Biol Chem 269: 30682–30687

u H, Schultz RM (1990) Relationship between secreted urokinase plasminogen activator activity and metastatic potential in murine B 16 cells transfected with human urokinase sense and antisense genes. Cancer Res 50: 7623–7633

Intervention in Receptor Tyrosine Kinase-Mediated Pathways: Recombinant Antibody Fusion Proteins Targeted to ErbB2

W. WELS[1], B. GRONER[1], and N.E. HYNES[2]

1 Introduction

The *erbB2* gene encodes a 185-kDa transmembrane glycoprotein that is a member of the type I family of receptor tyrosine kinases (RTK) which also includes epidermal growth factor (EGF) receptor, ErbB3, and ErbB4. Amplification and/or overexpression of *erbB2* is observed in human tumors arising at many sites, including the breast and the ovaries, where it correlates with an unfavorable patient prognosis. Its role in cancer development and its accessible location on the cell surface make ErbB2 a target for directed therapy. For reviews related to the biology of ErbB2 as well as to its mechanism of activation,

[1]Institute for Experimental Cancer Research, Tumor Biology Center, P.O. Box 1120, 79011 Freiburg, Germany
[2]Friedrich Miescher Institute, P.O. Box 2543, 4002 Basel, Switzerland

the reader is referred to two recent publications (Hynes and Stern 1994; Dougall et al. 1994).

ErbB2 offers advantages as a target for tumor cell-directed therapy. The differential expression of the receptor, high on the surface of tumor cells and low in normal epithelial tissues, makes it possible to design toxic molecules which preferentially "attack" tumor cells. Uptake of therapeutic molecules bound to the extracellular domain of the receptor can be facilitated either through normal receptor turnover or ligand-induced internalization. The extracellular accessibility of the ErbB2 protein and its growth promoting activity and involvement in signal transduction processes make different strategies for a targeted interference with tumor cell growth possible. First, cytotoxic molecules and effector cells can be engineered to specifically bind to the surface domain of the ErbB2 receptor, enabling them to recognize and selectively kill tumor cells. Second, inhibiting the function of ErbB2 as a signal-transducing molecule might directly affect the growth of tumor cells. ErbB2 antisense oligonucleotides have been shown to inhibit the growth of tumor cells with amplified *erbB2* gene copies (Colomer et al. 1994). Transformation of *erbB2* expressing cells can be reverted by dominant negative variants of the ErbB2 receptor (Messerle et al. 1994).

This review presents three different therapeutic strategies directed at tumors expressing ErbB2. They are based upon the availability of ErbB2-specific monoclonal antibodies (mAb) which have been used in the production of recombinant immunotoxins, in the construction of a novel antibody, CD3 ζ-chain chimera, supplying cytotoxic T cells with tumor cell specificity, and in the inhibition of the transforming ability of ErbB2 via "intracelluar immunization".

2 Recombinant Single-Chain Antibodies Specific for ErbB2

Monoclonal antibodies directed against the extracellular domain of the ErbB2 protein have been described by different groups. Some of these antibodies inhibit the in vitro or in vivo growth of tumor cells when tested alone (Hudziak et al. 1989; Stancovski et al. 1991; Harwerth et al. 1992) or in combination (Kasprzyk et al. 1992; Harwerth et al. 1993). Recent advances in the field of molecular immunology have made it possible to utilize antibody-derived binding domains for the target cell-specific delivery of therapeutic effector functions (Winter and Milstein 1991). From a panel of mAb which specifically recognize the ErbB2 protein (Harwerth et al. 1992), we have derived recombinant single-chain (sc) Fv molecules. They consist of the variable domains of the antibody heavy (V_H) and light (V_L) chains connected via a flexible linker sequence (Wels et al. 1992a). RNA isolated from hybridoma cells was used to prime first-strand cDNA synthesis of the L chain and H chain transcripts. Subsequently, the cDNA were amplified by polymerase chain reaction (PCR) using oligonucleotides designed to match consensus sequences deduced from a collection of known immunoglobulin cDNA sequences at the

5'- and 3'-ends of the rearranged (V_L) and (V_H) regions. The amplified cDNA sequences were joined into one open reading frame using a linker coding for the 15 amino acids (Gly, Gly, Gly, Gly, Ser)$_3$ (WELS et al. 1992a). This amino acid linker allows the correct folding of the antigen-binding domain in the recombinant molecule. The resulting scFv-encoding gene fragments were introduced into a vector suitable for the expression of recombinant proteins in *Escherichia coli.* Recombinant antibodies purified from bacterial extracts are able to bind to ErbB2 with high affinity. This was demonstrated by immunoprecipitation of scFv–ErbB2 complexes from a mixture of scFv molecules derived from mAb FRP5 and FWP51 with tumor cell lysates containing the ErbB2 protein (WELS et al. 1992a). In order to test the capacity of the scFv domain to direct an effector function specifically to ErbB2-expressing tumor cells, the FRP5-derived antibody gene fragment was fused to the bacterial alkaline phosphatase gene *phoA*. The resulting phosphatase fusion protein carrying the scFv domain at the N terminus is enzymatically active and binds specifically to ErbB2-expressing cells, indicating that this molecule is bifunctional (WELS et al. 1992c). Due to its superior activity, the scFv(FRP5) molecule was chosen as a binding domain in three different approaches aimed at the targeted interference with tumor cell growth via the ErbB2 protein.

3 A Bifunctional Antibody–Toxin Specific for ErbB2-Expressing Tumor Cells

3.1 Construction and Functional Characterization

Protein toxins of bacterial and plant origin have been chemically coupled with antibodies and other ligands, and the conjugates have been used for targeted cancer therapy. The elucidation of the molecular structure and the functional domains of bacterial toxins such as diphtheria toxin from *Corynebacterium diphtheriae* and exotoxin A from *Pseudomonas aeruginosa* has led to novel, biotechnologically produced molecules with improved features. Proteins with novel target cell specificity are produced by replacing the original cell-binding domain of such toxins with peptide hormones, growth factors, cytokines, or single-chain antibodies (PASTAN and FITZGERALD 1991).

Pseudomonas aeruginosa exotoxin A (ETA, PE) requires efficient internalization and activation along a multistep pathway for cell killing (ZDANOVSKY et al. 1993). Growth factor receptors such as ErbB2 present suitable targets for ETA-derived fusion toxins. Molecules binding to the extracellular domain of such receptors can be internalized via normal receptor turnover. A truncated ETA gene, encoding amino acids 252–613 and lacking the original cell-binding domain Ia, was fused to the 3'-end of the scFv(FRP5) gene in a bacterial expression vector, as shown in Fig. 1A (WELS et al. 1992b). The bacterially expressed 67-kDa scFv(FRP5)–ETA selectively and efficiently inhibits the in vitro growth of human tumor cell lines

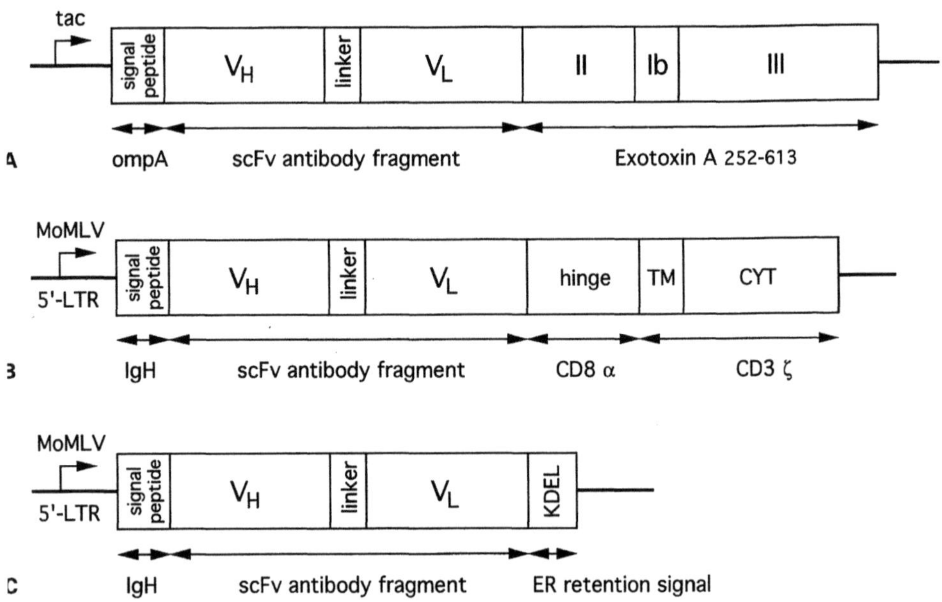

Fig. 1A–C. Single-chain (sc)Fv constructs. **A** Plasmid for the expression of the antibody–toxin scFv(FRP5)–exotoxin A (ETA) in *Escherichia coli* under the control of an inducible tac promoter (*tac*). cDNA sequences encoding the variable domains of the heavy *(V_H)* and light chain *(V_L)* of the antibody are connected via a synthetic linker sequence (*linker*) and fused to a truncated *Pseudomonas* exotoxin A gene encoding the translocation domain (*II*) and the enzymatic domain (*III*) of the toxin (amino acids 252–613). *ompA, E.coli* ompA signal peptide directing expressed protein to the periplasmic space. **B** Retroviral construct for the expression of a scFv CD3 ζ-chain fusion protein in T cells under the control of the 5'-long terminal repeat (LTR) promoter of the Moloney murine leukemia virus (*MoMLV*). The scFv sequences are connected to the transmembrane (*TM*) and cytoplasmic portion (*CYT*) of the CD3 ζ-chain via a flexible hinge region derived from the CD8 α-chain. *IgH*, immunoglobulin heavy chain signal peptide. **C** Retroviral construct for the intracellular expression of ErbB2-specific scFv molecules. A synthetic sequence encoding the amino acid stretch KDEL is fused to the scFv portion in order to achieve the retention of expressed protein in the endoplasmic reticulum (*ER*)

Table 1. Cytotoxic activity of single-chain toxins on human tumor cells in vitro

Cell line	Receptor expression[a]		IC_{50} of single-chain toxin (ng/ml)[c]		
	EGFR	ErbB-2	scFv(FRP5)–ETA[b]	scFv(225)–ETA[d]	TGFα–ETA[d]
SKBR3	+	+++++	34	>1000	72
MDA-MB468	++++	–	>1000	112	5.2
A431	+++++	+	33	5.8	7

EGFR, epidermal growth factor receptor; scFv, single-chain Fv; ETA, exotoxin A; TGF, transforming growth factor.
[a]Receptor expression was determined by quantitative western blot analysis (WELS et al. 1995a).
[b]Specific for ErbB2.
[c]IC_{50} was determined in a cell viability assay (WELS et al. 1992b, 1995a).
[d]Specific for the EGFR.

expressing ErbB2 and mouse cell lines transfected with human *erbB2* cDNA, with IC_{50} values ranging from 2 to 200 ng/ml (WELS et al. 1992b, 1995a). Cells which do not express ErbB2 are not affected by the toxin even at high concentrations, indicating that scFv(FRP5)–ETA toxicity is mediated via selective binding to ErbB2 (Table 1).

3.2 Factors Influencing Toxin Efficacy

The in vitro sensitivity of a target cell for a toxin directed at ErbB2 or other type I RTK family members is influenced by both the amount and the turnover of the receptor. Cells expressing high levels of ErbB2 are quite sensitive to scFv(FRP5)–ETA. Cells with low amounts are resistant (WELS et al. 1992b). We have also observed an increase in scFv(FRP5)–ETA toxicity in EGF-treated SKBR3 human breast carcinoma cells and EGF-treated HC11 mouse mammary epithelial cells expressing a transfected human *erbB2* cDNA (WELS et al. 1995a). This is likely due to the fact that EGF treatment of cells leads to the formation of activated ErbB2 and EGF receptor heterodimers which show an increased rate of internalization (KORNILOVA et al. 1992; DOUGALL et al. 1994). Similarly, HC11 mouse mammary epithelial cells transfected with a human *erbB2* cDNA carrying an activating point mutation in the transmembrane domain of the receptor are more sensitive to treatment with scFv(FRP5)–ETA than HC11 cell expressing normal human ErbB2 protein (WELS et al. 1992b). Activated ErbB2 has a markedly reduced half-life in HC11 cells when compared to its normal cellular counterpart, suggesting that the rapid internalization of toxin–receptor complexes leads to the increased cytotoxic activity of scFv(FRP5)–ETA on such cells.

Two recombinant toxins specific for the EGF receptor have also proven useful in analyzing factors which influence cellular sensitivity to a toxin. The mAb 225 competes with EGF for binding to the EGF receptor. A gene encoding scFv(225)–ETA was constructed as described above, and the recombinant protein was expressed and isolated from bacterial extracts. ScFv(225)–ETA displays potent in vitro cell-killing activity on cells which overexpress the EGF receptor (Table 1; WELS et al. 1995a). A growth factor toxin, transforming growth factor (TGF)-α–ETA, has also been examined for its toxicity on various tumor cell lines. SKBR3 cells express high levels of ErbB2 and moderate amounts of EGF receptor. The cells are sensitive to low concentrations of the ErbB2-specific toxin scFv(FRP5)–ETA and to TGFα–ETA, but resistant to high concentrations of scFv(225)–ETA (WELS et al. 1995a). These results can be explained when the "crosstalk" between the EGF receptor and ErbB2 is considered. TGFα–ETA induces the formation of activated EGF receptor–ErbB2 heterodimers. This leads to a stimulation of heterodimeric receptor internalization. ScFv(225)–ETA does not induce EGF receptor activation and is therefore restricted to passive internalization via normal EGF receptor turnover. This might not allow the accumulation of enough intracellular toxin sufficient for cell killing (M. SCHMIDT and W. WELS, unpublished results). A431 human squamous carcinoma cells overexpress EGF receptor and, as expected, are sensitive to low concentrations of both EGF receptor-directed toxins. Despite the fact that the A431 cells have only 10000–20000 ErbB2 molecules per cell, they are very sensitive to scFv(FRP5)–ETA. A431 cells produce TGFα, which is released into the culture supernatant and activates EGF receptor in an autocrine fashion (VAN DE VIJVER et al. 1991). It is likely that the TGFα-induced activation of ErbB2 through the formation of heterodimers with EGF receptor influences ErbB2 receptor turnover in these cells and may account for the high sensitivity of A431 cells towards the ErbB2-specific toxin.

3.3 Inhibition of Tumor Growth In Vivo

ScFv(FRP5)–ETA effectively inhibited the growth of established, ErbB2-expressing tumors in in vivo nude mouse tumor models. No nonspecific toxicity was observed. The tumors analyzed include SKOV3 human ovarian carcinoma xenografts (WELS et al. 1992b), A431 human squamous cell carcinoma xenografts (WELS et al. 1995a), NIH 3T3 mouse fibroblasts transformed by an activated human *erbB2* cDNA (WELS et al. 1995b), and MAXF1162 human breast carcinoma xenografts (M. SCHMIDT et al., unpublished results). Despite a short half-life in the circulation of mice of 30 min, daily treatment with low doses of scFv(FRP5)–ETA, ranging from 1 to 10 µg for a total of 8–10 days, led to a marked reduction in tumor size (Table 2). Different routes of application of the toxin were compared. Systemic treatment with scFv(FRP5)–ETA via intraperitoneal injection, continuous infusion from subcutaneously implanted miniosmotic pumps, and injection directly into the tumors were all effective.

In adults, ErbB2 is expressed at low levels in epithelial tissues, including the gastrointestinal, respiratory, reproductive, and urinary tract, skin, and breast (PRESS et al. 1990). The antibody–toxin scFv(FRP5)–ETA does not recognize the rodent ErbB2. The question of systemic toxicity of this molecule on normal tissues expressing low ErbB2 levels must be addressed in monkeys expressing

Table 2. Inhibition of tumor cell growth in vivo by single-chain (sc)Fv(FRP5)–exotoxin A (ETA)

Tumor	ErbB2 expression[c]	Treatment	Total dose (µg)	Tumor size (%)[d]
NIH/3T3 #3.7[a]	++	6 µg/day s.c. days 0–7 p.i	48	20
		6 µg/day s.c. days 4–11 p.i	48	55
SKOV3[b]	+++++	1 µg/day s.c. days 10–17 p.i	8	36
		6 µg/day s.c. days 10–17 p.i	48	13
A431[b]	+	10 µg/day i.p. days 6–15 p.i	100	44
MAXF 1162[b]	++	8 µg i.t. on days 25, 27,29, 33, 35 p.i.	40	23

s.c., continuous infusion from subcutaneously implanted miniosmotic pumps; p.i., postimplantation; i.p., intraperitoneal injection; i.t., intratumoral injection.

[a] 2×10^6 tumor cells were injected subcutaneously on day 0.

[b] Established tumor tissue was transplanted on day 0. Treatment was begun when tumor volumes reached approximately 100 mm^3.

[c] Determined by quantitative western blot experiments.

[d] Determined 10 days after the end of treatment as a percentage in comparison to mock-treated animals.

a cross-reacting ErbB2 protein or in rats using a rodent specific scFv–ETA fusion protein. In vivo experiments using a TGFα toxin (TGFα–PE40 PAI et al. 1991) and a heregulin toxin (HRGβ1–ETA) directed at the ErbB3/ErbB4 proteins have revealed that there is a dose-limiting toxicity due to effects on normal tissues expressing the target receptors (PAI et al. 1991; JESCHKE et al. 1995). A reduction of the toxin dose applied and/or the restriction to local treatment might allow a safe application of such reagents in humans (GOLDBERG et al. 1995).

4 Engagement of Cytotoxic T Cells for Directed Tumor Therapy

4.1 Cytotoxic Effector Cells in Tumor Therapy

Antibodies, antibody conjugates, and genetically engineered bifunctional proteins with tumor-specific recognition and toxin domains are emerging as valuable new tools in cancer therapy. In addition, cell-mediated immunity against cancer cells is a promising area for research. Cytotoxic T cells (CTL) are being examined for their ability to kill tumor cells. These cells are very effective in their ability to migrate and penetrate through microvascular walls. In contrast to biological molecules, CTL are designed to extravasate to reach their target cells. At the site of cytolytic action, they secrete cytokines and chemoattractants and activate additional non-specific effector cells. Cell-mediated immunity through the adoptive transfer of CTL clones specific for virally transformed tumor cells was effective in animal models (KAST et al. 1989). Activation of tumor-specific T cells is also effective against human tumors (CHANG and SHU 1992; ROSENBERG 1991). Different populations of lymphocytes have been employed. Lymphocytes from tumor patients were isolated and cultured in the presence of interleukin (IL)-2. These lymphokine-activated killer cells (LAK) killed freshly isolated tumor cells in culture. LAK cells consist of a heterogeneous population of nonspecific natural killer cells and T cells. Antitumor responses were found in patients treated with LAK cells and IL-2 (ROSENBERG et al. 1985). More specific and potent responses were obtained in lymphocytes which were isolated directly from tumor tissue. These tumor-infiltrating lymphocytes (TIL) are $CD8^{+}$T cells with an HLA class I restricted killing specificity. TIL isolated from tumor tissue, expanded in vitro in the presence of IL-2 and reinfused into the patient, caused tumor reduction (ROSENBERG et al. 1988).

4.2 Identification of Target Structures on the Surface of Tumor Cells

The demonstration of the antitumor effects of CTL is encouraging. Since the T cells possess "memory", they might be able to eliminate small numbers of dispersed metastatic cells or reoccurring tumor cells and thus provide long-term protection. A prerequisite for a T cell response, however, is the activation of the cells by foreign antigens. Interaction of these antigens with T and B cell receptors leads to proliferation and clonal expansion. CTL recognize short peptide sequences presented by HLA class I molecules on the target cell surface. Human tumor-associated antigens, however, are rare. The MAGE (melanoma antigen system; VAN DER BRUGGEN et al. 1991), the Pmel 17 (COX et al. 1994), and the MART (melanoma antigen recognized by T cells; KAWAKAMI et al. 1994) are examples of genes encoding cell surface antigens recognized by CTL. It appears that the majority of human tumors do not express cell surface antigens which are efficiently recognised by CTL and thus escape their elimination. For this reason we have resorted to the genetic manipulation of CTL to direct them against defined human tumor cells.

4.3 Structure of the T Cell Receptor

The possibility of genetically manipulating CTL and exploiting them for directed tumor cell therapy stems from recent insights into the structure and function of the T cell receptor. The T cell receptor consists of a variable α/β-heterodimer (Ti) which is responsible for the recognition of antigens. Invariant chains are associated with the α/β-heterodimer. The CD3γ, CD3δ, CD3ε, and the ζ-chains are integral parts of the T cell receptor. The ζ-chains are present as a homodimer in the T cell receptor complex. The early responses upon T cell receptor–antigen association have been studied in CTL which predominantly express the CD8 antigen. Tyrosine phosphorylation of several proteins has been observed. In addition, a phosphatidylinositol-specific phospholipase (phospholipase C, PLC) is activated, cytoplasmic free calcium is elevated, the guanosine triphosphate (GTP)-bound form of the p21 ras protein accumulates, and serine/threonine-specific kinases, including the mitogen-activated protein kinase, are induced. In the intracellular domains of the γ-, δ-, and ε-components of CD3 complex, as well as the ζ-component of the T cell receptor, there is a common sequence (antigen recognition activation motive, ARAM) which exhibits signaling function. Cross-linking of, for example, the ζ-molecule is sufficient to generate proximal and distal signal transduction events. Tyrosine phospshorylation of cellular proteins, PLC activation, calcium elevation, and IL-2 production as well as cytotoxic responses are triggered. This signaling is dependent on the presence of the ARAM, its tyrosine phosphorylation, and the recruitment of cytoplasmic proteins (e.g., ZAP 70) via their SH-2 domains (HOWE and WEISS 1995).

4.4 Chimeric T Cell Receptor Components

The ζ-chain plays a central role in T cell receptor signaling. The extracellular domain of the CD4 molecule and the ζ-molecule were linked (ROMEO and SEED 1991), leading to a recombinant receptor with four extracellular immunoglobulin-like domains linked to the transmembrane and cytoplasmic domains of the ζ-chain. When a chimeric gene encoding this fusion protein was introduced into T cells, it was possible to cross-link the extracellular part of the molecule with an antibody against CD4 and induce signaling. The CD4–ζ fusion receptor was able to direct a (MHC)-major histocompatibility complex-independent cytotoxic effect towards target cells which express the human immunodeficiency virus (HIV) glycoprotein (gp)120 molecule. The HIV envelope protein binds the CD4 molecule with high affinity and is recognized by the CD4–ζ fusion protein. These experiments suggested that the provision of an extracellular recognition domain to the ζ-component of the T cell receptor might be a strategy to direct cytotoxic T cells to predefined target cells.

Single-chain derivatives of mAb are valuable tools which can serve as recognition domains in fusion proteins. They can be used as extracellular recognition sequences for recombinant T cell receptor components. For the reasons discussed above, ErbB2-expressing tumor cells are a suitable model for directed therapy. Since overexpression of ErbB2 is observed in a high percentage of human adenocarcinomas and low expression is found in normal adult tissue, the extracellular domain of the receptor can be considered a tumor-associated antigen. A hybridoma cell line producing mAb FRP5 which recognizes the extracellular domain of ErbB2 was used to prepare cDNA encoding scFv(FRP5). A fusion gene was designed in which the scFv(FRP5) was linked to the transmembrane and cytoplasmic domain of the ζ-molecule. Sequences encoding an amino terminal hydophobic signal peptide derived from an mAb H chain were added to the 5'-end of the construct to direct the fusion protein to the plasma membrane. A schematic diagram of the construct is shown in Fig. 1B.

4.5 Introduction of a Single-Chain Fv-ζ Fusion Gene in T Cells

Retroviral gene transfer was chosen to introduce the fusion receptor gene in T cells. The gene shown in Fig. 1B was cloned into a retroviral vector, and high-titer retroviral supernatants were derived from packaging cell lines transfected with the construct (MORITZ et al. 1994). An established cytotoxic T cell line (CI96 cells) was infected with the recombinant retroviruses. G418-resistant, infected T cells, were examined for expression and cellular localization of the scFv–ζ fusion molecule. A protein comprising both the scFv and the ζ-portion of the molecule could be immunoprecipitated with a ζ- or scFv-specific antiserum. The inclusion of this chimeric protein into the cell membrane was shown by flow cytometry. Two recombinant proteins were made: in the first, the scFv portion of the molecule was fused via a short linker segment to the ζ-molecule; in the second, no linker segment was included. Only the molecule containing the linker showed efficient recognition of antigen by the scFv portion (MORITZ and GRONER 1995). In transduced T cells, the scFv–ζ molecule is present as a disulfide-linked dimeric complex containing either an endogenous ζ-chain or a second chimeric scFv–ζ. Affinity measurements of the scFv part of the molecule to its cognate antigen revealed only a slight difference between the parental antibody FRP5 (k_d, 0.8 × 10^{-9} M) and the scFv (FRP5)–ζ molecule (k_d, 1.5 × 10^{-9} M).

4.6 Signaling and Cytolysis Triggered via Recombinant Single-Chain Fv-ζ Receptors

The signaling function of the scFv–ζ chimera in CTL was investigated. Antigen-specific stimulation of the receptor induced proximal and distal T cell activation signals. ErbB2-mediated cross-linking led to an increase of intracellular calcium concentrations and to interferon-γ production (MORITZ et al. 1994). The scFv-ζ receptor behaves like an autonomous signal-transducing unit which stimulates the same intracellular chain of events as the association of the α/β-chain of the T cell receptor with MHC class I molecules and antigen. This is consistent with experiments in which antibody-mediated cross-linking of chimeric receptors has been investigated. Tyrosine phosphorylation of cellular substrates, phosphatidylinositol turnover, an increase in intracellular calcium concentrations, and the induction of cytokine production have been observed (ROMEO and SEED 1991; WEGENER et al. 1992).

Most importantly, CTL which express the scFv–ζ molecule recognize ErbB2 receptors on the surface of target cells and are able to lyse these cells. The cytolysis is independent of coreceptor functions and not restricted by the MHC. Similar results have been obtained with other recombinant T cell receptor components (ESHHAR et al. 1993). Transduced CTL are not only able to specifically recognize target cells in vitro, but are also able to exert cytolytic activity in vivo. The simultaneous injection of scFv–ζ-transduced CTL and ErbB2-expressing tumor cells resulted in growth inhibition of the tumor cells in nude mice. The in vivo and in vitro cytolotic effects were stricly dependent on the specifications of the CTL and the target cells: only cytotoxic T cells transduced with the scFv–ζ construct exerted lytic activity, and only target cells expressing human ErbB2 were lysed. In contrast to the in vitro experiments, in which 100% of the target cells could be lysed, not all tumor cells were eliminated in vivo. The efficiency with which distant tumor sites are detected by the cytotoxic lymphocytes might require higher T cell numbers for the quantitative elimination of the tumor cells.

The principle of genetically engineered cytotoxic T cells and their provision with grafted recognition specificity has been established. A major task for the future will be the adaptation of this principle to the situation found in a tumor patient. It will be necessary to transduce primary T cells with retroviruses at a high efficiency. This might be difficult in patients that have been subjected to chemotherapy and have low numbers of T cells. Genetically modified CTL can be directed towards many different surface targets. Tumor-directed toxins such as those described above rely on internalization and intracellular processing to exert their cytotoxic effects. It is conceivable that many tumor-associated antigens which could serve as targets for directed therapy do not mediate cellular internalization. For these targets, a killing mechanism similar to the one offered by the cytotoxic T cells would be advantageous. Since these targets, with the exception of viral antigens, do not represent real neoantigens, they are likely to be expressed on the surface of normal cells. To avoid symptoms of autoimmunity, extensive systemic toxicity studies will be required before genetically modified cytotoxic T cells can be used in human cancer therapy.

5 Intracellular Expression of ErbB2-Specific Single-Chain Antibodies

The intracellular expression of scFvs presents a novel approach to interfere with receptor function (BEERLI et al. 1994a,b; GRAUS-PORTA et al. 1995). We have used the approach to show the following: first, the importance of ErbB2 in transmitting signals from other members of the type I RTK family; second, the feasibility of reversing ErbB2-induced transformation via expression of scFvs. For expression in eukaryotic cells, the ErbB2-specific scFv(FRP5) cDNA was provided with an N-terminal immunoglobulin H chain-derived signal peptide which directs it to the secretory compartment of the cell, the same compartment through which ErbB2 passes on its way to the plasma membrane. At its C terminus, the scFv was provided with a KDEL peptide. This peptide is recognized by an endoplasmic reticulum (ER)-resident receptor which has been implicated in the retention of soluble proteins in the lumen of the ER (MUNRO and PELHAM 1987). A diagram of the scFv used for expression in eukaryotic cells is shown in Fig. 1C.

5.1 Activation and Crosstalk of Type I Receptor Tyrosine Kinases

Four members of the type I family of RTK have been identified: ErbB/EGF receptor, ErbB-2, ErbB3, and ErbB4. The four proteins are normally coexpressed in various combinations in diverse tissues excluding the hematopoietic system. Ligands binding to EGF receptor, ErbB3, and ErbB4 have been identified. Several growth factors, including EGF, TGFα, and amphiregulin, bind and activate the EGF receptor (reviewed in SALOMON et al. 1995). Recently, a number of peptide factors which were simultaneously identified by different groups were shown to induce tyrosine phosphorylation of ErbB2, and thus they were described as putative ligands for this receptor. The names reflect the source of isolation and include neu differentiation factor (NDF), heregulin (HRG), glial growth factor (GGF), and acetylcholine receptor-inducing activity (ARIA) (PELES and YARDEN 1993). However, it has recently been shown that ErbB2 does not directly bind NDF. Instead, ErbB3 and ErbB4 function as receptors for NDF (CARRAWAY and CANTLEY 1994; TZAHAR et al. 1994), and NDF-induced tyrosine phosphorylation of ErbB2 occurs only in the presence of either ErbB3 or ErbB4, presumably by heterodimerization and cross-phosphorylation. Similarly, activation of EGF receptor by EGF and other agonists stimulates tyrosine phosphorylation of ErbB2 by heterodimer formation (DOUGALL et al. 1994). Although ErbB2 alone cannot bind any of these ligands, it has been shown to modulate ligand affinities. ErbB2 confers high-affinity binding sites for EGF by heterodimerizing with EGF receptor (DOUGALL et al. 1994), as well as for NDF by heterodimerizing with ErbB3 (CARRAWAY and CANTLEY 1994). It is likely that ErbB2 also modulates intracellular signals elicited by EGF and NDF, since both factors cause its phosphorylation and activation.

In order to investigate the involvement of ErbB2 in EGF and NDF-induced signaling, intracellular expression of scFvs has been used to specifically and efficiently inactivate ErbB2 in vivo. ScFvs directed against the extracellular domain of ErbB2 have been expressed intracellularly in T47D human mammary carcinoma cells, which express moderate levels of all the known members of the type I RTK family. This leads to the specific retention of ErbB2 in the ER; the other members of the type I RTK are still present on the plasma membrane. In order to investigate intracellular signaling, two different kinases were examined: the mitogen-activated protein kinase (MAPK) and the ribosomal proteins S6 kinase ($p70/85^{S6K}$). EGF- and NDF-induced activation of both kinases and stimulation of growth are impaired in cells lacking cell surface ErbB2 (GRAUS-PORTA et al. 1995). These results shown that ErbB2 plays a central role in signaling by this family of receptors.

The strength of the signal leading to activation of both kinases is modulated by ErbB2 and directly correlates with the anchorage-independent proliferative response. These results may help to expalin why ErbB2 is overexpressed in many types of human tumors (HYNES and STERN 1994). Breast cancer cells often express ligands which bind the EGF receptor, including TGFα and amphiregulin (SALOMON et al. 1995). NDF has also been found in primary breast tumors (BACUS et al. 1993). Even when expressed at low levels, as in the T47D cells, ErbB2 has the ability to appropriate signals arising from other members of the receptor family. Its coexpression with other type I RTK leads to an increase in the transformed phenotype of T47D cells. It is possible that the high levels of ErbB2 expressed in primary tumors cause an enhancement of this phenomenon. It will be important to examine primary tumors for ErbB2 levels and for coexpression of ligands binding type I RTK. The scFv-mediated inactivation of growth factor receptors represents a novel way to study growth factor signaling. The application of such an approach is not limited to the study of type I RTK, but holds promise for other complex receptor systems.

5.2 Inhibition of the Transforming Potential of ErbB2

In order to test the potential of inhibiting transformed cell growth via intracellular expression of scFvs, ErbB2-transformed fibroblasts have been used as a model. The scFv(FRP5) was introduced into NIH/3T3 cells expressing an oncogenically activated form of the ErbB2 receptor that carries a single amino acid substitution (valine to glutamic acid) in the transmembrane domain. In these cells, the kinase activity of ErbB2 is constitutive and ligand independent. The cells show the typical phenotype of transformed fibroblasts as judged by morphology, focus formation, ability to undergo anchorage-independent growth, and tumor formation in nude mice. Expression of scFv(FRP5) in these cells leads to intracellular retention of ErbB2. Immunofluorescence and fluorescence-activated cell sorter (FACS) analyses confirmed that the receptor was not present on the plasma membrane. Expression of scFv(FRP5) did not affect the ER transit of other growth factor receptors, platelet-derived growth factor (PDGF) and fibroblast growth factor (FGF)-2 stimulated growth of the cells. However, retention of ErbB2 in the ER resulted in a drastic change in morphology of the ErbB2 transformed NIH/3T3 cells. The rounded shape of the transformed cells reverted to the more flattened appearance of normal fibroblasts. In addition, the scFv(FRP5) receptor expressing cells no longer formed foci. Finally, and most importantly, the anchorage-independent growth of the cells was reduced by 96% compared to the parental transformed cells (Beerli et al. 1994a). The mechanism leading to inhibition of transformation is likely due, at least in part, to the lower level of kinase activity displayed by the ER-retained ErbB2. In parallel, various intracellular substrates displayed lower levels of phosphotyrosine.

ScFv-mediated immunointerference represents a simple and powerful method to inhibit the in vivo function of receptors. The approach has shown the central role which ErbB2 plays in eliciting signals from the type I RTK family. It has also provided evidence that preventing the appearance of the ErbB2 protein on the plasma membrane of tumor cells may lead to reversion of the malignant phenotype.

6 Conclusions

The targeting strategies directed at *erbB2*-overexpressing tumor cells discussed in the review are based upon two characteristics of the protein. First, its extracellular accessibility provides a target for cytotoxic molecules or cells. The differential expression of the receptor, high on the surface of tumor cells and low in normal epithelial tissues, makes it possible to design toxic molecules which preferentially recognize tumor cells. However, before these molecules can be practically applied in the clinic, extensive preclinical toxicity studies are necessary as scFv–ETA or the engineered CTL might have unforeseen side effects in humans. Second, ErbB2 is a potent signal-transducing molecule whose activation stimulates numerous intracellular pathways which are likely to be important in tumor cell growth. We have shown that intracellular expression of scFvs can directly affect ErbB2 signaling. The utilization of these techniques will require the targeted delivery of the scFv-expressing gene to the tumor cell, an area which is dependent on the development of gene transfer vectors suited for systemic application.

The progress in such diverse areas of research as molecular oncology, tumor immunology, molecular biology, biotechnology, and gene therapy converges in the development of new therapeutic modalities. In the near future, it will become possible to "individualize" cancer therapy, i.e., to exploit the biochemical characteristics of the cancer cells of individual patients and administer drugs which act specifically on these cells. This might diminish the current side effects of cytotoxic drugs and increase the success rate of the treatment.

Acknowledgments. We would like to thank Roger Beerli, Dirk Moritz, and Diana Graus-Porta at the Friedrich Miescher Institute, Basel, and Mathias Schmidt at the Tumor Biology Center, Freiburg, for helpful discussions and their contributions to the work described in this review.

References

Bacus SS, Gudkov AV, Zelnik CR, Chin D, Stern R, Stancovski I, Peles E, Ben-Baruch N, Farbstein H, Lupu R, Wen D, Sela M, Yarden Y (1993) Neu differentiation factor (heregulin) induces expression of intercellular adhesion molecule 1: implications for mammary tumors. Cancer Res 53: 5251–5261

Beerli RR, Wels W, Hynes NE (1994a) Intracellular expression of single chain antibodies reverts ErbB-2 transformation. J Biol Chem 269: 23931–23936

Beerli RR, Wels W, Hynes NE (1994b) Autocrine inhibition of the epidermal growth factor receptor by intracellular expression of a single-chain antibody. Biochem Biophys Res Commun 204: 666–672

Carraway KL, Cantley LC (1994) A neu acquaintance for ErbB3 and ErbB4: a role for receptor heterodimerization in growth signaling. Cell 78: 5–8

Chang AE, Shu S (1992) Immunotherapy with sensitized lymphocytes. Cancer Invest 10: 357–369

Colomer R, Lupu R, Bacus SS, Gelmann EP (1994) ErbB-2 antisense oligonucleotides inhibit the proliferation of breast carcinoma cells with erbB-2 oncogene amplification. Br J Cancer 70: 819–825

Cox AL, Skipper J, Chen Y, Henderson RA, Darrow TL, Shabanowitz J, Engelhard VH, Hunt DF, Slingluff CL (1994) Identification of a peptide recognized by five melanoma-specific human cytotoxic T cell lines. Science 264: 716–719

Dougall WC, Qian X, Peterson NC, Miller MJ, Samanta A, Greene MI (1994) The neuoncogene: signal transduction pathways, transformation mechanisms and evolving therapies. Oncogene 9: 2109–2123

Eshhar Z, Waks T, Gross G, Schindler DG (1993) Specific activation and targeting of cytotoxic lymphocytes through chimeric single chains consisting of antibody-binding domains and the gamma or zeta subunits of the immunoglobulin and T-cell receptors. Proc Natl Acad Sci USA 90: 720–724

Goldberg MR, Heimbrook DC, Russo P, Sarosdy MF, Greenberg RE, Giantonio BJ, Linehan WM, Walther M, Fisher HAG, Messing E, Crawford ED, Oliff AE, Pastan IH (1995) Phase I clinical study of the recombinant oncotoxin TP40 in superficial bladder cancer. Clin Cancer Res 1: 57–61

Graus-Porta D, Beerli RR, Hynes NE (1995) Single-chain antibody-mediated intracellular retention of ErbB-2 impairs neu differentiation factor and epidermal growth factor signaling. Mol Cell Biol 15: 1182–1191

Harwerth IM, Wels W, Marte BM, Hynes NE (1992) Monoclonal antibodies against the extracellular domain of the erbB-2 receptor function as partial ligand agonists. J Biol Chem 267: 15160–15167

Harwerth IM, Wels W, Schlegel J, Müller M, Hynes NE (1993) Monoclonal antibodies directed to the erbB-2 receptor inhibit in vivo tumour cell growth. Br J Cancer 68: 1140–1145

Howe LR, Weiss A (1995) Multiple kinase mediate T-cell-receptor signaling. Trends Biochem Sci 20: 59–64

Hudziak RM, Lewis GD, Winget M, Fendly BM, Shepard HM, Ullrich A (1989) $p185^{HER2}$ monoclonal antibody has antiproliferative effects in vitro and sensitizes human breast tumor cells to tumor necrosis factor. Mol Cell Biol 9: 1165–1172

Hynes NE, Stern DF (1994) The biology of erbB-2/neu/HER-2 and its role in cancer. Biochim Biophys Acta 1198: 165–184

Jeschke M, Wels W, Dengler W, Imber R, Stöcklin E, Groner B (1995) Targeted inhibition of tumor cell growth by recombinant heregulin-toxin fusion proteins. Int J Cancer 60: 730–739

Kasprzyk PG, Song SU, DiFiore PP, King CR (1992) Therapy of an animal model of human gastric cancer using a combination of anti-erbB-2 monoclonal antibodies. Cancer Res 52: 2771–2776

Kast MW, Offringa R, Peters PJ, Voordouw AC, Meloen RH, van der Eb AJ, Melief CJM (1989) Eradication of adenovirus E1-induced tumors by E1A-specific cytotoxic T lymphocytes. Cell 59: 603–614

Kawakami Y, Eliyahu S, Delgado CH, Robbins PF, Sakaguchi K, Appella E, Yannelli JR, Adema GJ, Miki T, Rosenberg SA (1994) Identification of a human melanoma antigen recognized by tumor-infiltrating lymphocytes associated with in vivo tumor rejection. Proc Natl Acad Sci USA 91: 6458–6462

Kornilova ES, Taverna D, Hoeck W, Hynes NE (1992) Surface expression of erbB-2 protein is post-transcriptionally regulated in mammary epithelial cells by epidermal growth factor and by the culture density. Oncogene 7: 511–519

Messerle K, Schlegel J, Hynes NE, Groner B (1994) NIH/3T3 cells transformed with the activated erbB-2 oncogene can be phenotypically reverted by a kinase deficient, dominant negative erbB-2 variant. Mol Cell Endocrinol 105: 1–10

Moritz D, Groner B (1995) A spacer region between the single chain antibody and the CD3 ζ-chain domain of chimeric T-cell receptor components is required for efficient ligand binding and signaling activity. Gene Ther 2: 539–546

Moritz D, Wels W, Mattern J, Groner B (1994) Cytotoxic T lymphocytes with a grafted recognition specificity for erbB-2 expressing tumor cells. Proc Natl Acad Sci USA 91: 4318–4322

Munro S, Pelham HRB (1987) A C-terminal signal prevents secretion of luminal ER proteins. Cell 48: 899–907

Pai LH, Gallo MG, FitzGerald DJ, Pastan I (1991) Antitumor activity of a transforming growth factor α-Pseudomonas exotoxin fusion protein (TGF-α-PE40). Cancer Res 51: 2808–2812

Pastan I, FitzGerald D (1991) Recombinant toxins for cancer treatment. Science 254: 1173–1177

Peles E, Yarden Y (1993) Neu and its ligands: from an oncogene to neural factors. Bioassays 15: 815–824

Press MF, Cordon-Cardo C, Slamon DJ (1990) Expression of the HER-2/neu proto-oncogene in normal human adult and fetal tissues. Oncogene 5: 953–962

Romeo C, Seed B (1991) Cellular Immunity to HIV activated by CD4 fused to T cell or Fc receptor polypeptides. Cell 64: 1037–1046

Rosenberg SA (1991) Immunotherapy and gene therapy of cancer. Cancer Res 51: 5074s–5079s

Rosenberg SA, Lotze MT, Muul LM, Leitman S, Chang AE, Ettinghausen SE, Matory YL, Skibber JM, Shiloni E, Vetto JT et al (1985) Observations on the systemic administration of autologous lymphokine-activated killer cells and recombinant interleukin-2 to patients with metastatic cancer. N Engl J Med 313: 1485–1492

Rosenberg SA, Packard BS, Aebersold PM, Solomon D, Topalian SL, Toy ST, Simon P, Lotze MT, Yang JC, Seippe CA (1988) Use of tumor-infiltrating lymphocytes and interleukin-2 in the immunotherapy of patients with metastatic melanoma. A preliminary report. N Engl J Med 319: 1676–1680

Salomon DS, Brandt R, Ciardiello F, Normanno N (1995) Epidermal growth factor-related peptides and their receptors in human malignancies. Crit Rev Oncol Hematol 19: 183–232

Stancovski I, Hurwitz E, Leitner O, Ullrich A, Yarden Y, Sela M (1991) Mechanistic aspects of the opposing effects of monoclonal antibodies to the ERBB2 receptor on tumor growth. Proc Natl Acad Sci USA 88: 8691–8695

Tzahar E, Levkowitz G, Karunagaran D, Yi L, Peles E, Lavi S, Chang D, Liu N, Yayon A, Wen D, Yarden Y (1994) ErbB-3 and ErbB-4 function as the respective low and high affinity receptors of all neu differentiation factor/heregulin isoforms. J Biol Chem 269: 25226–25233

van der Bruggen P, Traversari C, Chomez P, Lurquin C, De Plaen E, Van den Eyde B, Knuth A, Boon T (1991) A gene encoding an antigen recognized by cytolytic T lymphocytes on a human melanoma. Science 254: 1643–1647

Van De Vijver MJ, Kumar R, Mendelsohn J (1991) Ligand-induced activation of A431 cell epidermal growth factor receptors occurs primarily by an autocrine pathway that acts upon receptors on the surface rather than intracellularly. J Biol Chem 266: 7503–7508

Wegener AM, Letourneur F, Hoeveler A, Brocker T, Luton F, Malissen B (1992) The T cell receptor/CD3 complex is composed of at least two autonomous transduction modules. Cell 68: 83–95

Wels W, Harwerth IM, Hynes NE, Groner B (1992a) Diminution of antibodies directed against tumor cell surface epitopes: a single chain Fv fusion molecule specifically recognizes the extracellular domain of the c-erbB-2 receptor. J Steroid Biochem Mol Biol 43: 1–7

Wels W, Harwerth IM, Müller M, Groner B, Hynes NE (1992b) Selective inhibition of tumor cell growth by a recombinant single-chain antibody-toxin specific for the erbB-2 receptor. Cancer Res 52: 6310–6317

Wels W, Harwerth IM, Zwickl M, Hardman N, Groner B, Hynes NE (1992c) Construction, bacterial expression and characterization of a bifunctional single-chain antibody-phosphatase fusion protein targeted to the human erbB-2 receptor. Biotechnology 10: 1128–1132

Wels W, Beerli R, Hellmann P, Schmidt M, Marte BM, Kornilova ES, Hekele A, Mendelsohn J, Groner B, Hynes NE (1995a) EGF receptor and $p185^{erbB\text{-}2}$ specific single-chain antibody-toxins differ in their cell killing activity on tumor cells expressing both receptor proteins. Int J Cancer 60: 137–144

Wels W, Moritz D, Schmidt M, Jeschke M, Hynes NE, Groner B (1995b) Biotechnological and gene therapeutic strategies in cancer treatment. Gene 159: 73–80

Winter G, Milstein C (1991) Man-made antibodies. Nature 349: 293–299

Zdanovsky AD, Chiron M, Pastan I, FitzGerald DJ (1993) Mechanism of action of Pseudomonas exotoxin. Identification of a rate-limiting step. J Biol Chem 268: 21791–21799

Interactions Between CD44 and Hyaluronic Acid: Their Role in Tumor Growth and Metastasis

M.S. Sy[1,2,3], D. Liu[1], R. Schiavone[1], J. Ma[1], H. Mori[1], and Y. Guo[1,3]

1 Introduction

Adhesion molecules enable lymphocytes to interact with antigen-presenting cells or target cells more effectively (Springer 1990; Butcher 1986; Stoolman 1989). Adhesion molecules also allow lymphocytes or monocytes to interact with endothelial cells or high endothelial venules (HEV). Adhesion molecules enable lymphocytes or monocytes to recirculate and home to specific anatomical sites during inflammation. Tumor metastasis is a complex phenomenon involving a sequence of events that remain poorly understood (Fidler 1978; Nicolson 1988; Kahn 1992; Stetler-Stevenson et al. 1993; Turley 1984). Tumor metastatic cascade involves tumor cell and host cell interactions and may also involve interactions among tumor cells. Tumor cells developed in different microenvironments may utilize different mechanisms for invasion and metastasis. This interpretation was supported by results obtained from experiments using orthoptopic implantation of human carcinoma cells in nude mice (Fidler and Radinsky 1990; Fidler et al. 1990). Human colon carcinoma cell lines do not metastasize unless they are

[1]Institute of Pathology, School of Medicine, Case Western Reserve University, Cleveland, OH 44106, USA
[2]Department of Dermatology, School of Medicine, Case Western Reserve University, Cleveland, OH 44106, USA
[3]Cancer Research Center, School of Medicine, Case Western Reserve University, Cleveland, OH 44106, USA

injected into the cecum or spleen of nude mice (FIDLER 1991). Subcutaneous injection of the same tumor cell lines results in a significantly lower frequency of metastasis formation. These observations have also been observed with human breast, stomach, pancreas, and prostate tumor cell lines (FIDLER and RADINSKY 1990; FIDLER 1990, 1991). In general, orthotopic transplants of human tumors into nude mice favors recapitulation of the metastatic patterns seen in patients with the same tumor. Tumor cells may use adhesion molecules for tumor cell–tumor cell or tumor cell–host cell interactions. These interactions may be essential for tumor growth and metastasis. Furthermore, tumor cells must be able to detach from the tumor mass in order to metastasize to other sites. Shedding or downregulation of adhesion molecules or expression of proteolytic enzymes by tumor cells may be one of the mechanisms that allows tumor cells to free themselves and migrate to other sites. Recent studies revealed that one of the adhesion molecules, namely CD44, may be important in the metastasis of tumor cells (HUGHES et al. 1981; TROWBRIDGE et al. 1982; HAYNES et al. 1989; CARTER and WAYNER 1988; LYNCH and CEREDIG 1988; JALKANEN et al. 1987; LESLEY et al. 1993). In this review, we summarize some of our recent work on the role of CD44 in tumor growth and metastasis.

2 CD44 Molecules

CD44 is a surface molecule originally thought to be involved in lymphocyte and HEV interaction (HUGHES et al. 1981; TROWBRIDGE et al. 1982; HAYNES et al. 1989; CARTER and WAYNER 1988; LYNCH and CEREDIG 1988; JALKANEN et al. 1987; LESLEY et al. 1993). Monoclonal antibodies against CD44 recognize an 80- to 90-kDa glycoprotein on human lymphoid cells. $CD44^+$ lymphocytes can adhere to the characteristic plump or "high" endothelium found lining the postcapillary venules. They also adhere to Peyer's patches and gut-associated lymphatic structures. CD44 may be associated with site-specific extravasation of lymphocytes into tissues. In addition, some anti-CD44 antibodies block T cell erythrocyte rosetting and stimulate T cell proliferation with other monoclonal antibodies specific for CD2 and CD3 antigens on T cells (HAYNES et al. 1989; HUET et al. 1989; SHIMIZU 1989; DENNING et al. 1990). Some monoclonal anti-CD44 antibodies have been shown to be able to inhibit T cell proliferation and cytokine production in vitro (ROTHMAN et al. 1991; GUO et al. 1993). Anti-CD44 antibodies also enhance natural killer (NK) cell activity (TAN et al. 1993). Cross-linking of CD44 on monocytes stimulates the production of interleukin (IL)-1 and tumor necrosis factor (TNF) (WEBB et al. 1990). Thus, CD44 may serve as a signal-transducing molecule. Expression of CD44 is enhanced on murine memory T cells and activated B cells (BUDD et al. 1987; CAMP 1991; MURAKAMI et al. 1990). Expression of CD44 is not restricted to lymphoid cells; it is also expressed in brain and on granulocytes, erythrocytes, fibroblasts, keratinocytes, and red blood cells (HUGHES et al. 1981;

TOWBRIDGE et al. 1982; HAYNES et al. 1989; CARTER and WAYNER 1988; LYNCH and CEREDIG 1988; JALKANEN et al. 1987; LESLEY et al. 1993).

The cDNA for CD44 has been cloned (NOTTENBURG et al. 1989; ZHOU et al. 1989; WOLFFE et al. 1990). Murine CD44 is located on chromosome 2. Human CD44 is located on the short arm of chromosome 11 (NOTTENBURG et al. 1989; ZHOU et al. 1989; WOLFFE et al. 1990). Southern blot analysis of DNA from various sources showed complex patterns, suggesting either the presence of multiple exons or a multigene family. In earlier studies, two isoforms of CD44 were identified based on their mRNA sizes and the molecular weights of the proteins (STAMENKOVIC et al. 1991; BROWN et al. 1991). The first form (CD44H) is expressed in hematopoietic cells with a protein product of 80–90 kDa. The second form (CD44E) is the epithelial protein of 130–160 kDa expressed weakly in formal epithelium, but highly expressed in carcinomas (STAMENKOVIC 1989, 1991). In addition to the epithelial form of CD44, carcinomas also express CD44H (STAMENKOVIC 1989, 1991). Recent data suggest that there are at least nine isoforms of CD44 generated by differential splicing of the mRNA (LESLEY et al. 1993; SCREATON et al. 1992). The hematopoietic form of CD44 in lymphoid cells represent the basic standard unit of CD44 proteins. This isoform has also been referred to as CD44S – the standard CCD44 isoform. Other isoforms are created by alternative splicing of the mRNA and the addition of new exons into the extracellular domain near the transmembrane region of the hematopoietic form of CD44. Each exon has the potential to encode for a separate domain in the mature CD44 protein. At least ten distinguishable variable domains have been proposed (TOLG et al. 1993). At the present time, 20 different exons have been identified in the human CD44 gene (SCREATON et al. 1993). Since all carcinomas express multiple CD44 isoforms, the isoform of CD44 which is most important in the growth and metastasis of the carcinoma is not known. While many of the carcinomas examined express multiple CD44 isoforms, the CD44H form is always by far the most abundant (M.S. Sy, unpublished results). So far we have been unable to identify any tumor cell line that express only the variant forms of CD44 without the expression of the hematopoietic form of CD44.

Most of the studies investigating the role of CD44 in tumor biology were carried out at the mRNA level either by northern blot analysis or reverse transcriptase polymerase chain reaction (RT-PCR). Which of these isoforms actually were expressed on the cell surface has never been studied biochemically. The extracellular domain of the hematopoietic form of CD44 consists of 268 amino acids rich in potential glycosylation sites (STAMENKOVIS 1989; NOTTENBURG 1989; ZHOU et al. 1989; WOLFFE 1990). Whether or not glycosylation plays a role in the function of CD44 is not known. A small part of the CD44 is present in a proteoglycan form of about 180–200 kDa, which contains about 120 kDa of glycosaminoglycans. Proteoglycans are multiple glycosaminoglycan chains covalently attached to a core protein and serve as a structural components in several types of extracellular matrixes (ROUSLAHTI 1988; HARDINGHAM and HOSANG 1992). The role of proteoglycan on the function of CD44 is not clear. Proteoglycans are known to be able to bind growth factors (ANDRES et al.

1989; Saksela and Rifkin 1990; Kiefer et al. 1990). Binding of growth factors by proteoglycans can concentrate growth factors or protect growth factors from degradation. Recent studies suggest that the proteoglycans on CD44 may also bind growth factors (Tanaka et al. 1993; Bennett et al. 1995). Proteoglycans on CD44 may be involved in the interaction between CD44 and IV collagen (Knutson et al. 1995). The ability of proteoglycan modified CD44 to bind growth factors and collagen will add additional complexities to the function of the CD44 molecule.

All CD44 isoforms have a transmembrane domain of 21 amino acids. The cytoplasmic domain contains 72 amino acids and is highly conserved among all CD44 cloned from different species. The cytoplasmic domain interacts with cytoskeletal proteins, actin, and ankyrin (Lacy and Underhill 1987; Kalomiris and Bourguignon 1988a). The cytoplasmic domain of CD44 can be acylated with palmitic acid (Bourguignon et al. 1991; Guo et al. 1994a) and phosphorylated by protein kinase C (PKC; Kalomiris and Bourguignon 1988b). Covalent modification of the cytoplasmic domain of CD44 with palmitic acid may play a role in the interactions between CD44 and ankyrin (Bourguignon et al. 1991). Modification of CD44 with lipid also plays a role in signal transduction in human lymphocytes (Guo et al. 1994a). The cytoplasmic domain of CD44 has also been reported to bind guanosine triphosphate (GTP) and has GTPase activity (Lokeshwar and Bourguignon 1992).

Thirty percent of the amino-terminal portion of the extracellular domain of CD44 is similar to cartilage-linked proteins and to the linked B element of the proteoglycan core protein (Chandrasekhar et al. 1983; Miyake et al. 1989; Lesley et al. 1990; Deak et al. 1986). The sequence homology between CD44 and macromolecules of the extracellular matrix suggests a general role for CD44 in proteoglycan or collagen-mediated matrix adhesion. This hypothesis was supported by findings that one of the major ligands for CD44 is hyaluronic acid (HA) (Aruffo et al. 1990; Murakami et al. 1991). HA is a polymer consisting of repeating disaccharide units of *N*-acetyl-D-glucosamine and D-glucuronic acid (Knudson et al. 1989; Laurent and Fraser 1992). However, CD44 also binds to collagen, fibronectin, and chondroitin sulfate (Carter and Wayner 1988; Faassem et al. 1992; Jalkanen and Jalkanen 1992). The receptor for HA has been studied over the last two decades by developmental biologists and investigators interested in wound healing (Peacock 1984; Underhill and Dorfman 1978; Poole 1986; Reid and Flint 1974; Toole 1990). HA is a major constituent of the extracellular matrix and is believed to create a low-resistance matrix, allowing enhanced cell motility. A microenvironment in which cells can migrate is essential for embryonic development and wound repair and is probably important for lymphocyte homing and tumor metastasis. HA receptors are known to be present in a variety of epithelia including the skin, cheek, tongue, esophagus, vagina, intestines, oviduct, and bladder (Picker et al. 1989; Kennel et al. 1993). In our earlier studies, we reported that only the lymphoid form of CD44 binds HA. The epithelial form of CD44 does not bind HA in vitro (Sy et al. 1991). Binding of HA to $CD44^+$ murine T cell hybridomas resulted in cellular activation and production of IL-2. Therefore, binding of HA to CD44 may also induce signal transduction (Guo et al. 1993).

CD44 is not the only HA-binding molecule. Some of the molecules known to bind HA include hyaluronectin/versican (KRUSIUS et al. 1987), aggrecan (DOEGE et al. 1987), an HA receptor complex termed RHAMM (HARDWICK et al. 1992) and a TNF-inducible protein known as TSG-6 (LEE et al. 1992). However, we have not been able to identify any tumor cell line that can bind HA in vitro but did not express CD44. Binding of HA to all CD44-bearing tumor cells can be blocked with some monoclonal anti-CD44 antibodies. Therefore, the only receptor on tumor cells that is capable of binding soluble HA in solution is CD44.

2.1 Interactions Between CD44 and Hyaluronic Acid In Vitro

The exact requirements for the interactions between CD44 and HA have not been well defined and remain controversial. Based on homology to other HA-binding proteins, the HA-binding site in CD44 has been speculated to be located in a basic amino acid motif. However, recent studies suggested that the actual binding sites are located just outside of the conserved domain of HA-binding proteins (PEACH et al. 1993). Binding of HA to cell surface CD44 requires more than simply the expression of CD44 molecules. Not all CD44-bearing tumor cells can bind HA. Despite the expression of high levels of CD44, normal murine lymphocytes or monocytes did not bind HA in solution. Interestingly, the capacity to bind HA in murine lymphocytes or tumor cells can be induced with one particular anti-CD44 monoclonal antibody (LESLEY et al. 1992). Murine tumor cell lines transfected with a CD44 gene lacking the cytoplasmic domain have reduced binding capacity, suggesting a role for the cytoplasmic domain (LESLEY et al. 1992). In human lymphocytes, binding to HA via CD44 can also be induced with monoclonal anti-CD44 (LIAO et al. 1993). More recently, MURAKAMI et al. reported that the induction of HA binding in human lymphocytes appears not to require cytoskeleton association, but does require new protein synthesis (MURAKAMI et al. 1994). Therefore, CD44 on normal murine and human lymphocytes exists in an inactive form. Since activation of lymphocytes with different polyclonal activators such as concanavalin A (ConA), Phytohemagglutinin (PHA), or phorbol myristate acetate (PMA) failed to induce this binding capacity, the physiologic relevance of these findings remain to be determined. It was postulated that binding of monoclonal anti-CD44 antibody may alter the conformation of the CD44 molecules and thus render them able to bind HA (LESLEY et al. 1992; LIAO et al. 1993; MURAKAMI et al. 1994).

Interactions between the cytoplasmic domain of CD44 and cytoskeletal elements have also been reported to be essential for the binding of HA. The cytoplasmic domain of murine CD44 interacts with cytoskeletal elements (LACY and UNDERHILL 1987; KALOMIRIS and BOURGUIGNON 1988a). Studies with human tumor cell lines transfected with CD44H forms with truncations of the cytoplasmic domain revealed that the cytoplasmic COOH-terminal 52 amino acids were critical for HA binding to the CD44 extracellular domain (LIAO et al. 1993). However, a soluble chimeric human CD44 molecule without the cytoplasmic domain of CD44 has been reported to be able to bind HA in vitro (PEACH et al. 1993). It was postulated that this genetically engineered soluble CD44 molecule may be "frozen" in an active form, thus enabling the molecule to bind HA in vitro. More recent studies revealed that aggregation of CD44 on the cell surface is important in the binding of HA (PERSCHL 1995). In this study, it was reported that the cytoplasmic domain of CD44 is not essential for binding. Therefore, whether or not the cytoplasmic domain of CD44 is essential for the interactions between CD44 and HA remains unclear.

Interactions between CD44 and HA also depend on the nature of the CD44 isoforms. We have previously reported that only the human CD44H, but not the CD44E form, can bind HA (SY et al. 1991). More recently, LIAO and coworkers reported that neither CD44H nor CD44E-transfected Jurkat T cells constitutively bind HA (LIAO et al. 1993). However, treatment of transfected Jurkat T cells with PMA induced CD44H-bearing but not CD44E-bearing Jurkat T cells to bind HA. Treatment of CD44E-bearing Jurkat T cells with one particular monoclonal antibody did enable the T cells to bind HA. Therefore, there are intrinsic differences between CD44H and CD44E in their interactions with HA. In contrast to human CD44, murine tumor cells bearing CD44E isoforms have been reported to constitutively bind HA (LESLEY et al. 1990). The discrepancies between these findings are not clear. It should be noted that most of the binding studies used HA in solution or immobilized on tissue culture plates. HA present in extracellular matrix is known to interact with other proteins and thus may have a different conformation (KNUDSON et al. 1989; LAURENT and FRASER 1992). Therefore, interactions between CD44 and HA in solution and CD44 and HA in extracellular matrix may be different.

We conjugated fluorescein (F) to HA (F-HA) and used F-HA to study binding of HA to CD44. F was conjugated to HA using a protocol described earlier (GUO et al. 1994). F-HA can be used as an immunofluorescent reagent. Stained cells were analyzed in a fluorescence-activated cell sorter (FACS). The melanoma cell line

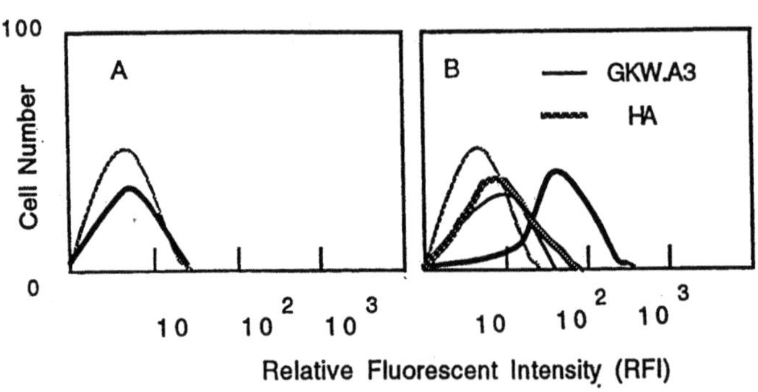

Fig. 1A,B. Binding of fluorescein–hyaluronic acid (*HA*) (F-HA) to CD44$^+$ melanoma cell line. SMMU-1 (**A**) or CD44$^+$ SMMU-2 (**B**) were stained with F-HA. For blocking of F-HA, cells were first incubated with unconjugated HA or monoclonal anti-CD44 antibody (GKW.A3). *Thick lines* are cells stained with F-HA alone. *Thin dotted lines* are negative controls with an isotype-matched monoclonal antibody

SMMU-1 did not express CD44 and did not bind F-HA (Fig.1A). Only the $CD44^+$ melanoma cell line SMMU-2 bound F-HA. Binding of F-HA to tumor cells can be completely blocked by two out of four (GKW.A3 and IM.7, but not Bu52 and P3H9) of the monoclonal anti-CD44 antibodies (Y.J. GUO, unpublished results). Therefore, only certain epitopes on CD44 are involved in the binding of HA. Binding of F-HA can also be blocked with unconjugated HA, but not with poly-L-lysine or heparan.

Recently, we investigated the binding of F-HA to different CD44 isoforms present on a panel of human tumor cell lines. These tumor cell lines are described in Table1. Seven of the human tumor cell lines expressed high levels of CD44. Two of the tumor cell lines, SMMU-1 and HepG2, did not express CD44, as demonstrated by immunofluorescent staining with a monoclonal anti-CD44 antibody GKW.A3. Melanoma SMMU-2, cervix carcinoma HeLa, and astrocytomas CRT and STT expressed the hematopoietic form of CD44. Colon carcinoma cell line HT29 expressed CD44 bearing domains V and VI (HOFMANN et al. 1991). Lung carcinoma Calu 6 was a double expressor and expressed CD44 variants carrying domains I–V and domains IV and V (HOFMANN et al. 1991). Keratinocyte line HeCat expressed domains I–V of CD44 (HOFMANN et al. 1991). Human tumor cell lines SMMU-2, Hela, CRT, HT29, and Calu 6 bind significant levels of F-HA (Table1). Binding of F-HA to these cell lines can be blocked with monoclonal anti-CD44 antibody GKW.A3 and unconjugated HA, indicating the specificities of the binding. Two tumor cell lines, one expressing the hematopoietic form of CD44 (STT) and the other expressing domains I–V of CD44 (HeCat), did not bind F-HA (Table 1). These two cell lines remained unable to bind F-HA even when the concentrations of F-HA used for staining were increased 50-fold (Y.J. GUO, results not shown). SMMU-1 and HepG2 did not express CD44 and did not bind F-HA. The levels of expression of CD44 in STT and HeCat are comparable to other tumor cell lines.

Since STT expressed the hematopoietic form of CD44 but was unable to bind F-HA, the correlation between expression of the hematopoietic form of CD44 and binding of HA is not absolute. This interpretation was further supported by

Table 1. Expression of CD44 isoforms in human tumor cell lines and their ability to bind fluorescein–hyaluronic acid (F-HA) in vitro

Tumor	CD44 isoform expressed	Binding to F-HA
SMMU-2	Standard 85–95 kDa	Yes
HeLa	Standard 85–95 kDa	Yes
CRT	Standard 85–95 kDa	Yes
STT	Standard 85–95 kDa	No
HT-29	Domains IV and V	Yes
Calu 6	Domains I–V Domains IV and V	Yes
HeCat	Domains I–V	No
SMMU-1	Negative	No
HapG2	Negative	No

another series of experiments using different human tumor cell lines transfected with the same CD44H gene. Human Burkitt's lymphoma Namalwa transfected with the hematopoietic form of the human CD44 gene was able to bind F-HA. In contrast, human lymphoma cell line Jurkat transfected with the identical CD44 gene was unable to bind F-HA. Therefore, the ability of the hematopoietic form of CD44 to bind F-HA may depend on some other cellular elements in addition to CD44. Binding of HA to CD44 may require another cellular protein or post-transitional modification that can only occur in some cell types. These observations are in good agreement with earlier findings in the murine and human systems. Transfection of the murine CD44 gene into some cell lines enables the transfectant to bind HA in vitro, while transfection of the same gene into other cell lines did not promote binding to HA.

Both tumor cells and normal cells can produce HA in vivo and in vitro (KNUDSON et al. 1989; PHILIPSON and SCHWARTZ 1984). Binding of endogenously produced HA to CD44 may render the CD44 molecules inaccessible to exogenous F-HA. Treatment of tumor cells with hyaluronate lyase should remove surface-bound HA, thus enabling cells normally unable to bind HA to become capable of binding HA. We first titrated the amount of hyaluronate lyase required to remove surface-bound HA. SMMU-2 tumor cells were first incubated with a concentration of HA known to be able to block the binding of F-HA. Cells were then washed with phosphate buffer saline (pH 7.4) to remove excess HA and incubated with various concentrations of hyaluronate lyase. After incubation, cells were then washed again with phosphate-buffered saline (PBS) and stained with F-HA. Tumor cells first incubated with HA and then treated with between 10 and 20 U hyaluronate lyase/ml were able to bind significant levels of F-HA. Tumor cells first incubated with HA and then incubated with medium alone remained unable to bind F-HA. Treatment of tumor cells with hyaluronate lyase did not alter the expression of CD44 molecules, as revealed by staining with a monoclonal anti-CD44 antibody. These results indicated that our treatment protocol is effective in removing surface-bound HA.

We next determined whether or not treatment with hyaluronate lyase of tumor cells that were unable to bind F-HA could enable them to bind F-HA. STT and HeCat cells were treated with 20 U hyaluronate lyase/ml. After enzymatic treatments, tumor cells were washed and stained with F-HA and anti-CD44 monoclonal antibody. Treatment of HeCat cells with hyaluronate lyase enabled the tumor cells to bind significant levels of HA. However, binding of F-HA to HeCat remained significantly lower compared to other CD44-bearing tumor cells. Increasing the concentrations of hyaluronate lyase did not further increase the binding of F-HA. Hyaluronate lyase may be unable to remove all surface-bound HA on HeCat. Alternatively, some of the CD44 molecules on HeCat may be constitutively unable to bind exogenous HA. STT tumor cells remained unable to bind F-HA after treatment with hyaluronate lyase. Treatment of $CD44^+$ SMMU-2 tumor cells with hyaluronate lyase did not enhance the binding of F-HA. $CD44^-$ tumor cell lines remained unable to bind HA after hyaluronate lyase treatment. Results from these experiments suggest that, in addition to CD44 isoforms, the nature of the cell type is also critical in determining the binding of F-HA. The inability of CD44$^+$ tumor cells to bind F-HA may be due to the endogenous production of HA by tumor cells or to other yet to be identified mechanisms.

In addition to HA, CD44 is known to bind collagen, chondroitin sulfate, and fibronectin (KNUDSON et al. 1989; LAURENT and FRASER 1992). We investigated whether binding of F-HA to CD44 on tumor cells can be inhibited with these other non-HA ligands. In contrast to unconjugated HA, which blocked binding of F-HA to SMMU-2 in a dose-dependent manner, none of the other ligands, including chondroitin sulfate A–C, type 1 collagen, heparin, and fibronectin, were able to block binding of F-HA (Y.J. GUO, results not shown). These results suggested that either CD44 does not bind to these other ligands in our in vitro binding conditions or that these ligands bind to different regions of the CD44 molecule distinct from the HA-binding site.

Binding of ligands to receptors may result either in the shielding or internalization of the ligand–receptor complexes. We incubated CD44$^+$ SMMU-2 tumor cells lines with F-HA and cultured the cells at 37°C for different lengths of time. Cells were then fixed and observed under a fluorescent microscope. Binding of F-HA to SMMU-2 first resulted in capping within 30 min, and then internalization of F-HA. Therefore, binding of HA to CD44 on the cell surface may also participate in the catabolism of HA in vivo. These findings are similar to earlier reports for human breast tumor cell lines (CULTY et al. 1992). Binding and internalization of HA may be relevant to tumor growth and metastasis in vivo. Breakdown products of HA have been reported to induce angiogenesis (WEST et al. 1985). Tumor cells may internalize and break down HA to create angiogenic HA fragments. Angiogenesis is a process that has been postulated to be essential for tumor growth and metastasis (FOLKMAN and KLAGSBUM 1987; FOLKMAN 1991). Angiogenesis usually involve sequential events, including endothelial cell activation, basement membrane disruption, endothelial cell migration into the interstitia space, endothelial cell proliferation proximal to the migrating tip, and lumen formation. The concept that tumors are angiogenesis dependent was based on observations that the growth rate of tumors is slow and linear before vascularization and rapid and nearly exponential after vascularization. Angiogenesis is not only important for tumor growth, but onset of angiogenesis also contributes to metastasis. Neo vascularization permits the shedding and migration of cells from the primary tumors to other sites.

As stated earlier, HA is a polymer consisting of repeat disaccharide units of *N*-acetyl-D-glucosamine and D-glucuronic acid. We used oligosaccharides of

Table 2. Inhibition of binding of fluorescein–hyaluronic acid (F-HA) by HA fragments

Disaccaride units	Concentration (mg/ml)	Inhibition(%)
4	1	0
8	1	0
15–20	1	50
15–20	0.5	30
15–20	0.1	20

defined disaccharide units to determine the minimum units of HA that can block the binding of intact F-HA in solution (Table 2). The human astrocytoma cell line CRT was first incubated with different concentrations of unconjugated HA or oligosaccharide of HA with defined disaccharide units. Tumor cells were then washed twice with PBS and stained with F-HA. Binding of F-HA was completely inhibited by unconjugated HA in a dose-dependent manner. Of the oligosaccharides tested, only oligosaccharides with 15–20 disaccharide units were able to block the binding of F-HA in a dose-dependent manner. Oligosaccharides smaller than 15–20 units of disaccharide were unable to block the binding of F-HA in the concentrations we have tested (up to 1 mg/ml). Blocking with the 15–20 disaccharide units is significantly less effective when compared to intact HA. It requires approximately 1 µg intact HA/ml to achieve 50% inhibition. In contrast, it requires approximately 1 mg/ml oligosaccharides with 15–20 disaccharide units to achieve 50% inhibition (results not shown). Our results are different from earlier studies indicating that the HA receptor present on the surface of fibroblasts was able to bind to smaller units of disaccharide (UNDERHILL and DORFMAN 1978; UNDERHILL 1982). The reasons for these differences are not known. One possibility is that in earlier studies radiolabeled oligosaccharides were used in direct binding assays, while in our studies oligosaccharides were used as inhibitors of binding of intact HA. Alternatively, binding of HA to fibroblasts may be different from binding of HA to human tumor cells.

2.2 Role of CD44 N-Linked Oligosaccharides in the Binding of Hyaluronic Acid

In the extracellular domain of human CD44, there are six potential N-glycosylation sites and five O-linked glycosylation sites (NOTTENBURG et al. 1989; ZHOU et al. 1989; WOLFFE et al. 1990). There are five potential N-glycosylation sites and ten O-linked glycosylation sites in murine CD44. Both complex-type and high mannose-type glycosylation are known to be present on CD44 molecules. Proper glycosylation is known to be essential for the expression and the function of some, but not all cell surface receptors.We investigated whether removal of the N-lined oligosaccharides on CD44 interfered with its ability to bind anti-CD44 monoclonal antibodies or HA. Murine T cell hybridoma 2C7 was cultured either with various concentrations of tunicamycin, an inhibitor of N-linked glycosylation, or with control medium for 18 h. Cells were then washed extensively and stained with monoclonal anti-CD44 antibodies (KM81 or KM201) or with F-HA. Tunicamycin did not significantly alter the staining profiles of the two different monoclonal anti-CD44 antibodies (KM81, KM201). Therefore, expression of CD44 does not require N-linked glycosylation. However, treatment with 5 µg, but not with 1 µg tunicamycin/ml significantly increased the binding of F-HA (Fig. 2). Treatment with tunicamycin reduced the molecular, mass of CD44 by about 4000–8000 kDa in two different experiments as revealed by surface iodination and immunoprecipitation with monoclonal anti-CD44 antibody. The enhancement ranged from a five-

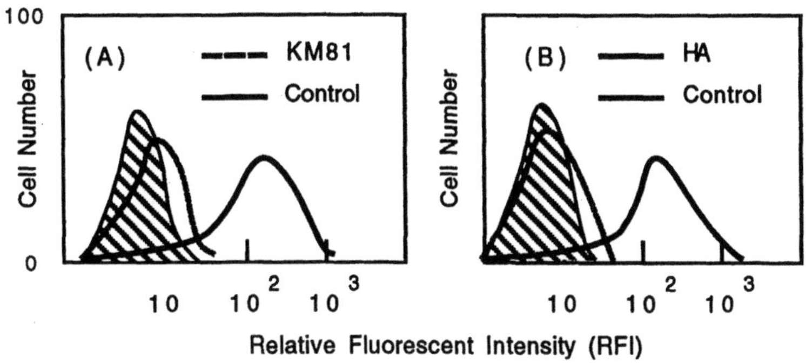

Fig. 2A,B. The effects of tunicamycin are specific for CD44 and hyaluronic acid (*HA*). Hybridoma 2C7 cells were treated with tunicamycin for 16 h as described. Cells were harvested and washed extensively. To block the binding of fluorescein–HA (F-HA), some of the samples were incubated on ice for 30 min with 25 µl KM81 monoclonal antibody (mAb) containing hybridoma culture supernatant (**A**) or 30 µl unconjugated HA (100 µg/ml; (**B**). After incubation, cells were washed twice with phosphate-buffered saline (PBS) to remove unbound mAb or HA. All samples were then stained with 40 µl F-HA (10 µg/ml) as described. After staining, cells were washed, fixed, and analyzed in a FACScan (fluorescence-activated cell sorter). *Hatched peaks* are results with the irrelevant antibody as control

to ten fold increase in the mean fluorescent intensities in five different experiments. Thus, N-linked oligosaccharides appeared to interfere with the binding of HA to CD44. Binding of F-HA to tunicamycin-treated cells can be completely blocked by unconjugated HA and by monoclonal anti-CD44 antibody KM81. When treated with tunicamycin under identical conditions, $CD44^-$ tumor cells remained unable to bind F-HA.

Sialic acid is the only sugar residue of glycoproteins that bears a net negative charge (Lloyd 1975; Schauer 1985). We also investigated whether removal of the sialic acid residues on CD44 enhanced the binding of HA. The T cell hybridoma 2C7 was treated in vitro with neuraminidase or PBS. Cells were then washed extensively and stained with monoclonal anti-CD44 antibodies (KM81 or KM201) or with F-HA. Neuraminidase treatment did not alter the staining profiles of monoclonal anti-CD44 antibodies, but increased the binding of F-HA (Fig. 3). Binding of F-HA to neuraminidase-treated cells is specific for HA and can be blocked by unconjugated HA, or by monoclonal anti-CD44 antibody KM81. In addition, treatment with neurominidase of a human tumor cell line stably transfected with the CD44 gene also enhanced the binding of F-HA. However, when treated with neuraminidase, the $CD44^-$ parental tumor cell line remained unable to bind to F-HA. These experiments provide additional support for the hypothesis that the presence of carbohydrate moieties may interfere with the binding of hyaluronic acid to CD44.

Murine CD44 contains both high mannose and complex-type N-linked glycans. To further investigate the role of N-linked oligosaccharides on the binding of HA to CD44, we obtained a murine T cell hybridoma that has an altered N-linked glycosylation pathway. This mutant is deficient in dolichol-phosphate-mannose (Dol-p-Man) biosynthesis (Thomas et al. 1991; DeGasperi et al. 1990). Truncated

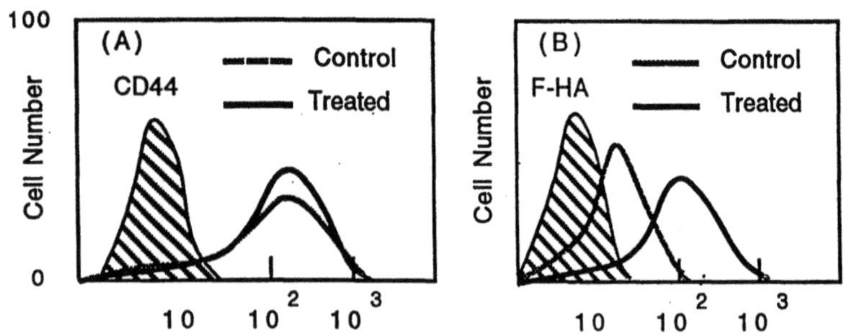

Fig. 3A,B. Treatment with neuraminidase enhanced the binding of fluorescein–hyaluronic acid (*F-HA*). Hybridoma 2C7 cells were treated with neuraminidase or control medium as described. Cells were harvested and washed extensively. **A** Samples containing 1×10^6 cells were incubated with 25 µl KM81 monoclonal antibody (mAb). An irrelevant rat mAb (53.6.72) was used as an isotype control. Cells were washed and then incubated for 30 min with 20 µl fluorescein isothiocyanate (FITC)-conjugated, affinity-purified goat anti-rat immunoglobulin (Ig). After the final incubation, cells were washed, fixed, and analyzed. **B** Other samples from control or treated groups containing 1×10^6 cells were incubated with 30 µl F-HA (10 µg/ml). After staining with F-HA, cells were washed, fixed, and analyzed in a FACScan (fluroscence-activated cell sorter). *Hatched peaks* are resuls with the irrelevant antibody as control

N-linked oligosaccharide precursors were synthesized and transferred to nascent proteins. They can be processed to become complex, but not high mannose glycans. The defects in this hybridoma have been extensively characterized (THOMAS et al. 1991; DEGASPERI et al. 1990). We investigated whether this mutant could bind more F-HA.

Both the parental hybridoma (YH16.33) and the mutant (M39/5) expressed comparable levels of CD44. However, the mutant cell line bound significantly more F-HA than the parental cell line. Treatment of tumor cells with PMA is known to increase the expression of CD44 molecules and the binding of F-HA. We treated the parental hybridoma and the mutant with PMA. The difference between the ability of the parental hybridoma and the mutant to bind F-HA was even greater when cells were first treated with PMA in vitro for 18 h (results not shown). These experiments provide strong evidence that CD44 on T cell hybridomas can bind more F-HA if the synthesis of high mannose is prevented. Mutant T cell hybridomas have been stably transfected with the yeast Dol-p-Man synthesis gene. N-linked glycosylation was either partially or completely corrected in some of these transfectants (DEGASPERI et al. 1990). We compared the binding of F-HA to one of the Dol-p-Man synthase transfectants with the original mutant and the wild-type parental cell line. All three cell lines express similar levels of CD44. However, the T cell hybridoma stably transfected with the yeast Dol-p-Man synthesis gene showed reduced ability to bind F-HA. The amounts of F-HA bound by the transfectant was similar to the original wild-type parental cell line, but was significantly lower than the mutant cells. Binding of HA to these tumor cells is specific for CD44–HA because binding can be completely inhibited by unconjugated HA or monoclonal anti-CD44 antibody KM81. These experiments provided conclusive evidence that N-linked glycans on CD44 affect the binding of HA.

The in vivo significance of our finding is currently not known. The amount of sialic acid present on a protein is developmentally regulated. For example, the amount of sialic acid present on the neural cell adhesion molecule (NCAM) found in embryonic brain tissue of the chicken is significantly higher than sialic acid on the NCAM in adult brain tissue (ROTHBARD et al. 1982). T lymphocyte differentiation is accompanied by an increase in sialic acid content of Thy-1 antigen (HOESSLI 1980). Both CD44 and HA are present in many cell types and in different tissues, as well as in the circulation. Modification of the number or structure of N-linked glycans or terminal sialic acid present on CD44 molecules may provide a means to avoid excessive interactions between CD44 and HA. However, in selective microenvironments or during certain stages of cellular activation, the N-linked glycans or terminal sialic acid on CD44 may be modified. CD44 on activated macrophages is known to be glycosylated differently than CD44 on nonactivated macrophages (CAMP et al. 1992). Activation of human neutrophils is known to cause the mobilization of sialidase from intracellular stores to the cell surface. Mobilization of sialidase to the cell surface also resulted in an increase in the adhesion responses of the cells (CROSS et al. 1991). Therefore, during cell–cell interactions, cell surface sialidase may be able to modify the terminal sialic acid on cell surface glycoproteins, resulting in the generation of a receptor with a higher affinity for its ligand. Sialic acid also plays an important role in tumor growth and metastasis (FOGEL et al. 1983; DENNIS et al. 1982). Surface sialic acid reduces the attachment of metastatic tumor cells to collagen and fibronectin (FOGEL et al. 1983; DENNIS et al. 1982). Differences in the amount of sialic acid on CD44 may affect the growth and metastatic potential of CD44-bearing tumor cells.

3 Relationship Between CD44 and Tumor Growth and Metastasis

To metastasize, malignant cells must be able to detach from the tumor mass, adhere to and penetrate endothelium of other organs, and establish a new tumor colony (FIDLER 1978; NICOLSON 1988; KAHN 1992; STETLER-STEVENSON et al. 1993; TURLEY 1984). Binding of tumor cells to endothelium is probably a prerequisite for metastatic colony formation. However, the invasive potential of tumor cells appears to correlate better with their ability to adhere to subendothelial basement membrane and matrix proteins than with their ability to adhere endothelium (FIDLER 1978; NICOLSON 1988; KAHN 1992; STETLER-STEVENSON et al. 1993; TURLEY 1984). HA is present in large quantities in the basement membrane. Elevated expression of CD44 on tumor cells may enhance the binding of tumor cells to the HA present on subendothelial basement membrane, resulting in enhanced tumor locomotion and invasiveness. Binding of HA to $CD44^+$ tumors may also result in signal transduction and the production of growth factors by tumor cells. These factors may be critical for tumor growth or metastasis. More recently, interactions between CD44 and hyaluronate has been directly implicated in the regulation of tumor development (BARTOLAZZI et al. 1994).

CD44 and HA may also interact with growth factors produced either by tumor cells or host cells. These growth factors may play important roles in the growth and metastasis of tumor cells. One of these growth factors is transforming growth factor (TGF)-β. Rous sarcoma virus (RSV) rapidly transforms many cell types in culture, but its ability to induce tumors in vivo is highly dependent on the tissue environment (MACLEOD et al. 1985). In normal young chicken hatchlings, sarcoma will only form at the site of RSV injection. However, wounding leads to tumor formation with 100% frequency in tissues that would otherwise remain tumor free. Subcutaneous injection of TGF-β could substitute for wounding in tumor induction in the chicken (SIEWEKE et al. 1990). Wound-related RSV tumorgenesis and the action of TGF-β may be related to CD44 and HA. RSV infection increases the expression of HA receptor; HA is important in wound healing; TGF-β stimulates the production of HA and TGF-β binds proteoglycans (WOODLAND and JONES 1988; RUOSLAHTI and YAMAGUSHI 1991). Interaction between CD44 and HA may affect the production of TGF-β or the expression of the receptor for TGF-β.

3.1 Upregulation of CD44 Expression During Tumor Growth and Metastasis

Prior to the cloning of the CD44 gene, many different lines of indirect evidence have already suggested a role for HA receptor and HA acid in tumor growth and metastasis. Transformation of fibroblasts with SV40 or RSV increased the expression of receptors for HA (UNDERHILL, B 1982). Carcinomas are frequently characterized by local accumulation of HA in vivo (KNUDSON et al. 1989; TAKEUCHI et al. 1981; HORAI et al. 1981). The ability of rabbit carcinoma to invade surrounding tissue correlates with its ability to induce formation of HA (TOOLE et al. 1979). Highly invasive human bladder carcinoma cell lines express high levels of HA receptors. Noninvasive bladder carcinoma cell lines express low levels of HA receptors (NEMEC et al. 1987). CD44 is expressed in the tumorigenic but not in the less tumorigenic variants of the murine *thymoma* SL12 (MACLEOD et al. 1985). More recent results suggest that high expression of the hematopoietic form of CD44 is associated with aggressive behavior, dissemination, and poor prognosis of human non-Hodgkin's lymphomas. Unlike the expression of CD44, expression of two other adhesion molecules, lymphocyte function-associated antigen (LFA)-1 and intercellular adhesion molecule (ICAM)-1 did not correlate with prognosis (HORST et al. 1990). In a panel of human breast cancer cell lines, in vitro invasiveness was positively correlated with CD44 expression and the ability of the tumor cell to bind and degrade HA (CULTY et al.1992). Both anti-CD44 antibodies and inhibition of CD44 expression by antisense oligonucleotides inhibited the invasion of glioma in vitro

(MERZAK et al. 1994). Human mesothelioma but not normal mesothelial cells express CD44 and bind HA (ASPLUND and HELDIN 1994).

Preferential expression of a CD44 isoform associated with tumor metastasis was first reported in the rat (GÜNTHERT et al. 1991). Later, a similar 230-kDa CD44 isoform was also found to be overexpressed in colorectal carcinomas and adenomatous polyps (HEIDER et al. 1993). Analysis of CD44 splice variants in tumor tissues from patients with colon, gastric, and breast cancer by polymerase chain reaction (PCR) revealed that there was gross overproduction of the alternatively spliced large molecular variants in all tumor tissues examined (TANAKE et al. 1993; MATSUMURA and TARIN 1993; WIELENGA et al. 1993). Only the mRNA encoding for the standard 94-kDa proteins were detected in the control samples. However, using monoclonal antibodies specific for various human cD44 isoforms instead of PCR, different CD44 isoforms were also found to be expressed in many normal tissues, including those from which the tumors arose (FOX et al. 1993, 1994). The reasons for these discrepancies are not known. Increased expression of V3, V6, and V10 exons have also been speculated to be involved in the blockage of myelopoiesis in acute myeloblastic leukemia (LEGRAS et al. 1995). More recently, increased and disorganized expression of CD44 variant exons was shown to occur in exfoliated tumor cells sedimented from the urine of bladder cancer patients. Interestingly, there appeared to be abnormalities in the splicing of the CD44 gene products in these bladder tumors, resulting in the accumulation of immature mRNA transcripts containing introns from this gene, as well as many abnormal variants (MATSUMURA et al. 1995). These observations also raised another important question as to whether or not some of these spliced variant isoforms were ever translated to produce functional proteins. The ability of tumor cells to preferentially express certain CD44 isoforms is reminiscent of an earlier report on abnormal alternative splicing of pre-mRNA in malignant human tumors (OYAMA et al. 1989). Alternative splicing of fibronectin pre-mRNA at the EIII-A region has been shown to be regulated in a tissue- and developmental stage-specific manner (SCHWARZBAUER et al. 1987). OYAMA and coworkers investigated the splicing pattern at the EIII-A region in malignant and nonmalignant human liver tissues and found that the relative population of the fibronection mRNA containing the EIII-A sequence is markedly increased in malignant liver tumors (OYAMA et al. 1989). These results suggest that the putative molecular machinery governing alternative RNA splicing of cellular proteins may be deregulated in malignant human tumors. In addition to tumor cells, CD44 variants are also present in activated normal human lymphocytes (KOOPMAN et al. 1993). Murine CD44 variant transcripts are present in many different normal tissues including colon, small intestine, stomach, kidney, and lung (SCREATON et al. 1993).

CD44 is important in the migration of melanoma cells in vivo (THOMAS et al. 1992). Metastatic melanomas express higher levels of CD44 (HART et al. 1991; BIRCH et al. 1991). We establish two melanoma cell lines from one patient with melanoma (GUO et al. 1994b). One of the tumor cell lines (SMMU-1) was established from the primary tumor. The other line (SMMU-2) was established from a metastatic lymph node. Both SMMU-1 and SMMU-2 grew in the subcutaneous tissue of athymic, nude mice, but the growth rate of SMMU-2 was higher. When injected subcutaneously, SMMU-2 but not SMMU-1 cells spontaneously formed metastatic tumors in approximately half of the injected animals. Only the metastasizing melanoma cell line SMMU-2 expressed the hematopoietic form of CD44 and bound fluorescent conjugated HA (F-HA) in vitro. Moreover, in vivo growth and metastasis of CD44$^+$ SMMU-2 tumor cells was inhibited with monoclonal anti-CD44 antibody. These experiments suggest that expression of CD44 is not essential for tumor growth, since both CD44$^+$ SMMU-2 and CD44$^-$ SMU-1 tumor cells were able to form tumors in nude mice. However, interactions between CD44 on tumor cells and its ligands in vivo may promote growth and metastasis.

More direct evidence that CD44 is important in tumor growth and metastasis came from gene transfection experiments. A rat carcinoma cell line which did not metastasize acquired metastatic properties, when transfected with a CD44 gene encoding for a high molecular weight 230-kDa CD44 isoform (Asplund and Heldin 1994). Expression of domain 6 (V6) of the CD44 variant protein appeared to be important in the metastasis of the rat tumor cell line. However, this rat tumor cell line expresses the standard CD44 isoform in addition to the CD44 V6 variant. We reported earlier that introduction of the human hematopoietic form of the CD44 gene into a CD44$^-$ tumor cell line, Namalwa, resulted in enhancement of tumor growth and metastasis in vivo (Sy et al. 1991). Moreover, the in vivo growth and metastasis of CD44$^+$ Namalwa can be blocked with soluble CD44 molecules (Sy et al. 1992). Therefore, interactions between CD44 and its ligands in vivo are important in the growth and metastasis of human melanomas. Not all human tumor cells express CD44 variants. A significant number of human lung carcinomas, melanomas, astrocytomas, and ovarian carcinomas express only the hematopoietic form of CD44. Since different CD44 isoforms may have different ligands, different CD44 isoforms may play different roles in tumor growth and metastasis. The role of different CD44 variant proteins may depend on species, the nature of the tumor cells, and/or the host microenvironments.

3.2 Downregulation of CD44 Expression in Some Human Tumors

Some human tumor cell lines do not express CD44 proteins at all. Human neuroblastomas do not express CD44 protein (Shtivelman and Bishop 1991). Shtivelman and Bishop reported that several upstream *cis*-acting elements contribute to the down-regulation of CD44 in neuroblastoma cells, the most prominent being a 120-bp DNA fragment located 450 bp upstream of the RNA initiation site (Shtivelman and Bishop 1991). Using antibodies against a synthetic peptide containing a sequence encoded by the V6 domain of CD44, Salmi et al. reported that human squamous cell carcinomas specifically downregulated the expression of the CD44 V6 variant protein during their malignant transformation (Salmi et al. 1993). Moreover, downregulation of the V6 variant protein correlates with good prognosis in patients with squamous cell carcinomas. Fujita et al. examined expression of CD44 in normal and carcerous tissues of the endometrium as well as in cell lines established from patients with endometrial cancers. Variant forms of CD44 were expressed in 81% of normal endometria, where as only 17% of endometrial carcinomas showed expression of the variants. Furthermore, lymph-vascular space involvement of cancer cells was observed to be statistically significant in the CD44⁻ group as opposed to the CD44⁺ group (FUJITA et al. 1994). Therefore, human endometrial carcinomas downregulate their CD44 molecules. Reduced CD44 expression might be related to the metastasis of endometrial cancer cells through lymph-vascular space. Epithelial cancer of the ovary spreads by implantation of tumor cells onto the mesothelial lining of the peritoneal cavity. Binding of ovarian cancer cells to peritoneal mesothelium is partially mediated by the adhesion molecule CD44 (CANNISTRA et al. 1993). Whereas nearly 94% of human ovarian tumors derived either from primary sites or peritoneal implants express CD44, more than 75% of the free-floating tumor cells from ascites do not express CD44. Shedding or downregulation of CD44 on tumor cells may allow release of tumor cells from the mesothelial lining into the peritoneal space. Alternatively, reexpression of CD44 may allow tumor cells in the peritoneal space to adhere to the mesothelial lining. CD44 expression was also found to negatively correlate with N-myc amplification in neuroblastomas and small cell lung carcinomas (GROSS et al. 1994; PENNO et al. 1994).

The speculation that during metastatic cascade some tumor cells may down-regulate their CD44 molecules was further supported by recent findings in our studies on the expression of CD44 in human prostate carcinomas. The immuno-histochemical investigation of 74 prostate cancers from 50 patients revealed that there is an inverse correlation ($p < 0.05$) between histological differentiation and the expression of CD44 (NAGABHUSHAN et al. 1995). As a second measure of the aggressiveness of a tumor, we also examined local lymph node metastases and found that they were much less likely ($p < 0.0006$) to express high levels of CD44 than primary prostate cancers. Immunohistochemically, the expression of CD44 is sometimes exclusively membranous and sometimes both membranous and cytoplasmic.

The observed decreased expression of CD44 in the least-differentiated primary cancers (Gleason grades 8 and 9) and in lymph node metastases as compared with primary prostatic carcinomas is of particular interest. To our knowledge, this is the first cancer in which aggressive characteristics have been associated with decreased expression of CD44. The immunohistochemical approach made it possible to investigate the neoplastic cells specifically. This is particularly important, since the microvasculature in the metastatic tumors often showed intense expression of CD44. If we had used the PCR to study extracts of tissue, the observed differences might have been less dramatic or totally obscured, since the extracts would contain material from neoplastic cells as well as from the microvasculature. Tumor cells must be able to detach from the primary tumor mass in order to metastasize to other sites, Shedding or down-regulation of adhesion molecules by tumor cells may be one of the mechanisms that allow tumor cells to free themselves and migrate to other sites.

4 Soluble CD44 as a Marker for Tumor Growth and Metastasis

Low levels of soluble CD44 can be detected in the circulation of normal individuals and in the joints of patient with rheumatoid arthritis (HERLYN et al. 1987; LUCAS et al. 1989). BAZIL and HOREJSI reported that monoclonal anti-CD44 antibody can induce shedding of the CD44 molecules from normal human leukocytes (BAZIL and HOREJSI 1992). Tumor cells have been reported to shed many of their cell surface molecules (HERLYN et al. 1987). Shedding of cell surface molecules by tumor cells may have physiologic significance in tumor growth and metastasis. Shedding or downregulation of adhesion molecules on the tumor cells may enable tumor cells to migrate or detach from the primary tumor. Shedding of ICAM-1 may be important in the growth and metastasis of human melanomas. HANNING et al. (1991) reported an increase of serum ICAM-1 levels in 100% of patients with stage I melanoma. ALTOMONTE et al. (1993) reported an increase serum ICAM-1 levels in more than 85% of patients with stage I and stage II melanomas. However, recent studies by KAGESHITA et al. (1993) revealed that the levels of serum ICAM-1 were increased only in patients with stage III melanoma with detectable metastatic lesions and in patients with stage IV melanoma. The reasons for these differences are not known. If not due to technical reasons, the conflicting results may reflect the different characteristics of the patient population analyzed. TSUJISAKI et al. (1993) reported that high levels of ICAM-1 antigens could be detected in the serum of patients with liver metastasis. However, increases in serum soluble ICAM-1 levels are not restricted to patients with malignant diseases. Soluble ICAM-1 levels are also increased in patients with diabetes mellitus, impaired renal function, chronic ambulatory peritoneal dialysis, rheumatoid arthritis, sepsis, and hypertension.

We devised an enzyme-linked immunosorbent assay (ELISA) with two different monoclonal anti-human CD44 antibodies to detect soluble CD44 in the serum of cancer patients (GUO et al. 1994c). The concentrations of soluble CD44 were elevated in 16 out of 17 patients with metastatic gastric cancer. Significantly elevated soluble CD44 concentrations were detected in five of the eight patients without metastatic tumors. We determined the concentration of soluble CD44 in 15 patients with chronic inflammatory rheumatic diseases. The concentrations of soluble CD44 in these patients were comparable to those found in normal controls (2.2 ± 1.6 n*M*). These results provide the first evidence that the serum CD44 antigen level may be a useful marker to monitor tumor burden in cancer patients. The concentrations of soluble CD44 were also significantly elevated in 14 of the 15 patients with metastatic colon cancer. Therefore, the serum CD44 antigen may be a useful marker to monitor the growth and metastasis of at least two different tumors. Enhanced levels of soluble CD44 have also been reported in the circulation of tumor-bearing mice (KATOH et al. 1994). We also determined the concentration of CD44 in the ascitic fluids of five patients with metastatic colon cancer and ascites. Soluble CD44 concentration in the ascitic fluids from these patients exceeded 100 n*M*. The mean concentration of soluble CD44 in ascites from ten patients with cirrhosis was 2.2±1.5n*M*.

We also investigated whether the levels of soluble CD44 in the circulation of cancer patients correlate with tumor burden. Tumors from 19 patients with metastatic colon cancer were surgically removed. Three weeks after surgery, the levels of soluble CD44 were determined. Removal of the tumors resulted in significant decreases in soluble CD44 concentration in the sera of patients with high concentrations of soluble CD44 prior to surgery. These observations provide additional evidence that serum levels of CD44 are an indicator of tumor burden in these patients. Western blotting with two different anti-CD44 antibodies identified one prominent band with a molecular mass of between 130 and 190 kDa in the immunoaffinity-purified materials from the sea of colon cancer patients. Two less prominent bands with molecular masses of about 72 kDa and 80 kDa were also identified. These proteins were absent from western blots of proteins from normal donors probed with the anti-CD44 antibody. More recently, using a rabbit polyclonal antibody specific for a synthetic peptide derived from the V6 domain of human CD44, we found that one of the larger soluble CD44 molecules present in the sera of colon cancer patients does indeed bear the V6 domain.

Low levels of soluble CD44 are present in the serum of normal individuals. The concentration of soluble CD44 in the serum is significantly elevated in patients with advanced gastric and colon cancers and in some patients with other tumors. Therefore, the concentration of soluble CD44 in the circulation may be an indicator of tumor growth for many different kinds of tumors. Serum CD44 levels correlated with tumor metastasis and tumor burden. Surgical resection of tumors resulted in decreases of serum CD44 levels. By western blot analysis, anti-CD44 antibody reacted with a major protein with molecular mass of between 130 and 190 kDa. In addition, two proteins with a molecular mass of 72 kDa and 80 kDa were also identified. Therefore, different CD44 isoforms or proteolytic products of larger molecular weight CD44 may be present in the sera of cancer patients.

Soluble CD44 present in the circulation of patients most likely came from tumor cells rather than normal cells. Complete surgical resection of tumor masses from these patients resulted in a significant reduction in their serum CD44 concentration. Elevated serum CD44 levels in these patients may be due to active shedding of CD44 molecules by the tumor cells. Alternatively, dying tumor cells may release CD44. CD44 on tumor cells may be released into the circulation by proteolytic mechanisms. Multiple dibasic peptide (Arg-Arg) sites for potential proteolysis exist in the extracellular domain of human CD44 molecules. Since different CD44 isoforms bear distant dibasic peptide sites, different isoforms of CD44 may be differentially susceptible to proteolysis. Their existence is speculative at this point. Alternatively, it has been reported that in the V1 domain of human CD44 there is a stop codon after the 17th amino acid residue (Screaton et al. 1993). This stop codon has been detected in over 90 haplotypes. The inclusion of V1 in human CD44 transcripts could yield one of the smaller soluble isoforms.

More recently, Classen and coworkers examined sera from 63 patients with primary and metastasized breast cancer using CD44 domain-specific monoclonal antibodies (Classen et al. 1995). All patients with metastatic tumor have elevated soluble CD44 bearing V6 domains. Furthermore, concentrations of V6 could be correlated with responsiveness to hormone or chemotherapy; the majority of the patients with V6 overexpression failed to respond to chemotherapy or endocrine therapy. This confirmed and extended our earlier findings that monitoring the levels of soluble CD44 in cancer patients may be a valuable indicator of metastasis and tumor burden as well as prognostic marker for responsiveness to chemotherapy.

Use of monoclonal antibodies specific for different CD44 isoforms may significantly improve the sensitivity of our ELISA. The reasons that we were not able to identify patients with stage I or some of the stage II gastric or colon carcinomas with a higher frequency may be related to the monoclonal antibodies we used for our soluble CD44 ELISA. The two anti-CD44 antibodies react with the 84-KDa CD44 isoform and cannot distinguish different CD44 isoforms. The presence of soluble 82-KDa CD44 in normal serum may interfere with our ability to detect small increases in the levels of other tumor-associated CD44 isoforms. This approach may eventually allow us to detect a small increase in various CD44 isoform concentration in patients with early stages of malignant diseases.

References

Alho AM, Underhill CB (1989) The hyaluronate receptor is preferentially expressed on proliferating epithelial cells. J Cell Biol 108: 1557–1565

Altomonte M, Colizzi F, Esposito G, Maio M (1993) Circulating intercellular adhesion molecule 1 as a marker of disease progression in cutaneous melanoma. N Engl J Med 327: 959

Andres JL, Stanley K, Cheifetz K, Massague J (1989) Membrane anchored and soluble forms of betaglycan, a polymorphic proteoglycan that binds transforming growth factor B. J Cell Biol 109: 3137–3145

Aruffo A, Stamenkovic I, Melnick M, Underhill CB, Seed B (1990) CD44 is the principal cell surface receptor for hyaluraonate. Cell 61: 1303–1311

Asplund T, Heldin P (1994) Hyaluronan receptors are expressed on human malignant mesothelioma cells but not on normal mesothelial cells. Cancer 54: 4516–4523

Bartolazzi A, Peach R, Aruffo A, Stamenkovic I (1994) Interaction between CD44 and hyaluronate is directly implicated in the regulation of tumor development. J Exp Med 180: 53–66

Bazil V, Horejsi V (1992) Shedding of the CD44 adhesion molecules from leukocytes induced by anti-CD44 monoclonal antibody stimulating the effect of a natural receptor ligand. J Immunol 149: 747–753

Bennet KL, Jackson DG, Simon JC et al (1995) CD44 isoforms containing exon V3 are responsible for the presentation of heparin binding growth factor. J Cell Biol 128: 687–698

Birch M, Mitchell S, Hart IR (1991) Isolation and characterization of human melanoma variants expressing high and low levels of CD44. Cancer Res 51: 6660–6667

Bourguignon LYW, Kalomiris EL, Lokeshwar VB (1991) Acylation of the lymphoma transmembrane glycoprotein GP85, may be required for GP-85 ankyrin interaction. J Biol Chem 266: 1761–1765

Brown TA, Bouchard T, St. John T, Wayner E, Carter WG (1991) Human keratinocytes express a new CD44 core protein (CD44E) as a heparin sulfate intransic membrane proteoglycan with additional exons. J Cell Biol 113: 207–216

Budd RC, Cerottini C, Horvath C et al. (1987) Distinction of virgin and memory T lymphocytes: stable acquisition of the Pgp-1 glycoprotein concomitant with antigenic stimulation. J Immunol 138: 3120–3129

Butcher EC (1986) The regulation of lymphocytes traffic. Curr Top Microbiol Immunol 128: 85–124
Camp RL, Kraus TS, Birkeland L, Pure E (1991) High levels of CD44 expression distinguish virgin from antigen primed B cells. J Exp Med 173: 763–766
Camp RL, Kraus TA, Pure E (1992) Variations in the cytoskeletal interaction and post translational modifications of the CD44 homing receptor in macrophages. J Cell Biol 115: 1283–1292
Cannistra SP, Kanasa GS, Niloff J, DeFranzo B, Kim Y, Ottensmeier C (1993) Binding of ovarian cancer cells to peritoneal mesothelium in vitro is partly mediated by CD44 H. Cancer Res 53: 3830–3838
Carter WG, Wayner EA (1988) Characterization of the class III collagen receptor, a phosphorylated, transmembrane glycoprotein expressed in nucleared human cells. J Biol Chem 263: 4193–4201
Chandrasekhar S, Kleinman HK, Rassell JR (1983) Interaction of link protein with collagen. J Biol Chem 258: 6226–6231
Classen SD, Eierman W, Wolf H, Kopp R, Wilmans W (1995) CD44 variant in serum of breast cancer patients: a prognostic factor for clinical progresison? Proc Am Assoc Cancer Res 36: 209
Cross AS, Wright DG (1991) Mobilization of sialidase from intracellular stores to the surface of human neutrophils and its role in stimulated adhesion responses of these cells. J Clin Invest 88: 2067–2076
Culty M, Miyake K, Underhill CB (1992) Binding and degradation of hyaluronan by human breast cancer cell lines expresssion different forms of CD44; correlation with invasive potential. J Cell Physiol 160: 275–286
Deak F, Kiss I, Sparks KJ, Argraves WS, Hampikian G, Foetinck PF (1986) Complete amino acid sequence of chicken cartilage link protein deduced from cDNA clones. Proc Natl Acad Sci USA 83: 3766–3770
DeGasperi R, Thomas LJ, Sugiyama E et al (1990) Correction of a defect in mammalian GPI anchor biosynthesis by a transfected yeast gene. Science 250: 988–991
Denning SM, Lee PT, Singer KH, Haynes BF (1990) Antibodies against the CD44 p80, lymphocyte homing receptor molecule augment human peripheral blood T cell activation. J Immunol 144: 7–15
Dennis J, Waller C, Timpl R, Schirmacher V (1982) Surface sialic acid reduces attachment of metastatic tumor cells to collagen type IV and fibronectin. Nature 300: 274–276
Doege K, Sasaki M, Horigan E, Hassel J, Yamada Y (1987) Complete primary structure of the rat cartilage proteoglycan core protein deduced from cDNA sequences. J Biol Chem 262: 17757–17763
Faassem AE, Schrager JA, Klein DJ, Oegema TR, Couchman JR, McCarthy JB (1992) A cell surface chondroitin sulfate proteoglycan, immunologically related to CD44 is involved in type collagen mediated melanoma cell motility and invasion. J Cell Biol 116: 521–531
Fidler IJ (1978) Tumor heterogeneity and the biology of cancer invasion and metastasis. Cancer Res 38: 2651–2660
Fidler IJ (1991) Orthotopic implantation of human colon carcinomas into nude mice provides a valuable model for the biology and therapy of metastasis. Cancer Metastasis Rev 10: 229–243
Fidler IJ, Radinsky R (1990) Genetic control of cancer metastasis. J Natl Cancer Inst 82: 166–168
Fidler IJ, Naito S, Pathak S (1990) Orthotopic implantation is essential for the selection, growth and metastasis of human renal cell cancer in nude mice. Cancer Metastasis Rev 9: 149–165
Fogel M, Altevogt P, Schiramacher V (1983) metastatic potential severely altered by changes in tumor cell adhesiveness and cell surface sialylation. J Exp Med 157: 371–376
Folkman J, Klagsbum M (1987) Angiogenesis factors. Science 235: 442–447
Folkman J (1991) Tumor angiogenesis. In: Holland JF (ed) Cancer medicine. Lee and Febiger, Philadelphia
Fox SB, Gatter KC, Jackson DG et al (1993) CD44 and cancer screening. Lancet 342: 548–549
Fox SB, Fawcett J, Jackson DG et al (1994) Normal human tissues, in addition to some tumors, express multiple different CD44 isoforms. Cancer Res 54: 4539–4546
Fujita N, Yaegashi N, Ide Y et al (1994) Expression of CD44 in normal versus tumor endometrial tissues: possible implication of reduced expression of CD44 in lymph-vascular space involvement of cancer cells. Cancer Res 54: 3922–3928
Gross N, Beretta C, Eruisseau GM, Jackson D, Simmons D, Beck D (1994) CD44H expression by human neuroblastoma cells: relation to MYCN amplication and lineage differentiation. Cancer Res 54: 4238–4242
Gunthert U, Hofman M, Rudy S et al (1991) A new variant of glycoprotein CD44 congers metastatic potential to rat carcinoma cells. Cell 65: 13–24
Guo YJ, Ma J, Wong H et al. (1993) Monoclonal anti-CD44 antibody acts in synergy with anti-CD2 but inhibtis anti-CD3 or T cell receptor mediated signaling in murine T cell hybridomas. Cell Immunol 153: 186–191

Guo YJ, Ling SC, Wong, Sy MS (1994) Acylation of CD44 may interfere with CD3 mediated signaling in human lymphocytes. Int Immunol 6(2): 213–221
Guo YJ, Ma J, WAng JH et al (1994) Inhibition of human melanoma growth and metastasis in vivo by anti-CD44 monoclonal antibody. Cancer Res 54: 1561–1565
Guo YJ, Liu GG, Wang XN et al (1994) Soluble CD44 in the serum as an indicator of tumor growth and metastasis in patients with gastric or colon cancer. Cancer Res 54: 422–426
Hanning R, Mainolfi E, Bystryn JC, Henn M, Merluzzi VJ, Rothlein R (1991) Serum levels of circulating intercellular adhesion molecule 1 expression in human malignant melanoma. Cancer Res 51: 5003–5005
Hardingham TE, Hosang AJ (1992) proteoglycans: many forms and many functions. FASEB J 6: 861–870
Hardwick C, Hoare K, Owens R et al (1992) Molecular cloning of a novel hyaluronan receptor that mediates tumor cell motility. J Cell Biol 117: 1343–1350
Hart I, Birch M, Marshall JF (1991) Cell adhesion receptor expression during melanoma progression and metastasis. Cancer Metastasis Rev 10: 115–135
Haynes BF, Telen MJ, Hale LP, Denning SM (1989) CD44 a molecule involved in leukocyte adherence and T cell activation. Immunol Today 10: 423–427
Haynes BF, Hale LP, Patton KL, Martin ME, McCallum RM (1991) Measurement of an adhesion molecule as an indicator of inflammatory disease activity. UP-regulation of the receptor for hyaluronate (CD44) in rheumatoid arthritis. Arthritis Rheum 34: 1434–1441
Heider K-H, Hofmann M, Hors E et al. (1993) A human homologue of the rat metastasis associated variant of CD44 is expressed in colorectal carcinomas and adenomatous polyps. J Cell Biol 120: 227–233
Herlyn M, Rodeck U, Koprowski H (1987) Shedding of human tumor associated antigens in vitro and in vivo. Adv Cancer Res 49: 189–221
Herrlich P, Rudy W, Hofmann M et al. (1993) CD44 and splice variants of CD44 in normal differentiation and tumor progression. In: Hemler M, Mihich E (eds) Cell adhesion molecules. Plenum, New York, pp 265–288
Hoessli DC (1980) T lymphocyte differentiation is accompanied by increase in sialic acid content of Thy-1 antigen. Nature 283: 576–577
Hofmann U, Rudy W, Zooler M et al (1991) CD44 splice variants confer metastatic behavior in rats: homologous sequences are expressed in human tumor cell lines. Cancer Res 51: 5292–5297
Horai T, Nakamura N, Tateishi R, Hattori S (1981) Glycosaminoglycans in human lung cancer. Cancer 48: 2016–2021.
Horst E, Meijer CJ, Radaszkiewicz T, Oseloppele GJ, Van Krieken JH, Pals ST (1990) Adhesion molecules in the prognosis of diffuse lage-cell lymphoma: expression of a lymphocyte homing receptor (CD44), LFA-1, (CD11a/18) and ICAM-1 (CD54). Leukemia 4: 595–599
Huet SH, Groux H, Caillou B, Valentin H, Prieur AM, Bernard A (1989) CD44 contributes to T cell activation. J Immunol 143: 798–801
Hughes EN, Mengod G, August JT (1981) Murine cell surface glycoproteins. Characterization of a major components of 80,000 daltons as a polymorphic differentiation antigen of mesenchyma cells. J Biol Chem 256: 7023–7027
Jalkanen S, Bargatze RF, del los Toyos J, Butcher EC (1987) Lymphocyte recognition of high endothelium: antibodies to distinct epitopes of an 85–95-kD glycoprotein antigen differentially inhibit lymphocyte binding to lymph node, mucosal, or synovial endothelial cells. J Cell Biol 105: 983–990
Jalkanen S, Jalknen M (1992) Lymphocyte CD44 binds the COOH-terminal heparin binding domain of fibronectin. J Cell Biol 116: 817–825
Kageshita T, Yoshii A, Kimura T et al (1993) Clinical relevance of ICAM-1 expression in primary lesions and serum of patients with malignant melanoma. Cancer Res 53: 4927–4932
Kahn P (1992) Adhesion protein studies provide new clue to metastasis. Science 257: 614–615
Kalomiris EL, Bourguignon LYW (1988a) Mouse T lymphoma cells contain a transmembrane glycoprotein (GP85) that binds ankyrin. J Cell Biol 106: 319–327
Kalomiris EL, Bourguignon LYW (1988b) Lymphoma protein kinase C is associated with the transmembrane glycoprotein gp85, and may function in gp85 ankyrin binding. J Biol Chem 264: 8113–8120
Katoh S, McCarthy JB, Kincade PW (1994) Characterization of soluble CD44 in the circulation of mice: levels are affected by immune activity and tumor growth. J Immunol 152: 3441–3449
Kennel SJ, Lankford TK, Foote LJ, Shinpock SG, Stringer C (1993) CD44 expression of murine tissues. J Cell Sci 104: 373–382

Kiefer MC, Stephans JC, Crawford K, Okino K, Barr PJ (1990) Ligand affinity cloning and structure of a cell surface heparan sulfate proteoglycan that binds basic fibroblast growth factor. Proc Natl Acad Sci USA 87: 6985–6989
Knudson W, Biswas C, Li X-Q, Nemec RE, Toole BP (1989) The role and regulation of tumor associated hyaluronan. The biology of hyaluronan. Ciba Found Symp 143: 150–160
Knutson JR, Fields GB, Lida J, Miles AJ, McCarthy JB (1995) A type IV collagen derived synthetic peptide, IN-H1, interacts with human melanoma CD44 chondroitin sulfate proteoglycan and inhibitis invasion of basement membranes. Proc Am Assoc Cancer Res 36: 68
Koopman G, Heider K-H, Horst E et al. (1993) Activated human lymphocytes and aggressive non-Hodgkin's lymphomas express a homologue of the rat metastasis associated variant of CD44. J Exp Med 177: 897–904
Krusius T, Gehlsen KR, Ruoslahti E (1987) A fibroblast chondroitin sulfate proteoglycan core protein contains lectin like and growth factor like sequence. J Cell Biol 262: 13120–13125
Lacy BE, Underhill CB (1987) The hyaluronate receptor is associated with actin filaments. J Cell Biol 105: 1395–1404
Laurent TC, Fraser JRE (1992) Hyaluronan. FASEB J 6: 2397–2404
Lee TH, Wisniewski H-G, Vilcek J (1992) A novel secretory tumor necrosis factor inducible protein is a member of the family of hyaluronate binding proteins closely related to the adhesion receptor CD44. J Cell Biol 116: 545–557
Legras S, Günthert U, Li Y et al. (1995) CD44 alternative splicing in normal human myelopoiesis and its deregulation in acute myeloblastic leukemia (AML). Proc Am Assoc Cancer Res 36: 465–465
Lesley J, Schulte R, Hyman R (1990) Binding of hyaluronic acid to lymphoid cell lines is inhibited by monoclonal antibodies against Pgp-1. J Exp Cell Res 187: 224–229
Lesley J, Q H, Miyake K, Hamann A, Hyman R, Kincade PW (1992) Requirements for hyaluronic acid binding by CD44: a role for the cytoplasmic domain and activation by antibody. J Exp Med 172: 257–266
Lesley J, Hyman R, Kincade PL (1993) CD44 and its interaction with extracellular matrix. Adv Immunol 34: 271–335
Liao HX, Levesque MC, Patton K et al (1993) Regulation of human CD44H and CD44E isoform binding to hyaluronan by phorbol myristate acetate and anti-CD44 monoclonal and polyclonal antibodies. J Immnol 151: 6490–6499
Lloyd CW (1975) Sialic acid and the social behavior of cells. Biol Rev 50: 325–350
Lokeshwar VB, Bourguignon LYW (1992) The lymphoma transmembrane glycoprotein GP85 (CD44) is a novel guanine nucleotide binding protein which regulates GP85 (CD44)-ankyrin interaction. J Biol Chem 267: 22073–22078
Lucas MG, Green AM, Telen MJ (1989) Characterization of the serum In (Lu)-related antigen: identification of a serum protein related to erythrocyte p80. Blood 73: 596–600
Lynch F, Ceredig R (1988) Lyt24 (Pgp-1) expression by thymocytes and peripheral T cells. Immunol Today 9: 7–10
MacLeod CL, Weinroth SE, Sterifinger C, Glaser SM, Hays EF (1985) SL12 murine T lymphoma: a new model for tumor cell heterogeneity. J Natl Cancer Inst 74: 875–879
Matsumura Y, Tarin D (1993) Significance of CD44 gene products for cancer diagnosis and disease evaluation. Lancet 340: 1053–1058
Matsumura Y, Smith S, Tarin JC (1995) Cancer diagnosis by detection of new abnormalities in splicing of CD44 gene products in bladder tumors and urine cell sediments. Proc Am Assoc Cancer Res 36: 283
Merzak A, Koocheckpour S, Pilkington GJ (1994) CD44 mediates human glioma cell adhesion and invasion in vitro. Cancer Res 54: 3988–3992
Miyake K, Underhill CB, Lesley J, Kincade PW (1989) Hyaluronate can function as a cell adhesion molecule and CD44 participates in hyaluronate recognition. J Exp Med 172: 69–75
Murakami S, Miyake K, June CH, Kincade PW, Hodes RJ (1990) IL-5 induces a Pgp-1 bright B cell subpopulation that is highly enriched in proliferative and Ig secretory activity and binds hyaluronate. J Immunol 145: 3618-3627
Murakami S, Miyake K, Bel R, Kincade PW, Hodes RJ (1991) Characterization of autoantibody secreting B cells in mice stimulatory chronic graft versus host reactions: identification of a CD44 high population that binds specifically to hyaluronate. J Immunol 146: 1422–1427
Murakami S, Shimabukuro Y, Miki Y et al (1994) Inducible binding of human lymphocytes to hyaluronate via CD44 does not require cytoskeleton association but does require new protein synthesis. J Immunol 152: 467–477
Nagabhushan M, Pretlow TG, Guo YJ, Amini SB, Pretlow TP, Sy MS (1995) Altered expression of CD44 in prostate cancer metastases (submitted)

Nemec RE, Toole BP, Knudson (1987) The cell surface hyaluronate binding sites of invasive human bladder carcinoma cells. Biochem Biophy Res Commun 149: 249–257
Nicolson GL (1988) Organ specificity of tumor metastasis: role of preferential adhesion, invasion and growth of malignant cells at specific secondary sites. Cancer Metastasis Rev 7: 143–188
Nottenburg C, Rees G, St. John T (1989) Isolation of mouse CD44 cDNA: structure features are distinct from the primate cDNA. Proc Natl Acad Sci USA 86: 8521–8525
Oyama F, Hiroshashi S, Shimosato Y, Titani K, Sekiguchi K (1989) Deregulation of alternative splicing of fibronectin pre-mRNA in malignant human liver tumors. J Biol Chem 264: 10331–10334
Payne LN, Pani PK, Weiss RA (1971) A domain epistatic gene which inhibits cellular susceptibility to RSV. J Gen Virol 13: 455–462
Peach RJ, Hollenbaugh D, Stamemkovic I, Aruffo A (1993) Identification of hyaluronic acid binding sites in the extracellular domain of CD44. J Cell Biol 122: 257–264
Peacock EEJ (1984) Wound repair, 3rd edn. Saunders, Philadelphia
Penno MB, August JT, Baylin SB et al (1994) Expression of CD44 in human lung tumors. Cancer Res 54: 1381–1387
Perschl A, Lesley J, English N, Trowbridge I, Hyman R (1995) Role of CD44 cytoplasic domain in hyaluronan binding. Eur J Immunol 25: 495–501
Philipson LH, Schwartz NB (1984) Subcellular localization of hyaluronate synthetase in oligodendrocytoma cells. J Biol Chem 259: 5017–5023
Picker LJ, Nakache M, Butcher EC (1989) Monoclonal antibodies to human lymphocytes homing receptors define a novel class of adhesion molecules on diverse cell types. J Cell Biol 109: 927–935
Poole AR (1986) Proteoglycans in health and disease: structure and functions. Biochem J 236: 1–14
Reid T, Flint MH (1974) Changes in glycosaminoglycan content of healing rabbit tendon. J Embryol Exp Morphol 31: 489–495
Rothbard JB, Brackenbury R, Cunnigham AB, Edelman GM (1982) Differences in the carbohydrate structures of neural adhesion molecules from adult and embryonic chicken brains. J Biol Chem 257: 11064–11069
Rothman BL, Blue M, Kelley KA, Wunderlich D, Mierz DV, Aune TM (1991) Human T cell activation of OKT3 is inhibited by a monoclonal antibody to CD44. J Immunol 147: 2493–2497
Rouslahti E (1988) Structure and biology of proteoglycans. Ann Rev Cell Biol 4: 229–255
Ruoslahti E, Yamagushi Y (1991) Proteoglycans as modulators of growth factor activities. Cell 64: 867–869
Saksela O, Rifkin DB (1990) Release of basic fibroblast growth factor-heparan sulfate complexes from endothelial cells by plasminogen activated mediated proteolytic activity. J Cell Biol 110: 767–775
Salmi M, Gron-Virta K, Sointu P, Grenman R, Kalimo H, Jalkannen S (1993) Regulated expression of exon V6 containing isoforms of CD44 in man: downregulation during malignant transformation of tumors of squamocellular origin. J Cell Biol 122: 431–442
Schauer R (1985) Sialic acids and their role as biological masks. Trends Biochem Sci 10: 357–360
Schwarzbauer JE, Patel RS, Fonda D, Hynes RO (1987) Multiple sites of alternative splicing of the rat fibronectin gene transcripts. EMBO J 6: 2573–2580
Screaton GR, Bell MV, Jackson DG, Cornelis FB, Gerth U, Bell JE (1992) Genomic structure of DNA encoding the lymphocyte homing receptor CD44 reveals at least 12 alternatively spliced exons. Proc Natl Acad Sci USA 89: 12160–12164
Screaton GR, Bell MV, Bell JI, Jackson D (1993) The identification of a new alternative exon with highly restricted tissue expression in transcripts encoding the mouse Pgp-1 (CD44) homing receptor. J Biol Chem 268: 12235–12238
Shimizu Y, Seventer GA, Siraganian R, Wahl L, Shaw S (1989) Dual role of the CD44 molecule in T cell adhesion and activation. J Immunol 143: 2457–2463
Shtivelman E, Bishop MJ (1991) Expression of CD44 is repressed in neuroblastoma cells. Mol Cell Biol 11: 5446–5453
Sieweke MH, Thompson NL, Sporn MB, Bissell MJ (1990) Mediation of wound related Rous sarcoma virus tumorigenesis by TGFB. Science 248: 1656–1660
Springer TA (1990) Adhesion receptors of the immune system. Nature 346: 426–434
Stamenkovic I, Aminot M, Pesando JM, Seed B (1989) A lymphocyte molecule implicated in lymph node homing is a member of the cartilage link protein family. Cell 56: 1057–1062
Stamenkovic I, Aruffo A, Aminot M, Seed B (1991) The hematopoietic and epithelial forms of CD44 are distinct polypeptides with different adhesion potentials for hyaluronate bearing cells. EMBO 10: 343–348
Stetler-Stevenson WG, Aznavoorian S, Liotta LA (1993) Tumor cell interactions with the extracellular matrix during invasion and metastasis. Annu Rev Cell Biol 9: 541–573

Stoolman LM (1989) Adhesion molecules controlling lymphocytes migration. Cell 56: 907–910
Sy MS, Guo YJ, Stamenkovic I (1991) Distinct effects of two CD44 isoforms on tumor growth in vivo. J Exp Med 174: 859–866
Sy MS, Guo YJ, Stamenkovic I (1992) Inhibition of tumor growth in vivo with a chimeric CD44-immunoglobulin molecule. J Exp Med 176: 623–627
Takeuchi J, Sobue M, Sato E, Yoshida M, Uchibori N, Miura K (1981) A high level of glycosaminoglycan synthesis of squamous cell carcinoma of the parotid cland. Cancer 47: 2030–2035
Tan PHS, Santos EB, Rossbach HC, Sandmaier BM (1993) Enhancement of natural killer activity by an antibody to CD44. J Immunol 150: 812–820
Tanabe KK, Ellis LM, Saya H (1993) Expression of CD44R1 adhesion molecule in colon carcinomas and metastasis. Lancet 341: 725–726
Tanaka Y, Adams DH, Hubscher S, Hirano H, Siebenlist U, Shaw S (1993) T-cell adhesion induced by proteoglycan-immobilized cytokine MIP-1B. Nature 361: 79–82
Thomas LJ, DeGasperi R, Sugiyama E et al (1991) Functional analysis of T cell mutants defective in the biosynthesis of glycosyl phosphatidylinositol anchor. J Biol Chem 266: 23175–23184
Thomas L, Byers HR, Vink J, Stamenkovic I (1992) CD44H regulates tumor cell migration of hyaluronate coated substrate. J Cell Biol: 971–991
Tolg C, Hofmann M, Herrlich P, Ponta H (1993) Splicing choice from ten variant exons establishes CD44 variability. Nucleic Acids Res 21: 1225–1229
Toole BP (1990) Hyaluronan and its binding proteins, the hyaladherins. Curr Opin Cell Biol 2: 839–844.
Toole BP, Biswas C, Gross J (1979) Hyaluronate and invasiveness of the rabbit V2 carcinoma. Proc Natl Acad Sci USA 76: 6299–6303
Trowbridge IS, Lesley J, Schulte R, Trotter J (1982) Biochemical characterization and cellular distribution of a polymorphic, murine cell surface glycoprotein expressed on lymphoid tissues. Immunogenetics 15: 299–312
Tsujisaki M, Imai K, Hirata H et al (1991) Detection of circulating intercellular adhesion molecule 1 antigen in malignant diseases. Clin Exp Immunol 85: 3–8
Turley EA (1984) Proteoglycans and cell adhesion: their putative role during tumorigenesis. Cancer Metastasis Rev 3: 325–339
Underhill CH, Dorfman A (1978) The role of hyaluronic acid in intercellular adhesion of cultured mouse cells. Exp Cell Res 117: 155–164
Underhill CB, Toole BP (1982) Transformation-dependent loss of the hyaluronate containing coats of cultured cells. J Cell Physiol 110: 123–128
Underhill CB (1982) Interaction of hyaluronate with the surface of simian virus 40-transformed 3T3 cells: aggregation and binding studies. J Cell Sci 56L: 177–189
Webb DSS, Shimizu Y, Van Seventer GA, Shaw S, Gerrard TL (1990) LFA-3, CD44 and CD45: Physiologic triggers of human monocyte TNF and IL-1 release. Science 249: 1295–1297
West DC, Sattar A, Kumar S (1985) Angiogenesis induced by the degradation products of hyaluronic acid. Science 228: 1324–1326
Wielenga VJM, Heider K-H, Offerhaus GJA et al. (1993) Expression of CD44 variant proteins in humal colorectal cancer is related to tumor progression. Cancer Res 53: 4754–4756
Wolffe EJ, Cause WC, Pelfrey CM, Holland SM, Steinberg AD, August JT (1990) The cDNA sequence of mouse Pgp-1 and to human CD44 cell surface antigen and proteoglycan core/link proteins. J Biol Chem 265: 341–352
Woodland H, Jones L (1988) Growth factors in amphibian cell differentiation. Nature 332: 113–115
Zhou DF, Ding JF, Picker LJ, Bargatze RF, Butcher EC, Goeddel DV (1989) Molecular cloning and expression of Pgp-1. The mouse homolog of the human H-CAM (Hermes) lymphocyte homing receptor. J Immunol 143: 3390–3395

Does Soluble CD44 Reflect the Clinical Behavior of Human Cancer?

R. RISTAMÄKI[1,2], H. JOENSUU[3], and S. JALKANEN[1]

1 Introduction

Adhesion molecules are intimately involved in mediation of proper cell–cell contacts and communications in a wide variety of biological functions. During malignant trasformation, cells often lose precisely regulated contacts to the mircoenvironment. This may result in an increase of the amount of soluble adhesion molecules in blood circulation. In this review, we will present data on soluble forms of one adhesion molecule, CD44, and discuss the possibilities of using CD44 analyses in clinical practice.

2 Multifunctional Nature of CD44

The history of CD44 started in 1980, when it was first reported as a brain-granulocyte-T lymphocyte antigen (DALCHAU et al. 1980). During the following

[1]National Public Health Institute, and MediCity Research Laboratory, Turku University, Tykistökatu 6, 20520 Turku, Finland
[2]Department of Oncology and Radiotheraphy, Turku University, Kiinamyllynhatu 4-8, 20520 Turku, Finland
[3]Department of Oncology and Radiotheraphy, Helsinki University, Haartmaninkatu 4, 00290 Helsinki, Finland

years, several laboratories described expression of an 80- to 90-kDa molecule on various cell types and its diverse functional and structural properties. This molecule was also a B cell antigen and a maturation marker for T cells; it participated in cell motility, adherence of cells to extracellular matrix molecules (fibronectin, collagen), and lymphocyte binding to the vascular endothelium (HAYNES et al. 1983; JACOBSON et al. 1984; LETARTE 1985; JALKANEN et al. 1986; CARTER and WAYNER 1988; PALS et al. 1989a; reviewed in HAYNES et al. 1989). Cross-precipitation studies using different antibodies and cloning and sequencing of the cDNA encoding CD44 showed that all these characteristics mentioned above were associated with CD44 (OMARY et al. 1988; PICKER et al. 1989; GOLDSTEIN et al. 1989; STAMENKOVIC et al. 1989). The enormous complexity of the CD44 molecule was fully appreciated after genomic cloning of the molecule, which revealed ten alternativeley spliced variant exons within the region coding the membrane-proximal portion of the extracellular domain (SCREATON et al. 1992). In addition, the CD44 molecule is a target for extensive post-translational modifications such as N- and O-glycosylation and addition of glycosaminoglycan side chains (JALKANEN et al. 1988; BROWN et al. 1991; JACKSON et al. 1995).

After cloning and sequencing the gene encoding CD44, new functional properties have been described for this molecule. It is a hyaluronan receptor that also serves as a signaling molecule through which lymphocyte function-associated antigen-1 (LFA-1) can be triggered to an active conformation and cytokine release from macrophages can be induced (MIYAKE et al. 1990; ARUFFO et al. 1990; KOOPMAN et al. 1993; WEBB et al. 1990). Such cytokines include tumor necrosis factor(TNF)-α, interleukin(IL)-1, and macrophage colony-stimulating factor(M-CSF). Furthermore, particular CD44 molecules containing the V3 exon have glycosaminoglycan side chains that are able to bind growth factors (BENNETT et al. 1995).

Participation of CD44 in malignant transformation and formation of metastases has been described both in animal models and in the clinical setting (PALS et al. 1989b; JALKNEN et al. 1991; GÜNTHERT et al. 1991; SY et al. 1991; HEIDER et al. 1993). For example, patients suffering from CD44$^-$ non-Hodgkin's lymphoma have a better prognosis than those with CD44$^+$ lymphoma, although the negative ones are more often morphologically highly aggressive (JALKANEN et al. 1991). Moreover, in many tumor types of epithelial origin, expression of certain variant forms of CD44 has been connected to metastatic spread of these malignancies (WIELENGA et al. 1993; LI et al. 1993; KAUFMANN et al. 1995). In contrast, in special cell types, expression of CD44 diminishes during malignant transformation and certain highly metastatic cell lines contain very low levels of CD44 (SALMI et al. 1993; JACKSON et al. 1994). The function of CD44 may be dependent on the cell type, and elucidation of the role of CD44 in clinical behavior of different types of human cancer warrants further study.

3 Structural Characteristics of Soluble CD44

DALCHAU et al. suggested the presence of CD44 in the serum as early as 1980. This was later confirmed by others (TELEN et al. 1983; LUCAS et al. 1989; PICKER et al. 1989). TELEN and her coworkers found the serum form to be of the same size (80 kDa) as the form detected on the surface of erythrocytes (TELEN et al. 1983; LUCAS et al. 1989). In addition to the major species of 60–80 kDa, larger forms of 100–150 kDa have also been detected (BAZIL and HORJESI 1992). Peptide maps have indicated that the 60- to 80-kDa and the 100- to150-kDa forms represent different molecular species of CD44. It is quite likely that the larger forms originate from the variant isoform(s), but it is also possible that the standard form decorated with chondroitin sulphate side chains is included in the larger forms (BAZIL and HORJESI 1992; RISTAMÄKI et al. 1994). Quite variable concentrations up to 5 µg/ml have been reported for circulating CD44 in normal individuals, but based on more recent publications the normal serum levels seem to vary between 10 and 200 ng/ml (LUCAS et al. 1989; GUO et al. 1994; RISTAMÄKI et al. 1994). Interestingly, CD44 expression is regulated by Lutheran inhibitory *In(Lu)* gene in such a way that individuals with the dominant form of Lu((a-b-) have lower levels of serum and surface CD44 on erythrocytes and leukocytes than others (TELEN et al. 1983). In the mouse, the concentration of soluble CD44 varies depending on the strain of the mouse and consists of a major 85- to 95-kDa species and of two minor, but larger components of 140 and 180 kDa in size (KATOH et al. 1994). Although the migration of the soluble 85- to 95-kDa form in the gel systems used is indistinguishable from the migration of the tissue-derived 85- to 95-kDa form, careful studies using antibodies against the cytoplasmic region of the molecule have shown the soluble form to lack the cytoplasmic tail. In addition, the difference in size found after deglycosylation of the serum and tissue forms of CD44 supports this assumption (KATOH et al. 1994). Circulating CD44 is monomeric, and it does not seem to be firmly associated with other proteins, and most notably not with the high molecular weight form of hyaluronan (KATOH et al. 1994). Certain mediators such as tumor necrosis factor-α, calcium ionophore A23187, and the chemoattractant peptide formyl-Met-Leu-Phe are able to decrease surface expression of CD44, but only that of neutrophils (CAMPANERO et al. 1991; BAZIL and HORJESI 1992), whereas treatment of neutrophils and lymphocytes with antibodies against CD44 has been shown to result in diminished expression of CD44 on both of these cell types (BAZIL and HORJESI 1992). Shedding rather than internalization seems to be the fate of CD44 after triggering the molecule, because subsequent to in vitro and in vivo modulation increased amounts of CD44 can be found in culture supernatants and in the serum, respectively (BAZIL and HORJESI 1992; CAMP et al. 1993). Although in in vitro conditions both lymphocytes and granulocytes shed CD44, the shed form originating from the granulocytes seems to represent only a minor component of the soluble CD44 isolated from the serum (BAZIL and HORJESI 1992).

Shedding of CD44 is most likely a consequence of endogenous protease activity, because in experimental settings protease inhibitors such as aprotinin and phenylmethylsulfonyl fluoride have been able to decrease the amount of CD44 shed from the cell surface (CAMPANERO et al. 1991). Metalloproteases and serine proteases are good candidates to be responsible for the cleavage of CD44, because their inhibitors, 1.10-phenanthroline and Nα-p-tosyl-L-lysine chloromethyl ketone, respectively, markedly inhibit the cleavage (BAZIL and STROMINGER 1994). However, the natural proteases responsible for the shedding remain to be characterized. The mechanism by which enzymatic cleavage is regulated is not known either. Modulation of the surface expression of CD44 with antibodies suggests that interaction between CD44 and its natural ligands may trigger the same phenomenon in the body. Binding to the natural ligand may change the conformation of CD44, exposing the cleavage site for one or more enzymes. Induced shedding by certain mediators suggests that signals transduced from intracellular compartments are also capable of activating the cascade, leading to shedding of CD44. Theoretically, the proteolytic enzymes can either be stored on the cell surface, perhaps anchored to other molecules such as glycosaminoglycans (known to be docking sites for certain enzymes and growth factors; D'AGOSTINO et al. 1989; RAPRAEGER et al. 1991), to be ready to attack the conformationally altered target or they can be generated from the proenzymes after stimulation.

4 Soluble CD44 in Human Cancer

4.1 Lymphomas

In general, there are as yet only a few reports on serum CD44 in human cancer. However, the results obtained with clinical series suggest that the level of circulating CD44 is markedly influenced by tumor growth (RISTAMÄKI et al. 1994). Patients suffering from lymphomas have significantly elevated levels of serum CD44 (median,510 ng/ml) at the time of the diagnosis as compared with the levels detected in healthy individuals (median, 15 ng/ml; Fig. 1). However, the concentration of circulating CD44 seems to reflect response to treatment; if a complete remission is obtained by chemotherapy, the CD44 serum level returns to the same range as is found in normal subjects (Fig. 2), whereas patients with progressive disease continue to have high levels of soluble CD44 in their sera. Elevation of serum CD44 in patients with lymphoma mainly occurs in the 70- to 80-kDa species of CD44. This was confirmed by gel analyses and also by measuring the level of the forms containing the variant exon V6 in dot blot analyses. Circulating CD44v6 levels were found to be only slightly higher in lymphoma patients than in the normal controls. Animal studies utilizing a mouse model have also shown that tumor growth increases the level of circulating CD44. These studies were also able to confirm that, in addition to normal lymphocytes, lymphoma cells can shed their surface CD44 in culture conditions (KATOH et al.

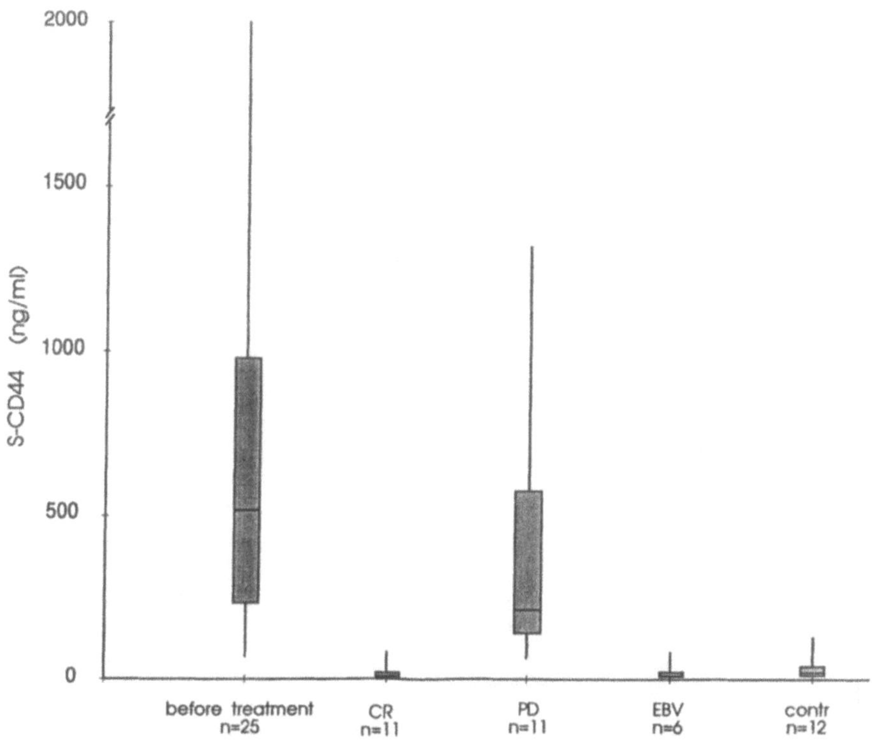

Fig. 1. Distribution of serum CD44 levels in lymphoma patients taken before treatment, in complete remission (*CR*) and progressive disease (*PD*), patients with Epstein-Bar virus infection (*EBV*, and healthy controls (*contr*). *Lines* represent the minimum and maximum levels, *boxes* contain values between the 25th and 75th percentiles, and *bars* indicate the medians

1994). Our recent experiments indicate that also in a clinical setting both normal and malignant lymphocytes can serve as a source of circulating CD44 (R. Ristamäki et al., submitted).

4.2 Cancers of Epithelial Origin

Regarding cancers of epithelial origin, the measurements of serum CD44 have so far concentrated on gastric and colon cancers (Guo et al. 1994). Patients with advanced gastric or colon cancer have approximately ten times higher serum levels of CD44 than normal controls. Patients with metastatic cancer tend to have higher levels than those who suffer from local tumor. In contrast to patients suffering from lymphoma, the prominent form of soluble CD44 in these patients is 130–190 kDa. In addition, minor components consisting of proteins with molecular masses of 80 and 72 kDa can be detected. This clearly illustrates the structural differences of circulating CD44 in different malignancies and supports the idea that in epithelial cancers the variant isoforms of CD44 predominate over the standard hematopoietic form. Moreover, it is highly likely that the elevated levels of serum CD44 of higher

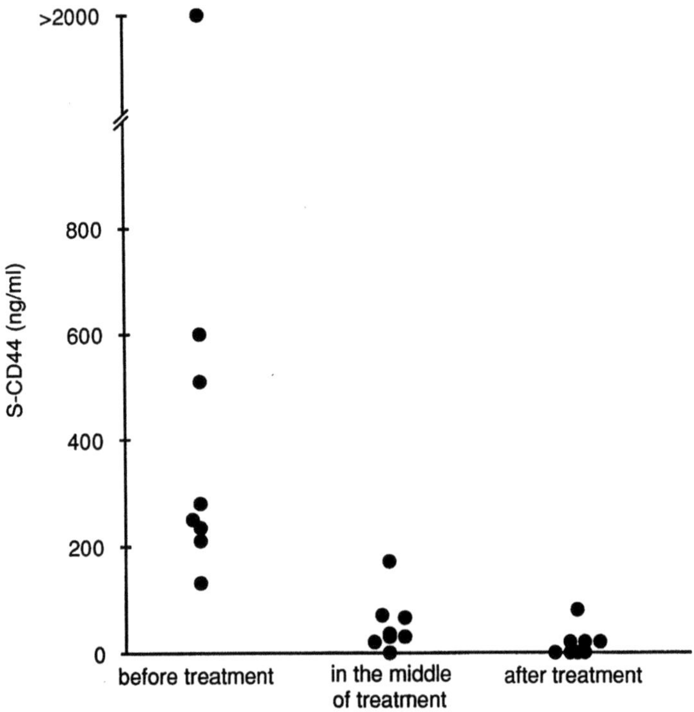

Fig. 2. Variations in the levels of serum CD44 of eight lymphoma patients achieving complete remission at the beginning, middle, and end of treatment

molecular weight originate from the tumor. A marked decrease in the serum concentrations of CD44 is achieved after surgical removal of the tumor.

5 Functional Significance of Soluble CD44

Circulating CD44 is functionally active. This has been shown unambiguously by analyzing its binding to hyaluronan and fibronectin (KATOH et al. 1994; R. RISTAMÄKI et al., submitted). An elevated level of serum CD44 might be one of the body's defense mechanisms to fight against cancer spread, because soluble CD44 may block the binding sites on the endothelial cell surfaces and the extracellular matrix and thus inhibit cancer cell attachment to these structures.

Transfection of the human gene encoding CD44 into lymphoma cells and subsequent transplantation of the transfected cells into severe combined immunodeficient (SCID) mice have provided strong evidence suggesting that soluble CD44 originates from lymphoma cells (R. RISTAMÄKI et al., submitted). In addition to cell adhesion and migration, CD44 may have a role in the control of cell proliferation. CD44 molecules are able to bind cytokines and growth factors (TANAKA et al. 1993; BENNETT et al. 1995). When present on the cell surface, CD44

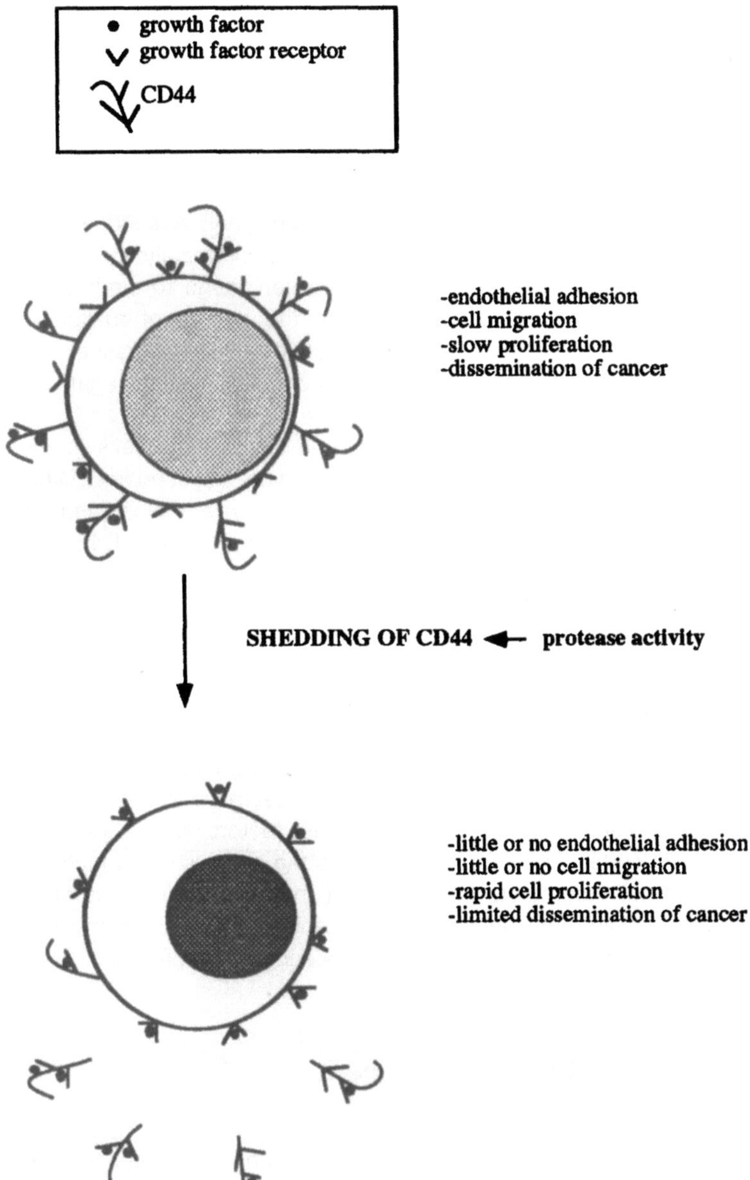

Fig. 3. Possible function of CD44 as a regulator of cell proliferation. See Sect. 5 for details

might bind growth factors and reduce their binding to their cognate receptors, which would result in slow or absent cell proliferation. When CD44 molecules are shed, growth factors could bind to their receptors on the cell surface in greater numbers, which, in turn, could provide a sufficient signal for cell proliferation (Fig. 3). This type of regulatory function has been suggested for another cell surface proteoglycan, syndecan (Mali et al. 1994).

There is some evidence to support this hypothesis. First, non-Hodgkin's lymphomas that stain negatively for CD44 have been found to often have a high grade of malignancy and to have larger S-phase fractions as measured by DNA flow cytometry than lymphomas that express CD44 (JALKANEN et al. 1991). On the other hand, the level of soluble CD44 has been found to be greater in high-grade, malignant non-Hodgkin's lymphomas as compared with lymphomas with a low grade of malignancy (R. RISTAMÄKI et al., submitted). Second, poorly differentiated squamous cell carcinomas proliferate more rapidly than well-differentiated ones, and expression of CD44 has been found to diminish in human squamous cell carcinoma in parallel with the degree of malignant transformation (SALMI et al. 1993). Third, in reactive lymphatic tissue, the germinal centers, which are the sites for rapid cell proliferation, stain negatively for CD44 (JALKANEN et al. 1986).

If we visualize CD44 simply as a molecule that anchors the cells to the extracellular matrix molecules, then it is essential for the cell to get rid of CD44 at least temporarily to be able to enter the metastatic pathways (the blood and the lymphatics). On the other hand, if binding to its ligand downregulates CD44 from the cell surface, active migration in the tissue needs rapid turnover of CD44, and the end result is seen as an elevated level of circulating CD44.

If we consider CD44 as a signaling molecule, shedding of CD44 from the cell surface may decrease the possibility to trigger LFA-1 via CD44, which may facilitate the tumor cells to escape from immunosurveillance. Triggering of CD44 by its ligand has also been shown to release certain cytokines from macrophages (WEBB et al. 1990). In this context, an increased amount of soluble CD44 may be effective in blocking cytokine secretion, leading to a decrease in the levels of tumor necrosis factor-α and interleukin-1. Lack of these factors may be beneficial for the tumor cells in certain circumstances. Due to the obvious multifunctional nature of CD44 but still incomplete knowledge of the precise functions of its various isoforms, the exact mechanisms of action and in vivo consequences of elevated levels of circulating CD44 remain purely speculative at present.

6 Soluble CD44 and Clinical Practice

It may be a challenging task to assess the clinical response of cancer to treatment. Currently, response evaluation may require extensive and expensive examinations, such as computed tomography scans and bone marrow biopsies, and it may be difficult to discriminate small cancer deposits from nonmalignant residual tumors without explorative surgery. Therefore, inexpensive and less cumbersome methods to analyze response to therapy would be welcome in clinical practice.

At present, thymidine kinase, lactate dehydrogenase, and β_2-microglobulin measurements are used in the follow-up of patients with lymphoma. However, the serum level of CD44 seems to better reflect the response to chemotherapy than the other laboratory investigations currently in use in patients with lymphoma. At the time of diagnosis, thymidine kinase and lactate dehydrogenase may have better prognostic significance regarding the overall outcome of the patient with lymphoma than soluble CD44 (R. Ristamäki et al., submitted), but serum thymidine kinase and lactate dehydrogenase levels may be influenced by cancer chemotherapy so extensively that they may be inferior to CD44 in the follow-up of treatment response in patients with lymphoma (Ristamäki et al. 1994). The levels of serum CD44 have not been systematically measured in different diseases, and, therefore, it is currently not known whether other diseases besides cancer can generate high levels of circulating CD44 and whether these conditions will interfere with the usefulness of serum CD44 levels in the follow-up of cancer patients. However, some information can be gathered from the few patients used as controls in recent studies. These were patients with banal viral infections of the upper respiratory tract, mononucleosis, or chronic rheumatic diseases (Ristamäki et al. 1994; Guo et al. 1994). None of these patients, not even those with fairly aggressive lymphoproliferative disease, have had elevated levels of circulating CD44. However, certain patients with severe and active rheumatoid arthritis may have increased levels of serum CD44 (Ristamäki et al. 1994). Cancer patients with such a disease are most likely easily distinguished, and their increased levels of soluble CD44 should be judged with caution. Thus, at present monitoring of the CD44 level, at least in lymphomas and in gastric and colon cancer, appears to be valuable in patient care. Elevation of different forms of CD44 in these hematopoietic or epithelial malignancies suggest that certain CD44 isoform-specific antibodies could be used in the follow-up of epithelial cancers in the future. Whether the measurements of CD44 levels are also beneficial in patients with $CD44^-$ cancers (approximately 10% of patients with non-Hodgkin's lymphoma) and in the rare patients with the dominant form of Lu (a-b-) remains to be determined.

Levels of soluble adhesion molecules are now being measured in various diseases (Gearing and Newman 1993; Bazil 1995), and concentrations of certain circulating adhesion molecules have been found to be elevated in various cancers (Tsujisaki et al. 1991; Banks et al. 1993; Christiansen et al. 1994). For example, melanoma patients show elevated levels (two- to threefold increase) of intercellular adhesion molecule (ICAM)-1, and this also has prognostic significance (Harning et al. 1991). Moreover, the level of soluble L-selectin is increased in patients with acute leukemia, and the level reflects the clinical response to leukemia treatment (Spertini et al. 1994). How levels of soluble CD44 relate to levels of other adhesion molecules in the serum and which of the adhesion molecules is the best diagnostic/prognostic marker in a certain type of cancer are still open questions. Availability of commercial kits will certainly make it easier to evaluate the true value of monitoring the serum concentrations of soluble adhesion molecules in the clinical setting.

7 Conclusions

In lymphoma, serum CD44 may mostly originate from the lymphoma cells. It remains functionally active and is able to bind to hyaluronan and fibronectin. In healthy individuals, both the standard and the variant forms of CD44 are detectable in the serum, but only in small quantities. In patients with lymphoma, the prominent species of soluble CD44 is the 70- to 80-kDa form, whereas in patients with gastric or colon cancer larger forms of 130–190kDa predominate.

The function and clinical significance of soluble CD44 are still mostly unsettled. According to a small pilot study, soluble CD44 is elevated in patients with lymphoma at the time of the diagnosis and in recurrent disease, but patients who acheive a complete response have similar serum levels of soluble CD44 as healthy controls. In lymphoma, the serum level of CD44 appears to decrease in parallel with the clinical treatment response, and, therefore, measurement of the serum CD44 level may be a new and effective method to monitor treatment of lymphoma.

In gastric and colon cancers, the serum levels have been found to decrease after surgical removal of the tumor. Hence, measurement of serum CD44 may be a method to monitor cancer treatment in human adenocarcinoma, too, but no information is currently available regarding other adenocarcinomas, such as cancers of the breast or the prostate. Similarly, the serum level and species of soluble CD44 is unknown in squamous cell carcinomas and in melanoma and sarcoma. High serum levels of CD44 appear to be associated with high-grade lymphoma and poor prognosis, but in general, the prognostic value of serum CD44 is mostly unsettled.

Acknowledgments. Our work is financially supported by the Finnish Academy, the Finnish Cancer Society, the Maud Kuistila Foundation, and the Sigrid Juselius Foundation.

References

Aruffo A, Stamenkovic I, Melnick M, Underhill CB, Seed B (1990) CD44 is the principal cell surface receptor for hyaluronate. Cell 61: 1303–1313

Banks RE, Gearing AJ, Hemingway IK, Norfolk DR, Perren TJ, Selby PJ (1993) Circulating intercellular adhesion molecule-1 (ICAM-1), E-selectin and vascular cell adhesion molecule-1 (VCAM-1) in human malignancies. Br J Cancer 68: 122–124

Bazil V (1995) Physiological enzymatic cleavage of leukocyte membrane molecules. Immunol Today 16: 135–140

Bazil V, Horejsi V (1992) Shedding of the CD44 adhesion molecule from leuckocytes induced by anti-CD44 monoclonal antibody simulating the effect of a natural receptor ligand. J Immunol 149: 747–753

Bazil V, Strominger JL (1994) Metalloprotease and serine protease are involved in cleavage of CD43, CD44, and CD16 from stimulated human granulocytes. J Immunol 152: 1314–1322

Bennett KL, Jackson DG, Simon JC, Tanczos E, Peach R, Modrell B, Stamenkovic I, Plowman G, Aruffo A (1995) CD44 isoforms containing exon v3 are responsible for the presentation of heparin-binding growth factor. J Cell Biol 128: 687–698

Brown TA, Bouchard T, St. John T, Wayner E, Carter WG (1991) Human keratinocytes express a new CD44 core protein (CD44E) as a heparan-sulfate intrinsic membrane proteoglycan with additional exons. J Cell Biol 113: 207–221

Camp RL, Scheynius A, Johansson C, Pure E (1993) CD44 is necessary for optimal contact allergic responses but is not required for normal leukocyte extravasation. J Exp Med 178: 497–507

Campanero MR, Pulido R, Alonso JL, Pivel JP, Pimentel-Muinos FX, Fresno M, Sanchez-Madrid F (1991) Down-regulation by tumor necrosis factor-α of neutrophil cell surface expression of the sialophorin CD43 and the hyaluronate receptor CD44 through a proteolytic mechanism. Eur J Immunol 21: 3045–3048

Carter WG, Wayner EA (1988) Characterization of the class III collagen receptor, a phosphorylated, transmembrane glycoprotein expressed in nucleated human cells. J Biol Chem 263: 4193–4201

Christiansen I, Gidlof C, Wallgren AC, Simonsson B, Totterman TH (1994) Serum levels of soluble intercellular adhesion molecule 1 are increased in chronic B-lymphocytic leukemia and correlate with clinical stage and prognostic markers. Blood 84: 3010–3016

D'Agostino L, Pignata S, Daniele B, Ventriglia R, Ferrari G, Ferraro C, Spaguolo S, Lucchelli P, Mazzacca G (1989) Release of diamine oxidase into plasma by glycosaminoglycans in rat. Biochim Biophys Acta 993: 228–234

Dalchau R, Kirkley J, Fabre JW (1980) Monoclonal antibody to a human brain-granulocyte-T lymphocyte antigen probably homologous to the W 3/13 antigen of the rat. Eur J Immunol 10: 745–749

Gearing AJH, Newman W (1993) Circulating adhesion molecules in disease. Immunol Today 14: 506–512

Goldstein LA, Zhou DF, Picker LJ, Minty CN, Bargatze RF, Ding JF, Butcher EC (1989) A human lymphocyte homing receptor, the Hermes antigen, is related to cartilage proteoglycan core and link proteins. Cell 56: 1063–1072

Guo Y-J, Liu G, Wang X, Jin D, Wu M, Jing M, Sy M-S (1994) Potential use of soluble CD44 in serum as indicator of tumor burden and metastasis in proteins with gastric or colon cancer. Cancer Res 54: 422–426

Günthert U, Hoffmann M, Ruby W, Reber S, Zöller M, Haussmann I, Matzku S, Wenzel A, Ponta H, Herrlich P (1991) A new variant of glycoprotein CD44 confers metastatic potential to rat carcinoma cells. Cell 65: 13–24

Harning R, Mainolfi E, Bystryn J-C, Henn M, Merluzzi VJ, Rothlein R (1991) Serum levels of circulating intercellular adhesion molecule 1 in human malignant melanoma. 51: 5003–5005

Haynes BF, Harden EA, Telen MJ, Hemler MR, Strominger JL, Palker TJ, Scearce RM, Eisenbarth GS (1983) Differentiation of human T lymphocytes. I. Acquisition of a novel human cell surface protein (p80) during normal intrathymic T cell maturation. J Immunol 131: 1195–1200

Haynes BF, Telen MJ, Hale LP, Denning SM (1989) CD44-A molecule involved in leukocyte adherence and T-cell activation. Immunol Today 10: 423–428

Heider KH, Hofmann M, Horst E, van den Berg F, Ponta H, Herrlich P, Pals S (1993) A human homologue of the rat metastasis-associated variant CD44 is expressed in colorectal carcinomas and adenomatous polyps. J Cell Biol 120: 227–233

Jackson DG, Schenker T, Waibel R, Bell JI, Stahel RA (1994) Expression of alternatively spliced forms of the CD44 extracellular-matrix receptor on human lung carcinomas. Int J Cancer 8: 110–115

Jackson DG, Bell JI, Dickinson R, Timans J, Shields J, Whittle N (1995) Proteoglycan forms of the lymphocyte homing receptor CD44 are alternatively spliced variants containing the v3 exon. J Cell Biol 128: 673–685

Jacobson K, O'Dell D, Holifield B, Murphy TM, August JT (1984) Redistribution of a major cell surface glycoprotein during cell movement. J Cell Biol 99: 1624–1633

Jalkanen S, Bargatze R, Herron LR, Butcher EC (1986) A lymphoid cell surface glycoprotein involved in endothelial cell recognition and lymphocyte homing in man. Eur J Immunol 16: 1195–1202

Jalkanen S, Jalkanen M, Bargatze R, Tammi M, Butcher EC (1988) Biochemical properties of glycoproteins involved in lymphocyte recognition of high endothelial venules in man. J Immunol 141: 1615–1623

Jalkanen S, Joensuu H, Söderström K-O, Klemi P (1991) Lymphocyte homing and clinical behavior of non-Hodgkin's lymphoma. J Clin Invest 87: 1835–1840

Katoh S, McCarthy JB, Kincade PW (1994) Characterization of soluble CD44 in the circulation of mice. Levels are affected by immune activity and tumor growth. J Immunol 153: 3440–3449

Kaufmann M, Heider K-H, Sinn H-P, von Minckwitz G, Ponta H, Herrlich P (1995) CD44 variant exon epitopes in primary breast cancer and length of survival. Lancet 345: 615–619

Koopman G, van Kooyk Y, de Graaff M, Meijer CJLM, Figdor CG, Pals ST (1990) Triggering of the CD44 antigen on T-lymphocytes promotes T-cell adhesion through the LFA-1 pathway. J Immunol 145: 3589–3593

Letarte M, Iturbe S, Quackenbush EJ (1985) A glycoprotein of molecular weight 85,000 on human cells of B-lineage: Detection with a family of monoclonal antibodies. Mol Immunol 22: 113–119

Li H, Hamou MF, de Tribolet R, Jaufeerally M, Diserens AC, Van Meir EG (1993) Variant CD44 adhesion molecules are expressed in human brain metastases but not in glioblastomas. Cancer Res 53: 5345–5349

Lucas MG, Green AM, Telen MJ (1989) Characterization of the serum In(Lu)-related antigen: identification of a serum protein related to erythrocyte p80. Blood 73: 596–600

Mali M, Andtfolk H, Miettinen HM, Jalkanen M (1994) Suppression of tumor cell growth by Syndecan-1 ectodomain. J Biol Chem 269: 27795–27798

Miyake K, Underhill CB, Lesley J, Kincade PW (1990) Hyaluronate can function as a cell adhesion molecule and CD44 participates in hyaluronate recognition. J Exp Med 172: 69–75

Omary MB, Trowbridge IS, Letarte M, Kagnoff MF, Isacke CM (1988) Structural heterogeneity of human Pgp-1 and its relationship with p85. Immunogenetics 27: 460–464

Pals ST, Hogervorst F, Keizer GD, Thepen T, Horst E, Figdor CG (1989a) Identification of a widely distributed 90-kDa glycoprotein that is homologous to the Hermes-1 human lymphocyte homing receptor. J Immunol 143: 851–857

Pals ST, Horst E, Ossekoppele G, Figdor CG, Scheper RJ, Meijer CJLM (1989b) Expression of lymphocyte homing receptor as a mechanism of dissemination in non-Hodgkin's lymphoma. Blood 73: 885–888

Picker LJ, de los Toyos J, Telen MJ, Haynes BF, Butcher EC (1989) Monoclonal antibodies against the CD44 (In(Lu)-related p80), and Pgp-1 antigens in man recognize the Hermes class of lymphocyte homing receptors. J Immunol 142: 2046–2051

Rapraeger A, Krufka A, Olwin B (1991) Requirement of heparan sulfate for bFGF-mediated fibroblast growth and myoblast differentiation. Science 252: 1705–1708

Ristamäki R, Joensuu H, Salmi M, Jalkanen S (1994) Serum in malignant lymphoma: an association with treatment response. Blood 84: 238–243

Salmi M, Grön-Virta K, Sointu P, Grenman R, Kalimo H, Jalkanen S (1993) Regulated expression of exon v6 containing isoforms of CD44 in man: downregulation during malignant transformation of tumors of squamocellular origin. J Cell Biol 122: 431–442

Screaton GR, Bell MV, Jackson DG, Cornelis FB, Gerthe U, Bell JI (1992) Genomic structure of DNA encoding the lymphocyte homing receptor CD44 reveals at least 12 alternatively spliced exons. PNAS 89: 12160–12164

Spertini O, Callegari P, Cordey AS, Hauert J, Joggi J, von Fiedner V, Schapira M (1994) High levels of the shed form of L-selectin are present in patients with leukemia and inhibit blast cell adhesion. Blood 84: 1249–1256

Stamenkovic I, Amiot M, Pesando JM, Seed B (1989) A lymphocyte molecule implicated in lymph node homing is a member of the cartilage link protein family. Cell 56: 1057–1062

Sy MS, Guo YJ, Stamenkovic (1991) Distinct effects of two CD44 isoforms on tumor growth in vivo. J Exp Med 174: 859–866

Tanaka Y, Adams DH, Hubscher S, Hirano H, Siebenlist U, Shaw S (1993) T-cell adhesion induced by proteoglycan-immobilized cytokine MIP-1β. Nature 361: 79–82

Telen MJ, Eisenbarth GS, Haynes BF (1983) Human erythrocyte antigens. Regulation of expression of a novel erythrocyte surface antigen by the inhibitor Lutheran IN(Lu) gene. J Clin Invest 71: 1878–1886

Tsujisaki M, Imai K, Hirata H, Hanzawa Y, Masuya J, Nakano T, Sugiyama T, Matsui M, Hinoda Y, Yachi A (1991) Detection of circulating intercellular adhesion molecule-1 antigen in malignant diseases. Clin Exp Immunol 85: 3–8

Webb DSA, Shimizu Y, van Seventer GA, Shaw S, Gerrard TL (1990) LFA-3, CD44, and CD45: Physiologic triggers of human monocyte TNF and IL-1 release. Science 249: 1295–1297

Wielenga VJ, Heider K-H, Offerhaus GJA, Adolf GR, van den Berg FM, Ponta H, Herrlich P, Pals ST (1993) Expression of CD44 variant proteins in human colorectal cancer is related to tumor progression. Cancer Res 53: 4754–4756

Signal Transduction As a Therapeutic Target

R. ALESSANDRO, J. SPOONSTER, R.P. WERSTO, and E.C. KOHN

1 Introduction

Cancer therapy is entering a critical period. Conventional approaches for cancer treatment, such as chemotherapy, radiotherapy, or surgery, have obtained results only in lymphoproliferative neoplasms and in a small percentage of solid tumors. The cause of this limited success can be ascribed to several factors, including the poor penetration of drug into tumor masses, biological heterogeneity of cancer cells in primary neoplasms and secondary foci, development of drug resistance, and the ability of malignant cells to invade and metastasize early in the tumorigenic process, often long before the primary tumor is detectable.

On the other hand, cancer research during the past 10 years has gained new insights into the genetic and biochemical mechanisms that are altered in malignant cells. To date, more than 100 proto-oncogenes have been identified and most encode components of the complex signal transduction pathways that in

Signal Transduction and Prevention Unit, Laboratory of Pathology, National Cancer Institute, 10 Center Drive, MSC1500, Bethesda, MD 20892-1500, USA

normal cells coordinate and regulate the effects of extracellular factors (e.g., hormones, growth factors, and adhesion molecules) on cell proliferation, differentiation, and cell death. In malignant cells containing the activated form of proto-oncogenes, some of these signaling pathways may become constitutively altered, causing the cells to acquire an invasive phenotype.

The components of the signaling cascade from the cell membrane to the nucleus represent new targets for anticancer drug design. Some of these drugs have already entered clinical trials, while others have shown promising preclinical results. This review introduces some of the key elements involved in signal transduction, focusing on transmembrane signal transduction that initiates the cascade of second messengers leading to modulation of gene expression. A relationship between oncogenes and signal transduction elements will be demonstrated, and in the last section novel antisignaling agents that are under development for cancer treatment will be discussed.

2 Transmembrane Signal Transduction

The cellular response to growth factor stimulation is dependent on a bidirectional communication between cell external environment and internal cytoplasmic compartments. Most of the molecules that initiate intracellular signaling bind specific cell surface receptors. This ligand–receptor interaction causes the production of a cascade of signal transduction pathways that ultimately elicit the modulation of gene expression and cellular activity. There are three general classes of transmembrane signal pathways through cell surface receptors: ion channel activity, guanine nucleotide-binding (G) proteins, and protein phosphorylation/dephosphorylation events.

2.1 Ion Channels

Membrane channels and transporters play a pivotal role in cellular development and homeostasis by altering intracellular ion concentrations that act as downstream effectors in signal transduction cascade. The importance of Ca^{2+} in the signal transduction cascade cannot be understated. Activation of ion channels, phosphorylation and/or dephosphorylation of second messengers, and activation of G proteins as well as biological properties such as cell adhesion and cytoskeletal organization are commonly regulated by Ca^{2+} homeostasis (Clapham 1995).

Four major types of mechanisms for mobilization of cytosolic Ca^{2+} have been identified: release from intracellular stores, receptor-operated calcium channels (ROCC), Ca^{2+} reuptake via depletion-operated (refilling) or capacitance channels (DOCC), and second messenger-operated calcium channels (SMOCC), as shown in Fig. 1 (Felder et al. 1994). ROCC constitute a diverse family of voltage-

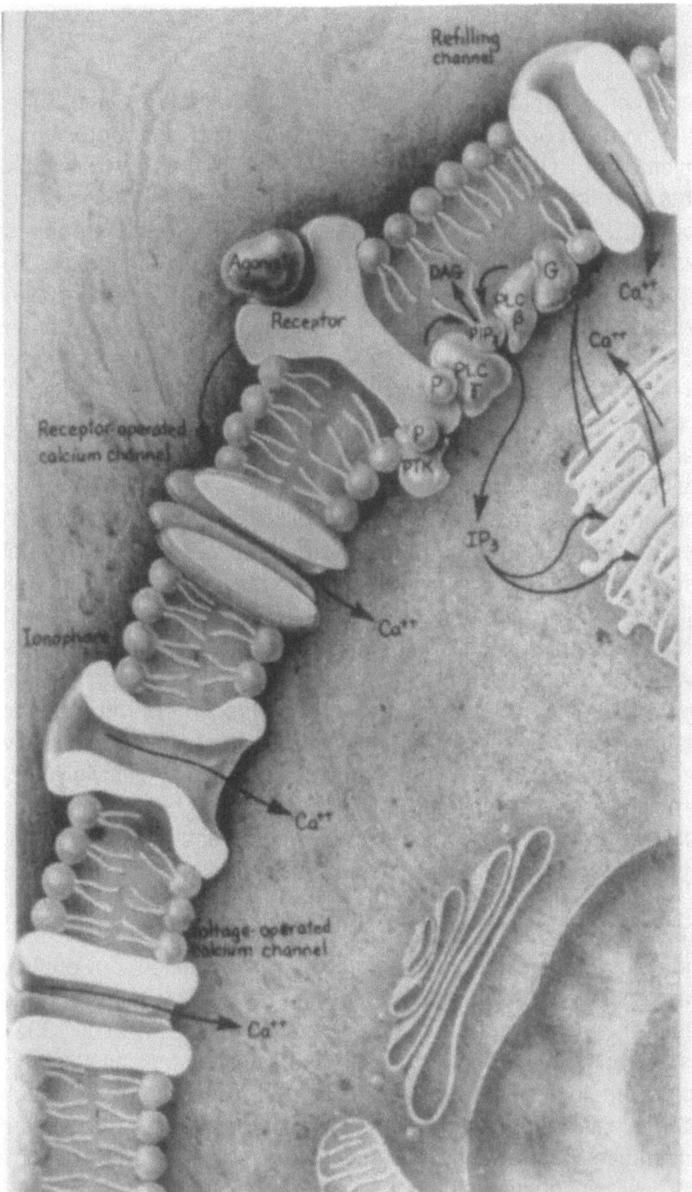

Fig. 1. Mechanism of transmembrane Ca^{2+} signal transduction. When a ligand binds to its receptor, it may activate phospholipase C (*PLC*) through either a G protein-dependent (PLC-β) or independent, tyrosine kinase-activated pathway (PLC-γ). This leads to the hydrolysis of phosphatidylinositol 4,5-bisphosphate (*PIP_2*) to generate inositol-1,4,5,-trisphosphate (*IP_3*) and 1,2-diacylglycerol (*DAG*). IP_3 can cause the mobilization of intracellular Ca^{2+} from internal stores, which secondarily stimulates release of Ca^{2+} influx factor (CIF). CIF, stimulates opening of the depletion-operated or refilling Ca^{2+} channels. Alternatively, increased Ca^{2+} influx may result from calcium influx through receptor-operated channels opened after ligand-receptor binding and independent of regulatory proteins. Ionophores are a laboratory method of opening Ca^{2+} channels which have counterparts in nature in the form of organic toxins. Lastly, in electrically active cells, voltage-operated Ca^{2+} channels (VOCC) may be opened and facilitate Ca^{2+} influx. These channels may be inhibited by the "classical" Ca^{2+} channel inhibitors diltiazem, verapamil, and nifedipine. (Original illustration by E.C. KOHN and J. SPOONSTER, artistic enhancement by K. SOMMERVILLE)

insensitive channels which directly induce an increase in Ca^{2+} influx without the requirement for involvement of second messengers. Inhibition of receptor function results in inhibition of ROCC function (FELDER et al. 1992). By contrast, the sequence of events by which intracellular Ca^{2+} pools are depleted and then refilled through activation of DOCC is more complicated. DOCC activation requires intracellular release of stored Ca^{2+}, in response to the generation of inositol-1,4, 5-trisphosphate (IP_3) from hydrolysis of phosphatidylinositol 4,5-bisphosphate (PIP_2) by phospholipase C (PLC) isozymes β or γ, and its binding to its receptor on the endoplasmic reticulum (FELDER et al. 1994). A putative soluble cytoplasmic factor, Ca^{2+} influx factor, has been suggested as the internal messenger that goes to the plasma membrane after secretion to stimulate opening of DOCC (RANDRIAMAMPITA and TSIEN 1993). The ability of this factor to open DOCC has been shown to be sensitive to phosphatase inhibition.

Another important ion channel that regulates cellular function is the Na^+/H^+ antiporter, an electroneutral plasma membrane-bound transporter. This channel not only regulates intracellular pH (pH_i) homeostasis, but also Na^+ transport, cell volume, and cell division (LAGARDE and POUYSSEGUR 1986). Polypeptide growth factors, lectins, hormones, and tumor-promoting agents such as phorbol esters can directly activate this antiporter through a number of divergent signal transduction pathways (SARDET et al. 1990). Some growth factor receptors with intrinsic tyrosine kinase activity have been shown to activate the Na^+/H^+ antiporter directly or via their downstream substrates. The PLC pathway products 1,2-diacylglycerol (DAG), which stimulates protein kinase C (PKC) activity, and IP_3, which mediates increase in intracellular Ca^{2+}, both also can control antiporter activity. By contrast, antiporter activation appears to be independent of intracellular cyclic adenosine monophosphate (cAMP) and not mediated by G proteins (GANZ et al. 1990). Importantly, the antiporter signal transduction pathway is receptor selective, cell type specific, and cell cycle specific. For example, pH_i is increased, not only by receptor-mediated regulation of the Na^+/H^+ antiporter, but also by increasing the number of cell–cell contacts and cell– substrate interactions (GALKINA et al. 1992).

Besides plasma membrane-bound channels and transporters, ion-selective channels are also present in nuclear membrances (MAZZANTI et al. 1990). Although they are speculated to balance the charge carried by macromolecules through the nuclear membrane, these channels may regulate cell proliferation, thereby becoming an important signal transduction step in the regulation of gene expression.

2.2 Guanine Nucleotide-Binding Proteins

Like Ca^{2+}, heterotrimeric G proteins are ubiquitous and have broad diversity in function (NEER 1995; SIMON et al. 1991). Localized to the cytoplasmic surface of the plasma membrane, G proteins share a common structural polypeptide motif and function in the signal transduction cascade as "on–off" switches coupling signals from cell surface receptors to membrane-bound intracellular effectors such as adenylyl cyclase, phospholipase A_2 (PLA_2), PLC-β, phosphodiesterases, and selected ion channels. Receptor binding initiates a complex series of events beginning with the release of guanosine diphosphate (GDP), which is tightly bound to the G protein α-subunit in its inactive ("off") state. GDP is then replaced by guanosine triphosphate (GTP). The GTP-bound α-subunit, now activated, dissociates from its associated $\beta\gamma$-protein subunits. Either or both of these components, α and $\beta\gamma$, can interact with coupled effector functions, activating production of second messengers. The protein is recycled to the "off" state by the GTP hydrolysis activity of the α-subunit, inactivating the α-subunit and permitting reassembly with the $\beta\gamma$-subunit. It has been shown that the $\beta\gamma$-subunit can independently regulate PLC-β, PLA_2, phosphoinositol 3'-kinase (PI 3'-kinase) and interact with *ras* to activate mitogen-activated protein (MAP) kinase (MAPK). Thus, while the G protein-mediated signal transduction pathway is complex, it allows multiple signals to converge on a single effector or multiple effectors to be activated by one transmembrane signal, adding redundancy, crosstalk, and amplification between signaling cascades.

2.3 Tyrosine Kinases

Tyrosine kinases, which phosphorylate specific tyrosine residues, are subdivided into two categories based upon the presence or absence of transmembrane regions which confer intracellular kinase activity. Receptor protein tyrosine kinases (RTK), including the receptors for colony-stimulating factor (CSF-1), fibroblast growth factors (FGF), epidermal growth factor (EGF), hepatocyte growth factor (HGF), platelet-derived growth factor (PDGF), and insulin-like growth factors (IGF), contain this domain and possess intrinsic tyrosine kinase activity activated by ligand binding. Protein tyrosine kinases (PTK) are cellular proteins with tyrosine kinase activity that do not have transmembrane domains. Many PTK are signaling molecules themselves and many, such as *src*, are proto-oncogene products. Tyrosine phosphorylation of target proteins at the cell membrane or in the cytoplasm can initiate several signaling pathways. Phosphorylation substrates include the signaling molecules PI3'-kinase, PLC-γ, GTPase-activating protein, and others (SCHLESSINGER and ULLRICH 1992). After activation by phosphorylation, these proteins may in turn activate downstream pathways, so that an initial stimulus may trigger a variety of responses and be propagated. For example, EGF binding to EGF receptor (EGFR) stimulates tyrosine phosphorylation of PLC-γ. This causes augmentation of the PLC-γ hydrolytic activity, producing IP_3 and DAG from PIP_2. Both compounds produced are second messengers that activate two subsequent signaling pathways. IP_3 stimulates the release of Ca^{2+} from intracellular stores and DAG binds to and activates PKC, a serine threonine kinase. The substrate, PIP_2, can also be a substrate for PI 3'-kinase, which is also phosphorylated in response to activation of the EGFR RTK.

RTK are coupled in different ways to the second category of kinases, the PTK. The PDGF receptor (PDGFR) as well as other RTK can directly phosphorylate PTK, such as members of the *src* family including *src*, *fyn*, and *yes*. Alternatively, autophosphorylation of RTK such as the EGFR can provide a phosphotyrosine for linkage to downstream pathways through binding of their phosphotyrosine to adaptor proteins such as Grb2. Grb2 is the doorway to the *ras/raf*/MAPK cascade of PTK and serine/threonine kinases (WILLIAMS and ROBERTS 1994). This cascade of kinases may branch at several points, resulting in the amplification and diversification of signal transmission. Several of the components of the kinase pathways are proto-oncogenes products, demonstrating how the aberrant activation of these signaling pathways may have the potential to cause malignant transformation and metastatic dissemination.

3 Oncogenes and Signal Transduction

The development and progression of a tumor is a complex process that requires multiple and sequential genetic changes (FIDLER and HART 1992; NOWELL 1976). The activation of several oncogenes, as well as the loss of tumor suppressor genes, is necessary to endow cells with the properties of transformation, invasiveness, and metastatic ability (BISHOP 1991; FEARON and VOGELSTEIN 1990). The identification and characterization of those genes involved in the malignant phenotype has led cancer biologists to consider cancer not simply as a consequence of unbridled cell proliferation, but as a disease of deranged signal transduction (COLE and KOHN 1994; GIBBS et al. 1994; KOHN and LIOTTA 1995). There are now a large number of well-defined oncogenes. Through the diversity of their actions, oncoproteins produce effects by activating different portions of the signaling cascade from the membrane to the nucleus. Their sites of action fall in four possible categories.

The first is where the oncogene products themselves act as growth or invasion-inducing factors. Examples of this category are c-*sis*, which encodes the B chain of PDGF (WATERFIELD et al. 1983) and the *int2/hst* gene product, which is a member of the FGF family (DELLI-BOVI et al. 1987). PDGF may also stimulate migration of a large number of cell types, including endothelial cells. Activation of these oncogenes can result in the inappropriate production of the encoded growth factor or a mutated growth factor. Activation may in turn cause the establishment of an autocrine loop function, leading to uncontrolled cell growth or motile function.

The second site of oncogene action in the signaling cascade is represented by oncoproteins which encode for the growth factor transmembrane receptors for growth factors. The oncogene *erbB1* was the first oncogene belonging to this class to be identified (DOWNWARD et al. 1984). It is highly homologous to the EGF receptor tyrosine kinase. Another oncogene, *neu*, also known as *erbB2*, has a close similarity to the EGF receptor and encodes for a 185-kDa protein with tyrosine kinase activity (BARGMANN et al. 1986). High levels of expression of both c-*erbB1* or c-*erbB2* can lead to transformation of NIH 3T3 cells (VELU et al. 1987). The overexpression of c-*erbB2* has been shown to represent a putative prognostic factor for breast and ovarian carcinomas (Dougall et al. 1994; Paterson et al. 1991). The oncogene *met* is another member of the RTK growth factor receptor family. The ligand for *met* is HGF, also known as scatter factor (SF) (Bottaro et al. 1991; Naldini et al. 1991). It has been shown that overexpression of *met* in NIH-3T3 cells induces a tumorigenic and metastatic phenotype after inoculation into nude mice (Iyer et al. 1990). Recent data have demonstrated that autocrine activation of *met* in NIH-3T3 cells enhances cellular motility, collagenase production, invasiveness in vitro, and metastasis formation in vivo (Rong et al. 1994). Other oncoproteins which resemble growth factor receptor-signaling molecules include *fms*, which encodes monocyte colony-stimulating factor (M-CSF/CSF-1) receptor (Sherr et al. 1985), *kit*, which is related to the PDGFR (Lev et al. 1994), and *ros*, which is related to the insulin receptor (Zong and Wang 1994). The proto-oncogene RTK form an important target for therapeutic development. Several new compounds have been designed to target the tyrosine kinase activity of selected growth factor receptors.

The third class of oncogenes encode cytoplasmic proteins that act as transducers of the signaling message from the membrane to the nucleus. These include *ras*, a family of small molecular weight G proteins found in colorectal adenomas and carcinomas, lung adenocarcinomas, pancreatic cancer, and other cancers (Bos 1990). Cytoplasmic tyrosine kinases such as *src*, *fyn*, *lck*, and *yes* can directly interact with the RTK through a specialized region called *src* homology-2 domains (SH2), which recognize phosphotyrosine residues in the context of a short sequence of amino acids (Koch et al. 1991). Binding through SH2 domains allows for one type of signal propogation. These PTK have a variety of downstream targets, thus providing a biochemical basis for the pleiotropic nature of signal transduction. It is also clear that variations in the second messenger response occur depending on the particular ligand–receptor complex and the specific cell type. Other cytoplasmic proto-oncogenes such as *mos*, *raf*, and *rel* encode kinases which phosphorylate proteins on serine and threonine residues (Macara 1989). *raf* is a kinase that can integrate signals received from the RTK or *ras* pathways. It can be activated by PDGF, EGF, or FGF receptors and therefore converts a signal from the RTK to a cytoplasmic serine/threonine kinase activity through subsequent activation of the MAPK cascade (Leevers and Marshall 1992). MAPK has been shown to take the message into the nucleus by phosphorylating transcription factors encoded by other proto-oncogenes, the fourth category (Chen et al. 1992; Davis 1993). The phosphorylation of nuclear transcription factors such as c-*jun* and c-*fos* regulates their function and subsequent expression. Activation of c-*jun* and c-*myc*-regulated transcription events can result in upregulation of the metastatic pathways, for example by inducing transcription of matrix metalloproteinses (Hennigan et al. 1994; Matrisian 1992; Sato et al. 1994).

The identification and study of the function of proto-oncogenes has shed light on the signaling scheme presenting messages from the cell surface to the nucleus. It is now clear that proto-oncogene products represent some of the critical components involved in the conversion of extracellular signals in intracellular events leading to cell division or invasion. Perturbation of signal transduction pathways by alteration of the activity or amount of the oncogenic proteins can result in cellular transformation and development of invasive behavior. Therapeutic modulation of the signaling pathways affected by oncogene expression may represent a new and alternative approach for cancer treatment.

4 Signal Transduction Therapy

The view of cancer as a disease of cell proliferation has been the ideological support for current antineoplastic therapies. Most of the present agents that are used in chemotherapy target DNA synthesis, DNA repair, and transcriptional events. Although these compounds have been successful in the cure of many hematopoietic malignancies, limitations have been observed in the treatment of solid tumors. As a consequence of their action at the DNA level, classical chemotherapeutic drugs may be extremely cytotoxic, and their continuous use during chemotherapy can result in unpleasant or toxic side effects (Bailar and Smith 1986). This demonstrates the need for a new generation of compounds that may be more specifically targeted against the malignant cell, especially for solid tumors.

With the advent of molecular biology and the identification and characterization of oncogene products, it has become clear that cancer is a disease of aberrant signal transduction (Cantley et al. 1991; Gibbs 1991; Kohn and Liotta 1995). The experimental evidence that many of the oncogenes encode proteins involved in transducing signals from the cell surface to the nucleus has lead researchers to focus on the inhibition of transmembrane and intracellular signaling pathways as a novel approach to cancer treatment (Cole and Kohn 1994; Kohn and Liotta 1995; Powis 1994). The consideration of targeting this ubiquitous signaling has generated numerous questions:

- What are the most opportune signaling pathways to be considered for intervention?
- Can selectivity for specific signal pathways be attained in tumor cells and not in normal cells?
- Will this therapeutic approach be less toxic than conventional chemotherapy?
- Will signal transduction therapy be cytotoxic or cytostatic in patients?
- What benefits may be obtained by combining signal transduction therapy with conventional therapy?

The proliferation of normal cells is regulated by the cooperation of different growth factors and cytokines that bind specific receptors on the cell surface (Deuel 1987). The interaction between growth factors and their receptors may activate multiple subsequent signaling pathways, as described previously, leading to the progression of cells through the cell cycle or into cell cycle arrest.

Signaling events in the cytoplasm or at the nucleus such as post-translational modification, altered gene expression, and alternative splicing increase the interconnections between the signaling pathways and provide cells with many regulatory checkpoints.

This degeneracy of signal transduction mechanisms has probably developed during evolution to give normal cells the ability to maintain homeostatic balance in response to chemical or genetic challenges (Fig. 2). The cell can compensate for challenges of lost or increased signaling events by modulating parallel pathways. In contrast, cancer cells accumulate molecular events such as deletion, mutation, and translocations during tumor progression and develop rigidity in their signaling

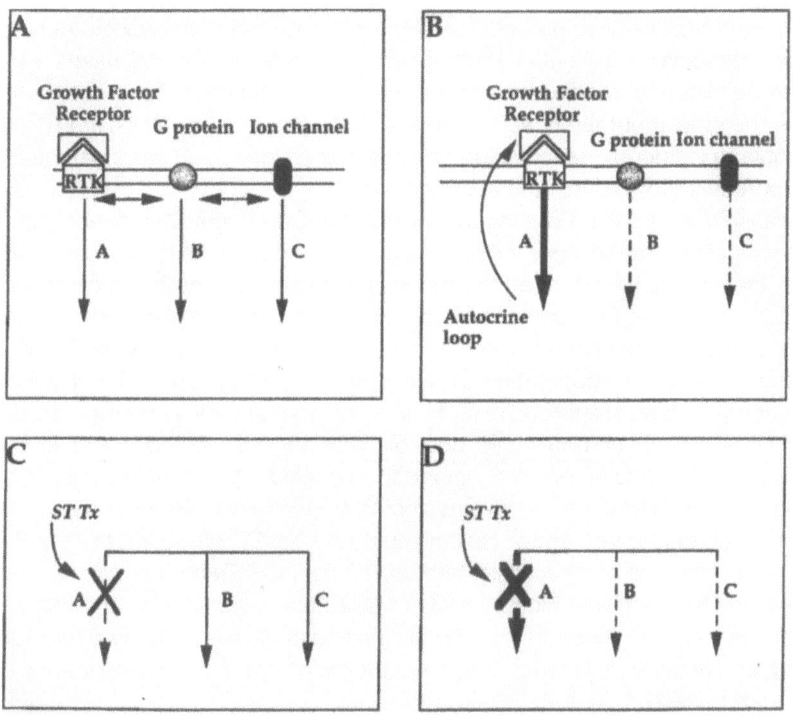

Fig. 2A–D. Altered signal transduction pathways in cancer cells represent new targets for therapeutic intervention. Interruption of the oncogene-activated pathway with signal transduction therapy should selectively block the invasive phenotype of malignant cells. In contrast, due to the redundancy of their signaling mechanisms, normal cells would not be effected by drug treatment and should therefore show a normal phenotype and gene expression. **A** Degenerate signaling pathways allow normal cells to maintain checks and balances in their signaling, resulting in a normal phenotype. *RTK*, receptor tyrosine kinase. **B** Accumulated molecular events such as oncogene overexpression, mutation, and deletion cause a rigidity in the signaling of the malignant cell. Autocrine loops through tyrosine kinase growth factor receptors represent one manifestation of this occurrence. **C** Intervention with signal transduction inhibition therapy (*ST Tx*) should not cripple normal cells due to the redundancy in signaling opportunities and should result in a normal molecular and cellular phenotype. **D** Signal transduction therapy (ST Tx) of rigidly signaling cells should reduce the activity of the predominant pathway, thereby modulating the aberrantly expressed signal and resulting in cytostasis and inhibition of the invasive phenotype

mechanisms. This can be manifested as overexpression of a specific pathway with the development of an autocrine loop or as development of constitutive activation of a signaling molecule that maintains an "on" message as is seen with mutated *ras*. This rigidity may then prevent a normal signaling balance from occurring. Exposure to an agent that targets the overactive signal pathway may allow the cell to reinitiate checks and tend towards a balanced state.

The interruption of a signal transduction pathway in nonmalignant cells in response to the exposure of a signal drug can be absorbed and overcome by the existence and action of redundant cellular signaling pathways; c-*src* knockout mice provide an example of this. Although c-*src* plays an important role in the propagation of growth factor-initiated signals (Kypta et al. 1990), c-*src* knockout mice showed normal development, with the only observed physiologic defect being osteoporosis (Soriano et al. 1991). In this case, other *src* family kinases may have compensated for the absence of the specific c-*src* function. Tumor cells, on the other hand, without the checks and balance found in normal cells, are more susceptible to loss or overexpression of signal transduction pathways, suggesting the potential utility of this therapeutic concept.

Because this new class of antitumor drugs does not target DNA directly, but selectively inhibits altered signaling pathways, thus restoring the previous cellular balance, they would be expected to produce a cytostatic rather than a cytotoxic action. Consequently, the schedule of administration of these agents should be adjusted to maintain continuous therapeutic pressure. A clinical end point for this cytostatic therapeutic intervention would then be anticipated to be disease stabilization. This type of approach may be used optimally in adjuvant treatment or chemoprevention. Over time, it is possible that the loss of cancer cells by apoptosis, chemotherapy, or immunological mechanisms may overbalance the arrested growth, causing a net regression of tumor mass. An ideal signaling inhibitor for chronic treatment of cancer must be a small compound that easily reaches its membrane or intracellular targets, is selective in action and site, is well tolerated, and has sufficient bioavailability to be administered readily. The following sections describe examples of signal transduction pathways available for targeting and antisignaling drugs that have preclinical promise or are already in clinical trial (Table 1).

4.1 Receptor Tyrosine Kinases

RTK play a regulatory role in signal transduction pathways initiated by the binding of a growth factor to its cell surface receptor (Schlessinger and Ullrich 1992). The demonstration that several oncogene products were growth factor receptors with tyrosine kinase activity (Bishop 1991) led to the search for RTK inhibitors as therapeutic agents. The most difficult aspect in the identification of useful inhibitors of tyrosine kinases is the selectivity required to block a specific RTK overexpressed or overactive in a particular tumor. While the kinase domains may be highly homologous, they have been demonstrated to be also highly selective.

Table 1. Drugs under development that modulate intracellular signaling pathways

Drug	Mechanism of action	Clinical status
Tyrphostins	Inhibit growth factor-stimulated tyrosine phosphorylation	Preclinical development
Staurosporine, sphingosine analogues	Inhibits PKC activity	In clinical trial
L-739 749	Inhibits farnesylation of *ras*	Preclinical development
Limonene	Inhibits isoprenylation of *ras*	In clinical trial
Edelfosine	Inhibits PLC, PI 3'-kinase activity	In clinical trial
Wortmannin	Inhibits PI 3'-kinase	Preclinical development
CAI	Inhibits select non-voltage-gated calcium influx	In clinical trial

PKC, protein kinase C; PLC, phospholipase C; CAI, carboxyamido-triazole; PI 3'-kinase, phosphoinositol 3'-kinase.

Most of the early RTK inhibitors were developed to compete for the receptor adenosine triphosphate (ATP)-binding site, a region that is conserved between tyrosine kinases. For example, the isoflavonoids that act at this site have a broad and nonselective inhibitory activity. Genistein, an isoflavonoid used as an inhibitor of tyrosine kinases, also has been shown to block the activity of DNA topoisomerase II (Akiyama et al. 1987; Markovits et al. 1989). Erbstatin, with some specificity for the EGFR tyrosine kinase, has been shown also to inhibit serine/threonine kinases (Umezawa et al. 1986).

An interesting series of synthetic compounds, the tyrphostins, has been derived from erbstatin. Some of them have demonstrated selectivity and are entering clinical trial (Levitzki 1994). These synthetic compounds were designed to compete with the RTK substrate and not with the ATP-binding site, yielding a high selectivity in their RTK-inhibitory activity. Different tyrphostins, synthesized with substitutions in their primary structure, can discriminate between the EGF and insulin receptors or between PDGFR and EGFR (Osherow and Levitzki 1994; Yaish et al. 1988). Some tyrphostins can distinguish between EGFR and the closely related product of the *neu* oncogene; this is extremely important, considering that a high percentage of tumors overexpress *neu* oncogene and a large number of normal cells have EGFR activity (Gazit et al. 1991; Levitzki 1994). Other tyrphostins have been directed to the tyrosine kinase activity of the PDGFR. They are effective in vitro against human glioblastoma cell lines overexpressing PDGFR and have shown efficacy in the inhibition of growth of human glioma xenografts in nude mice (Kovalenko et al. 1994).

A new inhibitor of EGFR tyrosine kinase which has been recently synthesized shows further potential for RTK inhibition. This drug, 4,5-dianilinophthalimide (DHAP-1), differs slightly from staurosporine, a nonselective inhibitor of (PKC) (Buchdunger et al. 1994). Selective activity has been shown against EGFR compared to PKC, *src, abl*, and protein kinase A (PKA), with a ten- to 1000-fold difference in potency (Buchdunger et al. 1994). Daily oral administration of DHAP-1 has been shown to inhibit EGFR-driven tumor growth in xenograft mouse models of A431 epidermoid and SK-OV tumors human ovarian carcinoma cell lines, respectively. No effects on proliferation of PDGF-driven tumors was observed (Buchdunger et al. 1994). This compound did not cause major toxicity in the treated animals.

4.2 Protein Kinase C

PKC consists of a multigene family of serine/threonine kinases that play a central role in many signal transduction pathways, leading to proliferation and invasion (Nishizuka 1992). PKC is activated in response to the binding of many growth factors to their receptors, with resultant production of the second messenger DAG. The activation of most PKC results in the translocation of the enzyme from the cytoplasm to the cell membrane. Other PKC isoforms are localized to the nucleus, where they may be involved in the phosphorylation of transcription factors leading to mitogenesis (Boyle et al. 1991). Several lines of evidence have linked PKC activity to the invasive phenotype of cancer cells. Tumor cells with increased PKC activity have an augmented ability to invade and metastasize (Isakov et al. 1991; Schwartz et al. 1990), and inhibition of PKC has resulted in decreased metastatic potential. Alterations in the expression of specific PKC isoforms has been found in several transformed cell lines (Goodnight et al. 1992; Mischak et al. 1991). Increased RNA expression of PKC-β1 has been demonstrated in gastric carcinoma cell lines (Schwartz et al. 1993). At present, there is indication that PKC-β1 may be an important factor in early stages of tumor development, whereas PKC-ζ and -θ are involved at later stages when the metastatic colonies are already formed (Albino et al. 1995). These findings suggest that PKC may be a potential target for therapeutic intervention.

Several inhibitors directed against the catalytic or regulatory domains of PKC have been developed and some are now in clinical trial (Nixon et al. 1992). Staurosporine, an antibiotic isolated from *Streptomyces* species, has strong and general protein kinase inhibitory activity, but has been studied primarily for its effects against PKC. Acting at the catalytic site of PKC, staurosporine inhibited invasion and metastases of human bladder carcinoma cells in in vitro models (Schwartz et al. 1993). UCN-01, a derivative of staurosporine with more selectivity for PKC than other kinases, has shown antitumor effect in vivo against three human xenografts models of epidermoid carcinoma (A431), fibrosarcoma (HT1080), and acute myeloid leukemia (Akinaga et al. 1991). Cationic lipids such as sphingosine and its derivatives are directed against the regulatory domain of PKC and can inhibit PKC activity. SPC100221, a sphingosine analogue, is highly specific for PKC and inhibits the invasion of human gastric cells without being cytotoxic to the cells (Albino et al. 1995).

Ether lipid analogues such as edelfosine, although developed as inhibitors of PLC and PI 3'-kinase (see below), also inhibit PKC (Powis 1994), suggesting that their activity as anticancer drugs may be due to inhibition of multiple signaling pathways. Tamoxifen a nonsteroidal antiestrogen compound used for the treatment and chemoprevention of breast cancer, has been shown to modulate multiple signaling pathways. Used at conventional doses, tamoxifen increases the production of transforming growth factor (TGF)-β, decreases the production of TGF-α, and reduces cell–matrix adhesion (MILLON et al. 1989), whereas high doses of tamoxifen inhibit PKC activity (O'BRIAN et al. 1988). The identification of compounds with high selectivity towards the different isoforms of PKC will help to dissect the signal transduction pathways in which this enzyme is involved in normal and tumor cell biology.

4.3 Ras

The *ras* proteins belong to the small molecular weight G protein family. They have been shown to play a key role in the regulation of cell proliferation and metastasis (BARBACID 1987). Oncogenic *ras* mutations are found in more than 30% of human tumors (BOS 1990). The predominant *ras* mutations detected in tumors yield protein that is unable to hydrolyze bound GTP to GDP. As a consequence, these mutant *ras* forms are locked into an active GTP-bound configuration, resulting in a constitutive stimulus to the downstream effector pathways. The *ras*-signaling pathway can be activated at several levels, including incoming signals from RTK and seven transmembrane heterotrimeric G protein-linked receptors (MCCORMICK 1993). Blockade of the signaling pathways which drive *ras* may therefore represent an opportune intervention point to target a wide variety of tumors.

After being synthesized in the cytoplasmic compartment, *ras* requires extensive post-translational modification for its localization to the inner face of the plasma membrane (GIBBS 1991). These modifications, such as farnesylation of the carboxy-terminal end of the protein, have been shown to be necessary for the transforming activity of the mutated *ras* (HANCOCK et al. 1989; SCHAFER and RINE 1992). Farnesyltransferase inhibitors have been developed which suppress the anchorage-independent growth of *ras*-transformed cells (GIBBS et al. 1994). Recently, KOHL et al. have demonstrated that a novel inhibitor of farnesyltransferase, L-739 749, inhibited the growth of *ras*-dependent tumors in nude mice (KOHL et al. 1994). The animals treated with L-739 749 did not show evidence of gross toxicity. Lovastatin is a competitive inhibitor of hydroxymethylglutaryl-coenzyme A (CoA) reductase, an enzyme involved in isoprenoid metabolism (ALBERTS et al. 1980). Administration of lovastatin has been shown to inhibit tumor formation by Ha-*ras*-transformed cells in nude mice (SEBTI et al. 1991). High doses of lovastatin have been tested in phase I and II clinical trials (A. THIBAULT et al., 1996 in press). The monoterpene limonene, which inhibits isoprenylation of p21 *ras*, was found to cause complete regression of chemically induced rat mammary carcinomas (HAAG et al. 1992). Limonene is in phase I clinical trials in the United Kingdom in patients with advanced breast and gastrointestinal cancer.

4.4 Phosphatidylinositol 3'-Kinase

PI3'-Kinase is an enzyme that catalyzes the phosphorylation of PI at the D-3 position of the myoinositol ring (FRY 1994). PI 3'-kinase consists of an heterodimer with 110-kDa and 85-kDa subunits. The p85 subunit has regulatory activity and links PI 3'-kinase to specific cellular targets through its SH2 and SH3 binding domains (MARGOLIS 1992). The 110-kDa subunit has catalytic activity and has been found to be homologous to a yeast protein involved in vacuolar protein sorting (SCHU et al. 1993). PI 3'-kinase has been found to be downstream in the signal transduction pathways of a variety of growth factors, including PDGF and EGF (FRY 1994).

Many studies have linked PI 3'-kinase to cell proliferation and tumorigenesis. Mutated tyrosine kinases that lack mitogenic or cell-transforming ability fail to associate with PI 3'-kinase (COUGHLIN et al. 1989; KAPLAN et al. 1987). Cellular levels of polyphosphoinositides are higher in transformed cells and in cells stimulated by growth factors than in their nonmal counterparts or in resting cells, respectively. In one study, the oncogenic form of *neu* was found permanently coupled to PI 3'-kinase, while a kinase-defective *neu* or a form with a carboxyl-terminal truncation did not associate constitutively with PI 3'-kinase (PELES et al. 1992).

The link between PI 3'-kinase and cellular transformation has led investigators to develop specific inhibitors of PI 3'-kinase activity. The tyrosine kinase inhibitors quercetin and genistein have been found to inhibit PI 3'-kinase weakly, probably due to their nonspecific inhibition of the ATP-binding site of the enzyme (MATTER et al. 1992). Other antisignaling compounds that have been tested, such as lovastatin, benzoylstaurosporine, or suramin, also have shown a nonspecific inhibitory activity against PI 3'-kinase. Recently, a class of ether lipid analogues has been found selectively to inhibit growth factor-stimulated polyphosphoinositide formation in normal cells and to result in inhibition of proliferation of cancer cells in culture and tumor formation in animals (BERGGREN et al. 1993; SEEWALD et al. 1990). Edelfosine is the prototype agent which is being proposed for preclinical development. Other specific and potent inhibitors of PI 3'-kinase activity include the D3-deoxy-substituted myoinositol analogues. These compounds inhibit the proliferation of v-*sis* transformed NIH-3T3 cells by competing with normally occurring myoinositol for incorporation in the phosphoinositides of the cellular membrane. Wortmanin is a secondary metabolite found in a variety of fungal species that has been shown to selectively and potently inhibit PI 3'-kinase. This compound was shown to specifically inhibit the PI 3'-kinase activity of Swiss 3T3 cells with an IC_{50} of 1.9 n*M*, while leaving other signaling enzymes unaffected (POWIS et al. 1994).

4.5 Phospholipase C

PLC isozymes play an important role in the mitogenic responses of cells to different growth factors, including PDGF, bombesin, and EGF. PLC catalyzes the hydrolysis of polyphosphatidylinositols, resulting in the production of two second messenger molecules, DAG and inositol polyphosphates. The PLC isoenzymes β and γ appear to be activated by different receptors through distinct mechanisms, as described above. The utility of PLC as a target has been suggested by several observations. Increased levels of inositol polyphosphates and DAG have been found in transformed cell lines (RILEMMA 1989; SMITH et al. 1990), NIH-3T3 cells are transformed by the addition of PLC γ- and β-activating receptors such as EGFR and neurotransmitter receptor subtypes, respectively. Microinjection of antibodies against PLC-γ into NIH-3T3 cells blocked *ras*-stimulated and serum-stimulated proliferation (SMITH et al. 1990). Interestingly, antibodies to PLC-β did not have the same effect, demonstrating selectivity between PLC isoenzymes. Higher immunoreactivity for PLC-γ has been detected in colorectal carcinomas or breast carcinomas when compared to normal tissues.

Several compounds have been described that inhibit PLC activity, and some have entered clinical trial. Alkyl-lysophospholipids, represented by Et-18-OCH3, have been shown to specifically inhibit PLC isozymes with an IC_{50} of 0.1–0.4 μ*M* (ARTEAGA et al. 1991; NOH et al. 1993; POWIS et al. 1992). This class of compounds was antimetastatic in numerous murine and rat tumors and models (BERDEL 1991). Et-18-OCH3 has been used as an antineoplastic agent in phase I and II clinical trials in Europe (ANDREESON 1988; KHANAVKAR et al., 1989). Derivatives of alkyl-lysophospholipids have been synthesized. Hexadecylphosphocholine, also known as miltefosine, is a competitive inhibitor of PLC and PKC (UBERALL et al. 1991; HILGARD et al. 1993) that has been shown to inhibit the motility and invasiveness of MO4 cells in vitro (SCHALLIER et al. 1991). Clinical trials with miltefosine have started, but gastrointestinal toxicity has been limiting (HILGARD et al. 1993). Although specificity of the various inhibitors on the different PLC isoenzymes has not been studied in detail, they have provided useful indications into the role of PLC in the malignant phenotype.

4.6 Ca^{2+} Signal Transduction and Carboxyamido-triazole

Ionized Ca^{2+} is the most common cellular signal transduction element. Originally, cells probably developed Ca^{2+} binding strategies to regulate and maintain cytosolic levels and later for more complete signal transduction roles (CLAPHAM 1995). A multitude of Ca^{2+} binding proteins, channels, and pumps have evolved to buffer and transport intracellular Ca^{2+} and to trigger signal transduction cascades. Intracellular Ca^{2+} may be mobilized from within the cell from endoplasmatic reticulum stores released in response to inositol trisphosphates, or it may enter the cell through voltage-gated or non-voltage-gated Ca^{2+} channels (KOHN et al. 1992; IRVINE 1992; FELDER et al. 1991). Ca^{2+} homeostasis is tightly regulated, with normal intracellular Ca^{2+} concentrations in the 10–100 n*M* range in comparison to the millimolar Ca^{2+} concentrations found in the extracellular environment. This 1000- to 10 000-fold concentration difference results in a Ca^{2+} concentration gradient that must be maintained carefully for cell survival.

Ca^{2+} is recognized as an important regulator of many normal and malignant cellular functions including cell proliferation, differentiation, adhesion, motility, and secretion (KOHN and LIOTTA 1990; KOHN et al. 1992; SAVARESE et al. 1992; TAYLOR and SIMPSON 1992). Intracellular Ca^{2+} is an important regulator of subsets of the three main transmembrane signal transduction categories: ion influxes, phosphorylation events, and G protein-mediated second messenger production. A review of signal transduction effector molecules involved in the processes of invasion and metastasis implicates Ca^{2+} as a key regulator in the biochemical steps of tumorigenicity and metastasis. Manipulation of intracellular Ca^{2+} influxes can alter cancer cell gene expression (KOHN et al. 1994b), inhibit cell cycle progression and proliferation (SHORT et al. 1993; TAYLOR and SIMPSON 1992), interfere in adhesion and motility (KOHN et al. 1995; KOHN and LIOTTA 1990; KOHN et al. 1992; LESTER and MCCARTHY 1992; SAVARESE et al. 1992), and produce programmed cell death (FURUYA et al. 1994). High levels of intracellular Ca^{2+} may yield increased activation of Ca^{2+} second messenger systems, resulting in the uncontrolled growth of malignant cells. Interruption of Ca^{2+}-dependent signaling pathways represents a novel target for the treatment and prevention of cancer development and dissemination (COLE and KOHN 1994; FELDER et al. 1993).

In view of a central role played by Ca^{2+} in the regulation of the biochemical steps of mitogenesis and metastases, strategies have focused on intervention at the level of Ca^{2+} signaling through inhibition of Ca^{2+} influx, blockade of Ca^{2+} efflux, and/or prevention of Ca^{2+} uptake/export by inhibition of Ca^{2+} release from internal stores (COLE and KOHN 1994; FELDER et al. 1991; REEVE 1991). It might therefore be possible to inhibit tumorigenicity and metastasis with an agent which could alter cellular Ca^{2+} homeostasis. Most epithelial cells, parental cells of carcinomas, are electrically neutral and lack voltage-gated Ca^{2+} channels as their primary method of Ca^{2+} entry. This defines non-voltage-gated Ca^{2+} influx as the primary regulator of Ca^{2+}-dependent cellular signaling pathways (KOHN and LIOTTA 1995). A two-pronged screen using inhibition of Ca^{2+}-sensitive tumor cell motility and inhibition of Ca^{2+} influx resulted in the identification of a novel non-voltage-gated Ca^{2+} influx inhibitor, carboxyamido-triazole (CAI; KOHN and LIOTTA 1990). The in vitro and in vivo experimental results with CAI suggested that intervention at the level of Ca^{2+} influx signaling may be a viable therapeutic option (FELDER et al. 1991; GUSOVSKY et al. 1993; KOHN et al. 1995; KOHN et al. 1994a; KOHN et al. 1994b; KOHN and LIOTTA 1990; KOHN et al. 1992).

The well-characterized signal transduction pathways of the muscarinic acetylcholine receptor system presented a direct model from which to study the biochemical effects of CAI. The m5 muscarinic acetycholine receptor is coupled to phosphoinositide metabolism through both PLC-β and PLC-γ, PLA_2-mediated release of arachidonic acid, adenylate cyclase-mediated production of cAMP, and the opening of receptor-operated calcium channels (FELDER et al. 1991; GUSOVSKY et al. 1993; KOHN et al. 1994a). Early studies using this Chinese hamster ovary (CHO)m5 model demonstrated that CAI inhibited agonist-induced non-voltage-gated Ca^{2+} influx in the concentration range of 1-10 μM (FELDER et al. 1991; GUSOVSKY et al. 1993). Downstream signaling pathways involving PLA_2, Ca^{2+}-sensitive PLC-γ, and PTK activity have been shown to be sensitive to CAI–Ca^{2+} influx inhibition (FELDER et al. 1991; GUSOVSKY et al. 1993; KOHN et al. 1994a). Biological studies demonstrated that CAI exposure inhibited human melanoma cell migration in response to Ca^{2+}-sensitive chemoattractants (KOHN and LIOTTA 1990; KOHN et al. 1992; LEHEL et al. 1994; SAVARESE et al. 1992). Similarly, adhesion of tumor cells to tissue culture plastic and collagen type IV have also been shown to be CAI sensitive (KOHN and LIOTTA. 1990). Exposure of tumor cells to CAI resulted in the inhibition of the production of gelatinase A at the level of gene expression (KOHN et al. 1994b).

Angiogenesis is a normal process in development and wound healing as well as a pathological condition required for the transformation from carcinoma in situ to invasive carcinoma (LIOTTA et al. 1991). The process of angiogenesis mirrors the steps of metastasis and can be considered a form of regulated invasion, requiring protease secretion to facilitate basement membrane remodeling, cell proliferation, and migration (FURCHT 1986; LIOTTA et al. 1991). CAI inhibited the proliferation, adhesion, motility to extracellular matrix components, and proteolytic activity of human umbilical vein endothelial cells in vitro. This inhibitory activity was confirmed in vivo with the chicken chorioallantoic membrane assay (KOHN et al. 1995). All effects, in vitro and in vivo, occured in the same concentration range as the inhibition of malignant invasion and the inhibition of Ca^{2+} influx requiring signaling pathways (KOHN et al. 1995; KOHN et al. 1994a).

Preclinical studies of CAI in mice demostrated oral bioavailability and the ability to attain targeted signal-inhibitory plasma concentrations of 1–10 μM without marked toxicity. Experimental metastasis assays using *ras*-transfected rat embryo fibroblasts and human colon cancer cell lines treats with CAI demonstrated a marked inhibitory effect to CAI on the formation of pulmonary metastases (KOHN et al. 1992). In vivo, oral administration of CAI resulted in growth arrest and decreased incidence of primary tumors and spontaneous metastases in human melanoma and human ovarian cancer xenografts in nude mice (KOHN et al. 1992). CAI plasma levels in the treated mice were at or above the concentrations demonstrated to be effective in the inhibition of proliferation and signal transduction in vitro (KOHN and LIOTTA 1990). No overt toxicity was observed at a gross or histologic level in these in vivo efficacy experiments.

4.7 Carboxyamido-triazole Clinical Trials

The in vivo efficacy and minimal toxicity observed in the animal toxicity studies in daily and weekly schedules led to the development of phase I clinical protocols. A sensitive, accurate, and reproducible high-performance liquid chromatography (HPLC) assay was developed for pharmacokinetic and metabolite analysis of blood and urine from patients participating in the phase I clinical trial (HOLMES et al. 1993). CAI has been administered orally in several formulations, including PEG-400 liquid, gelatin capsules, and micronized powder-containing capsules. Plasma CAI levels of 1-10 μM, in the range which inhibits signaling and invasion in vitro, have been demonstrated in all patients and was attained with an acceptable pattern of toxicity. Disease stabilization has been observed in several heavily pretreated patients lasting up to 7 months (KOHN et al. 1996). Phase II and III studies and chemoprevention strategies for oral CAI are under development.

5 Conclusions

The increased knowledge of the regulation of cell growth and the genetic and biochemical changes responsible for neoplastic development have opened new fields of investigations for cancer drug discovery. The traditional cytotoxic approach for tumor treatment may be supplanted by new therapeutic targets such as signaling pathways that result from and regulate altered gene expression. Although signal transduction therapy is still under early development, some antisignaling compounds have already entered clinical trials. It is important to emphasize that these new agents are cytostatic in their mode of action and show little toxicity compared to the conventional chemotherapeutic drugs. This requires a new mode of thinking about cancer biology and treatment. First, different biological assays for identification of antisignaling compounds need to be developed; secondly, disease stabilization and extended survival should be considered as a new clinical end point for cancer patient treatment. New paradigms for clinical investigation of these promising compounds are needed.

References

Akinaga S, Gomik K et al (1991) Antitumor activity of UCN-01, a selective inhibitor of protein kinase C in murine and human tumor models. Cancer Res 51: 4888–4892

Akiyama T, Ishida J et al (1987) Genistein, a specific inhibitor of tyrosine-specific protein kinases. J Biol Chem 262: 5592–5595

Alberts AW, Chen J et al (1980) Mevinolin: a highly potent competitive inhibitor of hydroxymethylglutaryl-coenzyme A reductase and α cholesterol lowering agent. Proc Natl Acad Sci USA 77: 3957–3961

Albino AP, Schwartz GK et al (1995) Possible roles of bFGF and PKC in the evolution of metastases in malignant melanoma and gastric adenocarcinoma. Bull Inst Pasteur 92: 276–284

Andreeson R (1988) Ether lipids in the therapy of cancer. Prog Biochem Pharmacol 22: 118–131

Arteaga CL, Johnson MD et al (1991) Elevated content of the tyrosine kinase substrates phospholipase C-γ1 in primary human breast carcinomas. Proc Natl Acad Sci USA 88: 10435–10439

Bailar JC, Smith E (1986) Progress against cancer? New Engl J Med 314: 1226–1232

Barbacid M (1987) Ras genes. Annu Rev Biochem 56: 779–827

Bargmann CI, Hung MC et al (1986) The neu oncogene encodes an epidermal growth factor-related protein. Nature 319: 226–230

Berdel WE (1991) Membrane-interactive lipids as experimental anticancer drugs. Br J Cancer 64: 208–211

Berggren M, Gallegos A et al (1993) Inhibition of the signalling enzyme phosphatidylinositol-3-kinase by antitumor ether lipid analogues. Cancer Res 53: 4297–4302

Bishop JM (1991) Molecular themes in oncogenesis. Cell 64: 235–248

Bos JL (1990) Ras gene mutations and human cancer. Elsevier, Amsterdam

Bottaro DP, Rubin JS et al (1991) Identification of the hepatocyte growth factor receptor as the c-met protooncogene product. Science 152: 802–804

Boyle WJ, Smeal T et al (1991) Activation of protein kinase C decreases phosphorylation of c-jun at sites that negatively regulate its DNA-binding activity. Cell 64: 573–584

Buchdunger E, Trinks U et al (1994) 4,5-Dianilinophtalimide: a protein-tyrosine kinases inhibitor with selectivity for the epidermal growth factor receptor signal transduction pathway and potent in vivo antitumor activity. Proc Natl Acad Sci USA 91: 2334–2338

Cantley LC, Auger KR et al (1991) Oncogenes and signal transduction. Cell 64: 281–302
Chen R, Sarnecki C et al (1992) Nuclear localization and regulation of erk- and rsk-encoded protein kinases. Mol Cell Biol 12: 915–927
Clapham DE (1995) Calcium signaling. Cell 80: 259–268
Cole KA, Kohn EC (1994) Calcium-mediated signal transduction: biology, biochemistry, and therapy. Cancer Metastasis Rev 13: 33–41
Coughlin SR, Escobedo JA et al (1989) Role of phosphatidylinositol kinase in PDGF receptor signal transduction. Science 243: 1191–1194
Davis RJ (1993) The mitogen-activated protein kinase signal transduction pathway. J Biol Chem 268: 14553–14556
Delli-Bovi P, Curatola AM et al (1987) An oncogene isolated by transfection of Kaposi's sarcoma DNA encodes a growth factor that is member of the FGF family. Cell 50: 729–737
Deuel TF (1987) Polypeptide growth factors: roles in normal and abnormal cell growth. Annu Rev Cell Biol 3: 443–492
Dougall WC, Qian X et al (1994) The neu-oncogene. 9: 2109–2123
Downward J, Yarden Y et al (1984) Close similarity of epidermal growth factor receptor and v-erbB oncogene protein sequences. Nature 307: 521–527
Fearon EN, Vogelstein BA (1990) Genetic model for colorectal tumorgenesis. Cell 53: 759–767
Felder CC, Ma AL et al (1991) The antiproliferative and antimetastatic compound L651582 inhibits muscarinic acetylcholine receptor-stimulated calcium influx and arachidonic acid release. J Pharmacol Exp Ther 257: 967–971
Felder CC, Poulter MO et al (1992) Muscarinic receptor-operated calcium influx in transfected fibroblast cells is independent of inositol phosphates and release of intracellular Ca^{2+}. Proc Natl Acad Sci USA 89: 509–513
Felder CC, MacArthur L et al (1993) Tumor-suppressor function of muscarinic acetylcholine receptors is associated with activation of receptor-operated calcium influx. Proc Natl Acad Sci USA 90(5): 1706–1710
Felder CC, Singer-Lahat D et al (1994) Voltage-independent calcium channels. Regulation by receptors and intracellular calcium stores. Biochem Pharmacol 48(11): 1997–2004
Fidler IJ, Hart IR, (1982) Biologic diversity in metastatic neoplasms – origins and implications. Science 217: 998–1001
Fry MJ (1994) Structure, regulation and function of phosphoinositide 3-kinases. Biochim Biopys Acta 1226: 237–268
Furcht L (1986) Critical factors controlling angiogenesis: cell products, cell matrix, and growth factor. Lab Invest 5: 505–509
Furuya Y, Lundmo P et al (1994) The role of calcium, pH, and cell proliferation in the programmed (apoptotic) death of androgen-independent prostatic cancer cells induced by thapsigargin. Cancer Res 54: 6167–6175
Galkina SI, Sud'Ina GF et al (1992) Cell-cell contacts alter intracellular pH. Exp Cell Res 200: 211–214
Ganz MB, Pachter JA et al (1990) Multiple receptors coupled to adenylate cyclase regulate Na-H exchange independent of cAMP. J Biol Chem 265: 8989–8992
Gazit A, Osherow N et al. (1991) Tyrphostins II: heterocyclic and α-substituted benzylide-nemalononitrile tyrphostins as potent inhibitors of EGF receptors and ErbB2/Neu tyrosine kinases. J Med Chem 34: 1897–1907
Gibbs JB (1991) Ras C-terminal processing enzymes – new drug targets? Cell 65: 1–4
Gibbs JB, Oliff A et al. (1994) Farnesyltransferase inhibitors: Ras research yields a potential cancer therapeutic. Cell 77(2): 175–8
Goodnight J, Kazanietz MG et al (1992) The cDNA sequence, expression pattern and protein characteristics of mouse protein kinase C-zeta. Gene 122: 305–311
Gusovsky F, Lueders JE et al (1993) Muscarinic receptor-mediated tyrosine phosphorylation of phospholipase C-gamma: alternative mechanism for cholinergic receptor-induced phosphoinositide breakdown. J Biol Chem 268: 7768–7772
Haag JD, Lindstrom MJ et al (1992) Limonene-induced regression of mammary carcinomas. Cancer Res 52: 4021–4026
Hancock JF, Magee AI et al (1989) All ras proteins are polyisoprenylated but only some are palmitoylated. Cell 57: 1167–1177
Hennigan RF, Hawker KL et al (1994) Fos-transformation activates genes associated with invasion. Oncogene 9: 3591–3600
Hilgard P, Klenner T et al (1993) Alkylphosphocolines: a new class of membrane-active anticancer agents. Cancer Chemother Pharmacol 32: 90–95

Holmes K, Chaffins AS et al (1993) Quantitation of human plasma levels of the anticancer agent carboxyamido-triazole by high performance liquid chromatography. J Chrom Biomed Appl 613: 317–325
Irvine RF (1992) Inositol phosphates and Ca^{++} entry: toward a proliferation or a simplification? Fed Am Soc Exp Biol J 6: 3085–3091
Isakov N, Gopas J et al (1991) Effect of protein kinase C activating tumor promoters on metastases formation by fibrosarcoma cells. Invasion Metastasis 11: 14–24
Iyer A, Kmieccik TE et al (1990) Structure, tissue-specific expression, and transforming activity of the mouse met-proto-oncogene. Cell Growth Differ 1: 87–95
Kaplan DR, Whitman M et al (1987) Common elements in growth factor stimulation and oncogenic transformation: 85 kD phosphoprotein and phosphatidylinositol kinase activity. Cell 50: 1021–1029
Khanavkar B, Ulbrich F et al (1989) Treatment of non-small cell lung cancer with the alkyl-lysophospholipid edelfosine. Contrib Oncol 37: 224
Koch CA, Anderson D et al (1991) SH2 and SH3 domains: elements that control interactions of cytoplasmic signaling proteins. Science 252: 668–674
Kohl NE, Wilson FR et al (1994) Protein farnesyltransferase inhibitors block the growth of ras-dependent tumors in nude mice. Proc Natl Acad Sci USA 91: 9141–9145
Kohn EC, Liotta LA (1990) L651582, a novel antiproliferative and antimetastasis agent. J Natl Cancer Inst 82: 54–60
Kohn EC, Liotta LA (1995) Molecular insights into cancer invasion: strategies for prevention and intervention. Cancer Res 55: 1856–1862
Kohn EC, Sandeen MA et al (1992) In vivo efficacy of a novel inhibitor of selected signal transduction pathways including calcium, arachidonate, and inositol phosphates. Cancer Res 52: 3208–3212
Kohn EC, Felder CC et al (1994a) Structure-function analysis of signal and growth inhibition by carboxyamido-triazole, CAI. Cancer Res 54: 935–942
Kohn EC, Jacobs W et al (1994b) Calcium influx modulates expression of matrix metalloproteinases-2 (72-kDa type IV collagenase, gelatinase A). J Biol Chem 269: 21505–21511
Kohn EC, Alessandro R et al (1995) Angiogenesis: role of calcium-mediated signal transduction. Proc Natl Acad Sci USA
Kohn EC, Reed E et al (1996) Clinical investigation of a cytostatic calcium influx inhibitor in patients with refractory cancers. Cancer Research (in press)
Kovalenko M, Gazit A et al (1994) Selective platelet-derived growth factor receptor kinase blockers reverse sis-transformation. Cancer Res 54: 6106–6114
Kypta RM, Goldberg Y et al (1990) Association between the PDGF receptor and members of the src family of tyrosine kinases. Cell 62: 481–492
Lagarde AE, Pouyssegur JM (1986) The Na+: H+ antiport in cancer. Biochim Biophys Acta 9: 1–14
Leevers SJ, Marshall CJ (1992) MAP kinase regulation – the oncogene connection. Trends Cell Biol 2: 283–285
Lehel S, Lasic Z et al (1994) Collagen type IV stimulates an increase in intracellular calcium in pancreatic acinar cells via activation of phospholipase C. Biochem J 299: 603–611
Lester BR, McCarthy JB (1992) Tumor cell adhesion to the extracellular matrix and signal transduction mechanisms implicated in tumor cell motility, invasion and metastasis. Cancer Metastasis Rev 11: 31–44
Lev SJ, Blechman M et al (1994) Steel factor and c-kit protooncogene: genetic lessons in signal transduction. Crit Rev Oncog 5: 141–168
Levitzki A (1994) Signal-transduction therapy. Eur J Biochem 226: 1–13
Liotta LA, Steeg PS et al (1991) Cancer metastasis and angiogenesis: an imbalance of positive and negative regulation. Cell 64: 327
Macara IG (1989) Oncogenes and signal transduction. Physiol Rev 69: 797–820
Margolis B (1992) Protein with SH2 domain: transducers in the tyrosine kinase signalling pathway. Cell Growth Differ 3: 73–80
Markovits J, Linassier C et al (1989) Inhibitory effects of the tyrosine kinase inhibitor genistein on mammalian DNA topoisomerase II. Cancer Res 49: 5111–5117
Matrisian L (1992) The matrix-degrading metalloproteinases. Bioessay 14: 455–462
Matter WF, Brown RF et al (1992) The inhibition of phosphatidylinositol 3-kinase by quercetin and analogs. Biochem Biophy Res Commun 186: 624–631
Mazzanti M, DeFelice LJ et al (1990) Ion channels in the nuclear envelope. Nature 343: 764–767
McCormick F (1993) How receptors turn Ras on. Nature 363: 15–16
Millon R, Nicora F et al (1989) Modulation of human breast cancer cell adhesion by estrogens and antiestrogens. Clin Exp Metastasis 7: 405–415

Mischak H, Kolch W et al (1991) Expression of protein kinase C genes in hemepoietic cells is cell-type and B-cell differentiation stage specific. J Immunol 147: 3981–3987

Naldini L, Weidner KM et al (1991) Scatter factor and hepatocyte growth factor are indistinguishable ligands for the MET receptor. EMBO J 10: 2867–2878

Neer EJ (1995) Heterotrimeric G proteins: organizers of transmembrane signals. Cell 80: 249–257

Nishizuka Y (1992) Intracellular signaling by hydrolysis of phospholipids and activation of protein kinase C. Science 258: 606–614

Nixon JS, Bishop J et al (1992) The design and biological properties of potent and selective inhibitors of protein kinase C. Biochem Soc Trans 20: 419–425

Noh D, Lee YH et al (1993) Elevated content of phospholipase C-γ1 in colorectal cancer tissues. Cancer 73: 36–41

Nowell PC (1976) The clonal evolution of tumor cell populations. Science 194: 23–28

O'Brian CA, Housey GM et al (1988) Specific and direct binding of protein kinase C to an immobilized tamoxifen analogue. Cancer Res 48: 3626–3629

Osherow N, Levitzki A (1994) Epidermal growth factor dependent activation of src family kinases. Eur J Biochem 225: 1047–1053

Paterson MC, Dietrich KD et al (1991) Correlation between c-erbB-2 amplification and risk of recurrent disease in node-negative breast cancer. Cancer Res 51: 556–567

Peles E, Lamprecht R et al (1992) Regulated coupling of the Neu receptor to phosphatidylinositol 3'-kinase and its release by oncogenic activation. J Biol Chem 267: 12266–12274

Powis G (1994) Recent advances in the development of anticancer drugs that against signaling pathways. Tumori 80: 69–87

Powis G, Bonjouklian R et al (1994) Wortmanin, a potent and selective inhibitor of phosphatidylinositol-3-kinase. Cancer Res 54: 2419–2423

Powis G, Seewald MJ et al (1992) Selective inhibition of phosphatidylinositol phospholipase C by citotoxic ether lipid analogues. Cancer Res 52: 2835–2840

Randriamampita C, Tsein R (1993) Emptying of intracellular Ca^{2+} stores releases a novel small messenger that stimulates Ca2+ influx. Nature 364: 809–814

Reeve J (1991) Mitogenesis and the search for new targets for anticancer drug design. Tumour Biol 12(1): 1

Rilemma JA (1989) Possible roles of phospholipase C in the regulation of cell division in normal and neoplastic cells. Med Hypothesis 29: 1–4

Rong S, Segal S et al (1994) Invasiveness and metastasis of NIH 3T3 cells induced by Met-hepatocyte growth factor/scatter factor autocrine stimulation. Proc Natl Acad Sci USA 91: 4731–4735

Sardet C, Counillo L et al (1990) Growth factors induce phosphorylation of the Na+/H+ antiporter, a glycoprotein of 110 kD. Science 247: 723–726

Sato H, Takino T et al (1994) A matrix metalloproteinase expressed on the surface of invasive tumor cells. Nature 370: 61–65

Savarese DMF, Russell JT et al (1992) Type IV collagen stimulates an increase in intracellular calcium: potential role in tumor cell motility. J Biol Chem 267: 21928–21935

Schafer WR, Rine J (1992) Protein prenylation: genes, enzymes, targets and functions. Annu Rev Genetics 30: 209–237

Schallier DK, Brunyneel EA et al (1991) Anti-invasive activity of hexadecylophosphocholine in vitro. Anticancer Res 11: 1285–1292

Schlessinger J, Ullrich A (1992) Growth factor signaling by receptor tyrosine kinases. Neuron 9: 383–391

Schu PV, Takegawa K et al (1993) Phophatidyl-inositol 3-kinase encoded by yeast VPS34 gene essential for protein sorting. Science 260: 88–91

Schwartz GK, Jiang J et al (1993) Protein kinase C: a novel target for inhibiting gastric cancer cell invasion. J Natl Cancer Inst 85: 402–407

Schwartz GK, Redwood SM et al (1990) Inhibition of invasion of invasive human bladder carcinoma cells by protein kinase C inhibitor staurosporine. J Natl Cancer Inst 82: 1753–1756

Sebti SM, Tkalcevic GT et al (1991) Lovostatin, a cholesterol biosynthesis inhibitor, inhibits the growth of human H-ras oncogene transformed cells in nude mice. Cancer Commun 3: 141–146

Seewald MJ, Olen RA et al (1990) Inhibition of growth factor-dependent inositol phosphate Ca^{2+} signaling by antitumor ether lipid analogues. Cancer Res 50: 4458–4463

Sherr CJ, Rettenmier CW et al (1985) The c-fms proto-oncogene product is related to the receptor for the mononuclear phagocyte growth factor, CSF-1. Cell 41: 665–676

Short AD, Bian J et al (1993) Intracellular Ca^{2+} pool content is linked to control of cell growth. Proc Natl Acad Sci USA 90: 4986–4990

Simon ML, Strathmann MP et al (1991) Diversity of G proteins in signal transduction. Science 252: 802–808

Smith MR, Liu YL et al (1990) Inhibition of serum- and Ras-stimulated DNA synthesis by antibodies to phospholipase C. Science 247: 1074–1077

Soriano PC, Montgomery R et al (1991) Targeted disruption of the c–src protoncogene leads to osteoporosis in mice. Cell 64: 693–702

Taylor JM, Simpson RV (1992) Inhibition of cancer cell growth by calcium channel antagonists in the athymic mouse. Cancer Res 52(9): 2413–2418

Thibault A et al (1995) Lovastatin in phase I and II clinical trials (submitted)

Thibault A, Samid D, Tompkins AC, Figg WD, Liang B, Patronas N, Hohl RJ, Cooper MR, Venzon DJ, Reed E, Myers CE (1996) A phase I study of lavastatin, in inhibitor of the mevalonate pathway, in patients with cancer. Clin Cancer Res (in press)

Uberall F, Oberhuber H et al (1991) Hexadecylphosphocholine inhibits inositol phosphate formation and protein kinase C activity. Cancer Res 51: 807–812

Umezawa H, Imoto M et al (1986) Studies on a new epidermal growth factor-receptor kinase inhibitor, erbstatin, produced by MH435-hF3. J Antibiot (Tokyo) 39: 170–173

Velu TJ, Beguinot L et al (1987) Epidermal-growth-factor-dependent transformation by a human EGF receptor proto-oncogene. Science 238: 1408–1410

Waterfield MD, Scrace GT et al (1983) Platelet-derived growth factor is structurally related to the putative transforming protei p28sis of simian sarcoma virus. Nature 304: 35–39

Williams NG, Roberts TM (1994) Signal transduction pathways involving the raf proto-oncogene. Cancer Metastasis Rev 13: 105–116

Yaish P, Gazit A et al (1988) Blocking of EGF-dependent cell proliferation by EGF receptor kinases inhibitors. Science 242: 933–935

Zong CS, Wang L (1994) Modulatory effect of the transmembrane domain of the protein kinase encoded by oncogene ros: biological function and substrate interaction. Proc Natl Acad Sci USA 91: 10982–10986

Immunotherapy of Metastases

V. SCHIRRMACHER, V. UMANSKY, and M. ROCHA

1 Introduction

Many aspects of tumor growth and metastasis are dependent on tumor cell–host cell interactions. The capacity of tumor cells to elicit or suppress various host tissue responses seems as important for development of solid tumors as for tumor cell proliferation in different organ microenvironments. The process of tumor angiogenesis, the consequences of which extend far beyond nutrient supply, can serve as an illustrative and important example (RAK et al. 1993). The

---

German Cancer Research Center, Division of Cellular Immunology (0710), Im Neuenheimer Feld 280, 69120 Heidelberg, Germany

importance of tumor–host interactions and of signals from the microenvironment for development of cancer metastases was pointed out as early as 1980 (Schirrmacher 1980), but has only recently attracted much attention and been studied at the molecular level (Schirrmacher 1994; Nicolson 1988,1991; Fidler 1990; Kerbel 1990; Gullino 1991; Heppner 1989). The way in which tumor cells react to signals from the microenvironment is determined by their reception via specific receptors, by the way these signals are transduced, and by the way these transduced signals are translated in the cellular response. Cellular calcium homeostasis is the result and the regulator of many signal transduction pathways and plays a central role of regulation of cell proliferation, invasion, and metastatic potential (Liotta et al. 1991). While the interaction of tumor cells with host endothelial cells and extracellular matrix (Ingber 1991; Ingber and Folkman 1989; Jain 1990) is being discussed by experts contributing to this series of articles, the interaction with cells of the immune system, in particular with antigen-specific immune mechanisms, will be the focus of this review. It is necessary at this stage to mention the existence of natural tumor resistance mechanisms (Vaage 1985; Janeway 1989; Schirrmacher 1985), which can be exerted by non-antigen-specific cells such as natural killer (NK) cells, (Kaminski and Auerbach 1988) macrophages, granulocytes, or even endothelial cells. Cytokines can influence natural resistance mechanisms as well as specific antitumor immune responses and have to be included when it comes to strategies of using components of the immune system such as antigens, antibodies, or immune cells for intervening with the process of metastases. This article attempts to review the rationale, the experimental approaches, and different strategies of immunotherapy of metastases. Some aspects will be exemplified by recent findings from our own group.

2 Defining Therapeutic Targets and Aims

Approaches for metastasis treatment can have different therapeutic targets, aims, and strategies (Schirrmacher 1985). All the metastasis-related molecules disussed in the previous volume of *Current Topics in Microbiology and Immunology* could be potential therapeutic targets for intervention with metastases. Two examples in this volume are the targeting to the CD44 variant form and to the EGF or erbB2 receptors. Therapeutic targets can of course also be tumor-associated genes (oncogenes, suppressor genes) and tumor antigens. Tumor-associated antigens can be targeted by either antibodies or T cell receptors, depending on whether humoral or cell-mediated immune responses and active, passive, or adoptive immunotherapies are being aimed for. All active immunization procedures involve the use of tumor-associated antigens or whole tumor cells as tumor vaccines or of anti-idiotypic antibodies as internal images of tumor antigens. To increase the immunogenicity of such tumor vaccines, various approaches are possible, including virus infection, gene transfer, and attachment of additional molecules such as adhesion molecules, costimulatory molecules, antibodies or bispecifics. Tumor cell-targeted gene transfer aims at introducing in tumor cells molecules of importance for antigen presentation. The aims of active immunization strategies with such modified tumor vaccines are to induce antitumor immune responses, activate already primed memory cells, and overcome host–tumor tolerance (SCHÖNRICH et al. 1991; FERBER et al. 1994). Other strategies of immunotherapy of metastases aim at eliminating host factors which are essential for aggressive tumor growth, such as angiogenesis factors, chemokines, and growth factors (PEKAREK et al. 1995). In advanced disease stages, important additional strategies would be to counteract immunosuppression (TURK and PARKER 1982; YE and MOKYR 1984) and possibly immune escape mechanisms (SCHREIBER 1993) by the tumor cells. It thus has to be defined whether the therapy aims at micrometastases, in which case a monotherapy may be sufficient, or whether it aims at macrometastases, in which case combination therapies should be considered in order to have a realistic chance of success.

In late disease stages in which the host is already severely affected by its tumor, active specific immunotherapy procedures (ASI) with cancer vaccines (SCHIRRMACHER 1993) may be useless unless they are combined with antisuppressive (TURK and PARKER 1982; YE and MOKYR 1984) and/or substitutive (TALLBERG et al. 1985) treatment. As we recently discovered in an animal model, even in such late stages, potent cellular therapies can still be useful and effective (SCHIRRMACHER et al. 1995). Transferred immune cells could improve therapeutic effectiveness by changing the milieu around primary tumors and around organ metastases. In addition, immune cell transfer could cause activation of additional tumor resistance mechanisms from the host such as production of nitric oxide (NO) by liver endothelial cells (ROCHA et al. 1995a) and Kupffer cells (UMANSKY et al. 1995).

Therapy aims and evaluation have to be discussed both for preclinical models and for clinical trials and have to include methods for therapy evaluation. Because clinical immunotherapy trials are difficult and expensive and the end points sometimes difficult, new methodologies are important for therapy evaluation. Without surrogate end points and some good ideas about how to refine these trials by careful choice of target populations, it will be very hard to develop such therapies. It would be important for progress, especially in adjuvant therapy of cancer, to have markers that may give an early end point of therapy evaluation. This may not be the final answer, which is recurrence-free survival (RFS) and overall survival, but could help monitoring therapies more quickly. New ways of immune monitoring of tumor vaccination effects have recently been reported (BERD et al. 1995; KOTERA et al. 1994; MUKHERJI et al. 1995).

3 Development of Preclinical Models

3.1 Visualization and Quantitation of Metastases

3.1.1 Micrometastases

The processes of tumor metastasis formation and its interplay with the host immune system have been extensively studied in the murine lymphoma system Eb/ESb/ESb-MP (SCHIRRMACHER et al. 1981, 1982, 1992; ALTEVOGT et al. 1984; SCHIRRMACHER and BARZ 1986; VON HOEGEN et al. 1987). The spontaneous highly metastatic lymphoma variant ESb is able to kill syngeneic animals in less than 12 days following s.c. inoculation of 10^5 cells, whereby the metastatic spread involves several visceral organs, in particular the liver.

Detection of micrometastases and also of low levels of dormant tumor cells (SCHIRRMACHER et al. 1981) has only recently been possible in this model by genetically tagging the tumor line with the bacterial *lac-Z* gene. This foreign gene codes for β-D-galactosidase (β-gal), whose enzymatic activity can be detected by staining with the chromogenic substrate X-gal. This results in a precipitate of an indigo blue reaction product which allows single ESb *lac-Z* tumor cells to be detected in infiltrated target organs such as lymph nodes, bone marrow, or liver (KRÜGER et al. 1994a). Despite expression of the *lac-Z* gene, the tumor cells were still tumorigenic, highly metastatic, unchanged in phenotype, and therefore comparable to the parental ESb cells. After spontaneous metastasis, whole organ staining revealed metastatic foci at the surface of the liver, while x-gal staining of frozen tissue sections revealed micrometastases in the form of clusters and diffusely disseminated single cells (KRÜGER et al. 1994b).

It was also possible to reisolate cells from metastases and to quantify all cells per organ that have metastasized into it via loading with the substrate fluorescein β-D-galactopyranoside (FDG), which allowed tumor cell fluorescence staining and quantitative FACScan fluorescence-activated cell sorter analysis. In a typical experiment, 12 days after intradermal tumor cell inoculation, 55% of the reisolated cells from the liver and 13% of the cells from the spleen were tumor cells (SCHIRRMACHER et al. 1995).

3.1.2 Macrometastases

With nuclear magnetic resonance (NMR) imaging, it was recently possible to detect the formation of focal metastases in the liver and kidneys noninvasively and without contrast agents. The NMR studies were performed at high magnetic field strength (7.0 Tessler) using a Bruker AM-300 spectrometer and a superconducting magnet with a 15-cm-diameter vertical bore. During the measurements, mice were anesthesized with chloral hydrate for 1-2 h in the case of 31P-NMR spectroscopy of primary tumors or with isofluorane inhalation narcosis for 2–3 h in the case of imaging and kept at a temperature of about 32°C. Metastases with a diameter greater than 0.3 mm could usually be detected by day

21–23 after ESb-MP inoculation. This noninvasive imaging procedure in live tumor-bearing mice enabled us to follow the growth of metastases in internal organs in individual animals. It also allowed us to follow the effect of immunotherapy (SCHIRRMACHER et al. 1995).

3.2 Models for Studying Tumor-Host Interactions

A new parameter for evaluating immune responses to metastases is the phenotype of metastasized organs. We recently reported that the pattern of liver metastases was dependent on and strongly influenced by the host immune status (KRÜGER et al. 1994b). While in immunocompromised mice primary tumor growth and metastases of ESb *lac-Z* lymphoma cells were progressive and metastases appeared as diffuse or focal, in immunocomponent syngeneic animals metastatic spread was to a large extent delayed and when it eventually appeared, a distinct mosaic-like metastasis pattern developed in involved livers. It turned out that the shift from a diffuse or focal pattern to a distinct mosaic-like pattern was a reliable indicator of an ongoing host immune response. Using whole organ staining, it was possible to distinguish these two phenotypes and thus to use the metastatic phenotype as a new parameter for evaluating whether or not animals experienced an antitumor immune response.

Most of the studies published so far about interactions of tumor and host cells were done in vitro, testing the ability of tumor cells to modify their properties (adhesion, invasiveness, cytokine production) when they interacted with different cell subpopulation. However, validation of such in vitro findings with the in vivo situation is often difficult. Organotropism and targeting of tumor cells to selective sites in metastasized organs are determined by seed and soil characteristics, as predicted by Stephen Paget as early as 1889 (PAGET 1889). The nature of these seed and soil characteristics can be analyzed directly using newly designed methods of ex vivo isolation and immunostaining of tumor and host cells from involved organs. We adapted a method allowing the separation of *lac-Z* tumor cells from sinusoidal cells from the liver with a high viability (more than 93%) and reproducibility (SCHIRRMACHER et al. 1995; KRÜGER et al. 1994a, b). This method, which can be used for other target organs, allows the direct characterization of the cells without further in vitro culture. In one such study we investigated the expression of major histocompatibility complex (MHC) class II and B7 molecules, in metastasized tumor cells and also of adhesion molecules during tumor growth and growth retardation or second expansion. It was possible to demonstrate dynamic expression changes in vivo of adhesion and costimulatory molecules and to demonstrate that these changes were associated with load and pattern of lymphoma liver metastases. The application of such ex vivo analysis also allowed us to demonstrate the participation of liver endothelial cells and Kupffer cells (UMANSKY et al., submitted) in T cell de pendent host resistance to lymphoma metastases by production of nitric oxide in vivo. The activity of these host cells could also be influenced by active immunization with tumor vaccines or by adoptive immunotherapy (ADI) following transfer of immune T lymphocytes (UMANSKY et al. 1995; ROCHA et al. 1995b).

3.3 Models for Therapy Evaluation

To evaluate the importance of different effector mechanisms in vivo, several classical methods have been developed: in the Winn assay, tumor cells are mixed with effector cells in vitro and the mixture is then injected s.c. into an animal to determine whether local tumor growth is prevented (SCHIRRMACHER and GRIESBACH 1994; SCHIRRMACHER and ZANGEMEISTER-WITTKE 1994). Another way to find out about the relative importance of certain lymphocyte subsets or cytokines is by in vivo depletion or elimination by treatment with antibodies specific for different lymphocyte subpopulations (SCHIRRMACHER and ZANGEMEISTER-WITTKE 1994) or cytokines (SCHIRRMACHER 1995). A third approach involves the adoptive transfer of effector cells, cytokines, or antibodies into normal or immune-incompetent animals challenged with tumor cells. Another more indirect way of determining the importance of immune effector mechanisms in vivo is to analyze tumor variants which have developed respective immune escape mechanisms (SCHIRRMACHER et al. 1982; BOSSLET and SCHIRRMACHER 1982). The phenotypic changes observed in these variants may indicate which effectors were responsible for immunoselection in vivo. In this way a variety of tumor immune escape mechanisms were discovered: (a) failure of the tumor to express a suitable target for immune recognition, e.g., lack of antigenic epitope (ALTEVOGT et al. 1984; BOSSLET and SCHIRRMACHER 1981), downregulation of MHC molecules (SEUNG et al. 1993), deficient antigen processing, antigenic modulation or antigenic masking; (b) failure to induce an effective immune response e.g., in addition to the above, lack of costimulatory signal (BASKAR et al. 1995), production of inhibitory substances by the tumor (TORRE-AMIONE et al. 1990), shedding of antigen and membrane vesicles (SCHIRRMACHER and BARZ 1986), tolerance induction and sneaking through (SCHREIBER 1993), and alterations in signal-transducing molecules (MIZOGUCHI et al. 1992; FINKE et al. 1993; NAKAGOMI et al. 1993) and transcription factors (GOSH et al. 1994) in tumor-infiltrating lymphocytes.

As a new method for therapy evaluation in individual live animals, we applied high-field NMR spectroscopy of phosphor metabolites of the primary tumor as a noninvasive method (SCHIRRMACHER et al. 1995). The [^{31}P] phosphor spectra were acquired in 40 min from anesthesized animals. A detailed analysis of the spectra revealed that tumor growth was associated with an increase of the total amount of phosphate metabolites and an increase in phosphor monoesthers (PME) and inorganic phosphate in both relative and absolute amounts. The average tissue pH determined from the chemical shift of inorganic phosphate showed a decrease with time (SCHIRRMACHER et al. 1995).

The application of this technique to analyze the effects of transferred immune T cells (i.e. adoptive cellulare immunotherapy or ADI) demonstrated that successful therapy (complete remission) was accompanied by a dramatic decrease in PME, a decrease in total phosphate metabolites (necrosis and cell death), an increase in the relative amount of phosphocreatine, and a return to neutral tissue pH. Changes associated with effective ADI therapy (increase in tissue pH, decrease in PME) were detectable within 5 days.

Therapy evaluation at the level of visceral metastases was also possible using proton-magnetic resonance imaging (^{1}H-MRI). Images obtained from the same mouse from virtually the same liver section on day 27, 1 day before completed ADI treatment, showed several light-grey foci of 0.3–2 mm diameter which were completely eradicated after therapy, a finding which was confirmed in images of seven contiguous liver slices (FICHTNER et al., in preparation).

The antimetastatic effect of adoptive cellular immunotherapy (ADI) was also documented at the single cell level using the *lac-Z*-transduced ESb lymphoma line. In untreated control animals injected intradermally with ESb *lac-Z* cells, the metastasized tumor cells were detected 4 weeks later, predominantly in the periportal areas of liver lobuli. In contrast, liver sections from animals which had been ADI-treated on day 16 did not contain any detectable x-gal-stainable tumor cells, and no damage could be detected later on in the normal liver tissue (SCHIRRMACHER et al. 1995).

These findings were further corroborated and quantified following a pre-established method of in situ liver perfusion in order to reisolate tumor cells and live liver sinosuidal cells (Kupffer cells and endothelial cells) from tumor-bearing controls or ADI-treated mice. For FACScan detection of *lac-Z*-positive cells, the reisolated cells were loaded with FDG, which is converted by β-galactosidase into a fluorescent dye. Approximately 50% of the liver cells were quantified in contour blots as *lac-Z* tumor cells, whereas after ADI treatment this distinct cell population was undetectable, while all other subpopulations of the normal liver could still be seen (SCHIRRMACHER et al. 1995). The ADI therapy effect was further evaluated quantitatively by testing the total tumor load from animals bearing intradermal ESb *lac-Z* tumors (day 28, 13-mm diameter) and macroscopic liver metastases. Disregarding the dead cells after reisolation and FDG loading, the minimal estimate of live cells was 6.6×10^7 *lac-Z* tumor cells per liver and 2.5×10^7 *lac-Z* tumor cells per primary tumor. When day 28 in situ perfused liver cells from ADI-treated animals were subjected to cytospin centrifugation on microscope slides, fixed, and stained with x-gal, no tumor cells could be detected in 10^6 cells, indicating complete tumor cell eradication after a single treatment with 2×10^7 immune cells (SCHIRRMACHER et al. 1995).

4 Antibody-Based Therapies

4.1 Monoclonal Antibodies

Monoclonal antibodies (mAb) have been very useful as diagnostic tools to identify tumor cells which have disseminated early from primary carcinomas to the bone marrow and also to obtain information about the phenotype and prognostic significance of these cells (Lindemann et al. 1992). They have also enabled identification of molecular changes occurring in primary tumor cells at the time they develop their metastatic potential. Data indicate that changes in adhesive properties of solid tumor cells, such as downregulation of desmosome proteins and neoexpression of molecules such as intercellular adhesion molecule (ICAM)-1 or others are important determinants of the metastatic capability of individual malignant cells. Cell adhesion molecules are developmentally and environmentally regulated molecules, and it seems likely that their expression pattern during tumor progression reflects a disturbance at the level of the molecular elements normally responsible for controlling their expression. Such expression changes may make new cell interactions possible, particularly in the vascular and lymphatic system.

Because of their isolated location and low frequency in early stages of carcinoma progression, tumor cells in the bone marrow of cancer patients have been regarded as ideal targets for immunotherapeutic approaches. A first clinical randomized trial with the mAb 17-1A has confirmed this insofar as treatment was effective in significantly reducing the 6-year mortality rate of patients with colorectal cancer (Riethmüller et al. 1994).

mAb might function as adhesion molecule blockers, but there is also the possibility of blocking with small molecules such as peptides (Renschler et al. 1995) or carbohydrates. Obviously, a great problem with antibodies is whether they can be used repeatedly without creating immune responses.

Since mAb can also react with cell surface receptors, they can interfere with signaling and cause cell cycle arrest and even apoptosis. Therapeutic mAb could thus be selected on their agonistic activity. A combination of signaling mAb could thus be developed for tumor cell cycle arrest and apoptosis (Vitetta 1995).

4.2 Combined Use of Monoclonal Antibodies and Cytokines

mAb effectiveness is closely tied to the level of antigen expression and to heterogeneity in expression of the target antigen. Antigenic heterogeneity is a property shared by two human tumor antigens, tumor-associated glycoprotein (TAG)-72 and carcinoembryonic antigen (CEA). Treatment by interferon (IFN)-α or γ was shown to cause increased tumor antigen expression from the tumor site, while the surrounding normal tissue remained negative. This allowed better tumor targeting, for instance by radiolabled mAb. A phase II trial with patients diagnosed with metastatic breast cancer revealed that a combination of ^{131}I-CC49 with IFN-α increased TAG-72 tumor antigen expression and enhanced ^{131}I-CC49 tumor localization. Of nine patients enroled, two partial and one mixed response were documented.

4.3 Monoclonal Antibodies and Prodrug Activation

New approaches to tumor targeting use antibody to deliver nontoxic molecules to the tumor and then target the toxic molecules separately. The general technology (*pretargeting*) involves using the antibody, e.g., to target streptavidin to the tumor and then selectively clearing it from the blood stream. This permits the antibody to switch "antigens on the cell surface", substituting streptavidin for the tumor antigen. Biotin, a small nontoxic vitamin, binds streptavidin in four sites with an affinity 1–10 million times stronger than the antibody for its antigen. This strong affinity results in nearly instantaneous uptake of biotinylated molecules to the avidinylated tumor or their rapid exit from the body. This pretargeting technique allows replacement of antibody targeting with small (drug-sized) molecule targeting (ABRAMS 1995).

Pretargeting can be a general technology platform for the selective delivery of molecules such as drugs, prodrugs, toxins, cytokines, genes, and superantigens to tumor cells. Toxic agents for direct tumor cell killing fall into three groups: protein toxins, low molecular drugs, and radioisotopes (ALVAREZ-DIEZ et al. 1995). Protein toxins are extremely potent, but tend to penetrate normal cells and are immunogenic. Low molecular weight drugs of similar potency have recently been discovered which may overcome these problems. Radioisotopes result in much greater killing, including possible antigen-negative bystander tumor cells, than protein toxins or drugs. Irradiation of the bone marrow by the circulating conjugate may, however, result in dose-limiting toxicity (ALVAREZ-DIEZ et al. 1995).

In antibody-dependent enzyme-mediated prodrug therapy (ADEPT), a nontoxic prodrug is converted into a cytotoxic drug by an enzyme linked to an mAb, which targets the conjugate to the tumor. For instance, a F(ab)'2 fragment of an anti-CEA antibody conjugated to the bacterial enzyme carboxy-peptidase was used to selectively convert a methyloxyethyl-chloroethyl-mustard benzoyl-glutamate prodrug to its corresponding benzoic acid mustard drug. In a colorectal carcinoma nude mouse xenograft model, this ADEPT schedule resulted in complete tumor regressions in about 50% of the tumors and in tumor growth delays of more than 30 days (SPRINGER 1995). A promising selective genetic strategy against cancer is *virally directed enzyme prodrug therapy* (*VDEPT*). This approach makes use of viral vector to carry a prodrug-activating enzyme gene into both tumor and normal cells. With the linking of the enzyme gene downstream of tumor-specific transcription units, tumor-specific prodrug activation is achieved.

Sequences within the first 500 base pairs of the $^{5'}$-flanking region of the *erbB2* oncogene drive specific transcription in many breast, pancreatic, and gastric tumors. A strategy for genetic therapy using VDEPT against cancers of the breast and pancreas is being developed by producing a chimeric minigene consisting of the *erbB2* promoter linked to either cytosine deaminase (converting the prodrug 5-fluorocytosine into cytotoxic 5-fluoruracil) or herpes simplex virus thymidine kinase (conferring sensitivity to ganciclovir) (ALI and LEMOINE 1995).

In *fusion protein-mediated prodrug activation* (*FMPA*), the prodrug can be activated by a recombinant fusion protein consisting, for instance, of humanized

CEA-specific binding region linked to human β-glucuronidase. This can be targeted to CEA expressing human colon xenografts in nude mice. Upon injection of a glucuronyl-doxorubicine prodrug, strong tumor growth-inhibitory effects were seen under conditions in which xenografts were completely resistant with the maximum tolerated dose (MTD) of doxorubicin. Such superior therapeutic effects were caused by five to ten times higher and two to five times lower drug concentrations in tumor and normal tissue, respectively, compared to standard treatment (BOSSLET et al. 1995).

4.4 Antibody Engineering and Bispecifics

The ability to manipulate immunoglobulin genes may enable mAb to be produced with properties better suited for immunodiagnosis and immunotherapy. Research has shown that several constructed immunoglobulin (Ig) forms have improved serum half-life characteristics as well as increased tumor penetration. Examples are single-chain Fv-binding regions (scFv) and domain-deleted antibodies.

Antibody engineering also allows for Ig class switching, humanization of animal-derived mAb, and construction of entirely new molecules such as bispecifics and janusins (GOTTER et al. 1995). Cure of human xenografted B cell tumors has recently been reported by treatment with human T lymphocytes and the bispecific mAb CD30-CD3 and CD30-CD28 (RENNER et al. 1994).

5 Tumor Vaccine Based Therapies

ASI procedures involve the use of tumor vaccines in an attempt to boost the function of the immune system in patients with cancer. Some tumors in experimental animals and probably also in the humans have tumor-specific antigens against which immune responses can be induced that can cause tumor injection (i.e., tumor rejection antigens). Antigens encoded by viral genes or mutated cellular oncogenes are of particular interest in view of both their specificity and association with a neoplastic phenotype. If antigens that are encoded by mutated suppressor genes can be found, they may likewise qualify as good targets. Essentially all neoplasms express normal differentiation antigens of various degrees of tumor selectivity. Some of these may induce immune responses which may be more damaging to tumor cells because they express more antigen than normal cells do. Even those neoplasms that express rejection antigens generally grow in immunocompetent hosts. Failure of tumors to provide proper costimulation of the immune response appears to be at least partially responsible.

A review of the relationship between the degree of immunosuppression and malignancy in patients on immunosuppressive drugs or immunosuppressed by human immunodeficiency virus (HIV) infection, postoperative blood transfusions, or pregnancy provides the most convincing evidence of the importance of intact T cell immunity in resistance to cancer. In addition, correlation of decreased MHC antigen expression with increased malignancy and diminished tumor-infiltrating lymphocyte generation provides the most convincing evidence that one factor necessary to ensure survival of most spontaneous tumors is mutation, which enables tumor cells to escape rejection by cytotoxic T lymphocytes (CTL) (OLIVER 1995). Other factors that may limit effective tumor immunity may have to do with T cell tolerance, clonal deletion, anergy, ignorance, and tumor-induced immunosuppression. Most modalities of conventional cancer therapy such as surgery, radiotherapy, and chemotherapy have suppressive effects on immune responses. Since defects of MHC antigen expression are less pronounced in early cancer and since the tumor burden is still relatively low, many ASI immunotherapy studies use active vaccination as adjuvant treatment in combination with surgery in high-risk groups of cancer patients.

5.1 Viral Vaccines

Viruses that cause cancer are subject to the same rules of immunology as those that cause acute infectious disease. Hence, present viral vaccines and those to be developed in the future will be expected to prevent the infections that lead to an increasing number and variety of cancers in humans and animals. Examples of vaccines which are effective against animal cancers caused by viruses are SV-40 (irradiated tumor preparations), Marek's disease virus (infected avian embryonic cells or cell-free dried virus), feline leukemia, and simian immunodeficiency disease virus. Examples of licensed and experimental vaccines against established or putative cancer-associated viruses in humans include hepatitis B (plasma-derived and recombinant, licensed), adenovirus (live and killed, licensed), retrovirus HIV-1 and HIV-2 acquired immunodeficiency syndrome (AIDS), human T lymphotropic virus (HTLV-1 and HTLV-2 leukemia), Epstein-Barr virus, papilloma virus, and hepatitis C.

Although viruses may cause no more than 10%–15% of neoplasms in humans, their role as primary factors and cofactors in cancer is being increasingly recognized and their importance emphasized. The importance of and potential for vaccines against cancer is illustrated by the remarkable sucess of a vaccine against hepatitis B, the world's first vaccine against cancer in man (HILLEMAN 1993).

5.2 Virus-Infected Tumor Cell Vaccines

One way of increasing tumor immunogenicity is to infect cells with viruses, such as influenza, vaccina, or Newcastle disease virus (NDV). Tumor antigen-specific lymphocytes that would recognize tumor antigens on such virus-infected tumor cells or in viral oncolysates might become activated and recruit cells that are involved in the defense against viruses. This is an approach that we in Heidelberg have developed over the last 10 years using NDV. This virus is well suited for application in cancer patients because it has few side effects, causes no neurotropic symptoms, and has in itself a number of antineoplastic as well as immune-stimulating properties (SCHIRRMACHER et al. 1993).

5.3 Cytokine Gene-Transfected Tumor Cell Vaccines and Other Methods of Immunogene Therapy

Increased knowledge in basic immunology has led to a variety of innovative and imaginative approaches for tumor-specific immunotherapy. One of these approaches is based on the premise that tumor cells do not normally stimulate an effective tumor-specific immune response, because they do not efficiently present tumor antigens to the relevant lymphocytes. To overcome such inadequate antigen presentation, it has been hypothesized that tumor cells can be genetically engineered to present tumor peptides directly to T lymphocytes in a way similar to antigen-presenting cells (APC) (ÖSTRAND-ROSENBERG 1994). In support of this it was shown that interleukin (IL)-2 transfection into tumor cells leads to local production of cytokine which bypasses T-helper function in the generation of an antitumor response (FEARON et al. 1990). Many cytokine genes have now been introduced in tumor cells, including IL-1_{α}, IL-2, IL-4, IL-5, IL-6, IL-7, IL-12, granulocyte-macrophage colony-stimulating factor (GM-CSF), IFN-α, IFN-γ, and tumor necrosis factor (TNF) (COLOMBO and FORNI 1994). These cytokine expression strategies are based on the hypothesis that the tumor cells themselves de facto present endogenously synthesized tumor peptides in the context of their MHC class I molecules and would therefore be capable of stimulating T cell responses if helper cytokines are provided. However, the target cells for the cytokine-transfected tumor cells are diverse, ranging from CD8 CTL to B lymphocytes, NK cells, eosinophils, and others. Vaccination with tumor cells that have been transduced with several of these cytokine genes does not result in systemic immunity, indicating that in some cases effective antigen presentation to T cells does not occur. If systemic antitumor immunity is induced, however, it is at least partially mediated by CD8 T cells.

A major direction of current gene therapy of cancer is directed towards generation of cancer vaccines. It involves the use of established cell lines, derived from either cancerous or normal cells, into which genes of interest are transferred. This can be achieved by direct DNA transfer or by virus injection using, for instance, retroviral or adenoviral gene expression vectors. The gene-transfected cell lines are then inactivated, most frequently by γ irradiation, and injected either alone or together with autologous patient-derived cancer cells as vaccines back into the patients in order to achieve active specific immunization and immunotherapy (ASI) effect (HELLSTRÖM and HELLSTRÖM 1993).

Whole cell vaccines can be distinguished from vaccines consisting of soluble tumor-associated antigens. Among the whole cell vaccines we can further distinguish between autologous and allogeneic vaccines and between unmodified, virus-infected, or genetically engineered vaccines. Clinical studies are also being carried out with tumor cell lysate vaccines, shed antigen vaccines, ganglioside vaccines, and anti-idiotype vaccines. Many of these include approaches for potentiating vaccine immunogenicity by using novel adjuvants or delivery systems. Interesting new techniques include the use of tumor-derived peptide-loaded heat-shock proteins (SHRIVASTAVA and UDONO 1994) or the DNA gun (KELLER et al. 1995), which allows particle-mediated gene transfer (PMGT) into tumors.

There is optimism that procedures will be developed that will make it possible to induce long-lasting immune responses to tumor antigens with the primary goal of destroying small foci of tumor cells that remain after other types of therapy.

5.4 Preclinical Studies

Major contributions to the understanding of the principles of tumor vaccination and ASI were made in the L10 guinea pig hepatocarcinoma (HANNA et al. 1982) and in the ESb mouse lymphoma (HEICAPPEL et al. 1986; SCHIRRMACHER et al. 1986; SCHIRRMACHER and VON HOEGEN 1993; SCHIRRMACHER 1992). A number of similarities existed between the optimum protocols developed in these two metastasizing tumors and the therapeutic results obtained. Whereas in the L10 tumor postoperative immunotherapy with irradiated autologous tumor cells admixed with *Bacillus Calmette-Guérin* (BCG) protected against lymph node and lung metastases, in the ESb tumor postoperative treatment with irradiated NDV-infected autologous tumor cells protected against hematogenous metastases. Important variables for optimum therapeutic effects were the time of operation of the primary tumor, the residual disseminated tumor burden, and the dose of virus added to a standard dose of 10^7 irradiated tumor cells (HEICAPPEL et al. 1986). The site of vaccination also appeared to be important: in the L10 guinea pig model, three vaccinations injected intradermally protected against lymphatic spread, whereas in the ESb mose model, intraperitoneal administration of the vaccine had the strongest protective effect against visceral metastases. In both models, postoperative vaccination with inactivated viable tumor cells without viral or bacterial adjuvants had no therapeutic effect.

The efficiency of the antimetastatic effect of vaccination was dependent on the size of the metastases, their vascularization, and the total tumor burden. In both tumor models it was shown that vaccination leads to a local inflammatory reaction which extends to the draining lymph nodes and eventually leads to a systemic immunity and inflammatory reaction at sites of metastases. In the guinea pig, the interconnections between local vaccine reactions and systemic immunity were evaluated by means of removal of the local reaction sites of the skin or of the draining lymph nodes at different times and testing the effects on postoperative survival. The stronger the local reaction, the better the extent of systemic immunity induced. The combination of autologous tumor cells and BCG in the vaccine had a synergistic effect both on the intensity of the local reaction and on the protective immunity. In both models it was documented that the viability of the inactivated tumor cells was of decisive importance (HANNA et al. 1982; SCHIRRMACHER and VON HOEGEN 1993) and that the cells had to be frozen under controlled conditions, otherwise their immunogenicity was greatly decreased. In the ESb tumor system, we found that simple shock freezing/thawing of the tumor cells completely destroyed their immunogenicity with regard to stimulating tumor-specific CTL in mixed lymphocyte–tumor cell (MLTC) cultures. This defect could not be restored by the addition of cytokines, such as IL-1, IL-2, IFN-γ (SCHIRRMACHER and VON HOEGEN 1993). The stronger immunogenicity of intact tumor cells compared with membrane preparations could be due to accessory adhesive and/or signaling molecules.

When we compared the two vaccines in the L10 model, L10/BCG and L10-NDV vaccine were found to have similar efficacy. A combination of the two types of vaccines did not lead to better protective immune effects. BCG induces long-lasting ulcers in animals and patients, whereas NDV does not. Another advantage of NDV may be its more precise localization at the tumor cell surface where immune cell–tumor cell interactions take place. When we analyzed the immune responses of animals inoculated in the pinna with either live ESb or live ESb-NDV cells, we found three levels at which NDV exerted an amplifying effect on the immune response: (1) an increased $CD4^+$ T cell-mediated helper response, (2) an increased frequency of tumor-specific CTL precursors, and (3) an increased stimulatory capacity of NDV-modified ESb cells for activating tumor-specific CTL precursors (CTLP) to mature CTL in vitro (VON HOEGEN et al. 1987, 1988, 1990).

Most experiments were performed with the NDV strain Ulster, which is a nonlytic avirulent strain. This negative stranded RNA virus possesses an external envelope with spike glycoproteins of two types: HN, a glycoprotein with hemagglutinating and neuraminidase activity which binds to ubiquitous sialic acid containing gangliosides on cell surfaces and also increases adhesive function (JURIANZ et al. 1995) and CTL activity (ERTEL et al. 1993), and F, a glycoprotein with membrane-fusing ability. In comparison with influenza virus, NDV is more restricted in its host range and much less hazardous for humans (AHLERT and SCHIRRMACHER 1990).

5.5 Strategies for Further Improvement of Autologous Tumor Vaccines

A therapeutic application of NDV-modified tumor vaccines was effective in the ESb model, but only against micrometastases and only in half of the treated mice (HEICAPPEL et al. 1986). Therefore, we are trying to improve on it. Treatment of ESb tumors with IFN-α was ineffective, but in combination with immune CD8 T cells IFN-α treatment was highly effective (KAIDO et al. 1994). This finding is in line with the observation that IFN-α has a costimulatory effect on CTL activation (VON HOEGEN et al. 1990). In the B16 melanoma model, the therapeutic application of NDV-modified tumor vaccines was more effective than the application of MHC class I-transfected tumor cells, but was ineffective against macrometastases (PLAKSIN et al. 1994).

We recently constructed a single-chain antibody for coupling ligands to tumor cells infected with Virus. A single-chain antibody (scFv) specific for the hemagglutinin-neuraminidase (HN) of NDV is used to anchor costimulatory molecules for T cell activation to NDV-infected autologous tumor cells in a vaccine for ASI. The anti-HN reagent contains a sulfhydryl group near its C terminus for coupling costimulatory proteins followed by five histidines for purification by metal chelate chromatography (IMAC). The single-chain antibody binds specifically to NDV particles with an affinity comparable to that of the parent antibody. The purified scFv and an scFv conjugate were able to bind to cells infected with NDV, as shown by FACS analyses (GOTTER et al. 1995).

One example of a costimulatory molecule that may attach to ATV-NDV via bispecifics is B-7, an immunoglobulin-like cell surface molecule of APC which interacts with CD28 receptors on T cells and thereby supplies costimulatory signals for T cell activation. It was found that for a helper T cell to produce sufficient IL-2 to allow autocrine-driven clonal expansion, so-called "costimulatory" or "accessory" signals are required in addition to T cell receptor ligation by antigen–MHC molecules (Ag-MHC). Activation of the T cell receptor in the presence of costimulatory signals result in T cell clonal expansion and in the induction of effector functions such as the production of lymphokines. By contrast, the interaction of T cells with Ag-MHC in the absence of costimulatory ligand is not a neutral event, but can lead to induced unresponsiveness or even to death by apoptosis.

Since the majority of human cancers are carcinomas derived from epithelial tissues, they are likely not to express B7, since these costimulatory ligands are characteristic only of APC such as activated lymphocytes, macrophages, and dentritic cells. With the help of genetic engineering, it is hoped and expected that carcinoma cell lines will be able to be converted into professional APC by transferring genes or molecules such as B7 for costimulation. Recent discoveries indicate that CD28 and B7 are each members of larger gene families and suggest that the regulation of costimulation is more complex than previously imagined (JUNE et al. 1994). Gene therapy will therefore not be simple, and it is likely that more than one gene is required to turn a cancer cell into an APC: cytokine genes, B7, and cell adhesion molecules (CAM) are just a few of these.

Now that human tumor-associated and peptide epitopes have been identified, new strategies are being developed to use host APC such as dendritic cells or EBV-transformed autologous B cells or macrophages (MUKHERJI et al. 1995) for their presentation during vaccination.

Reagents which have been used in cancer patients for many years for active immunization include intact irradiated autologous tumor cells (HOOVER et al. 1993; BERD et al. 1990), allogeneic fresh or cultured tumor cells of the same histological type or allogeneic tumor cell homogenates (MITCHELL et al. 1990), tumor oncolysates (CASSEL et al. 1983), and soluble tumor antigens (HOLLINSHEAD et al. 1987).

A variety of methods and procedures to modify cancer vaccines (SCHIRRMACHER 1993) have been developed to increase their immunogenicity: chemical and enzymatic modification, xenogenization with virsuses, modification by mutagens, transfection of genes coding for MHC class I antigens, upregulation of the expression of MHC antigens, integrins and tumor-associated antigens (TAA) by in vitro exposure to IFN-γ and/or TNF-α, use of the vaccine together with low doses of IL-1 or IFN-γ, and use of molecularly engineered tumor cells carrying genes for lymphokines. Potentiation of immune stimulation can also be expected with unmodified or modified tumor cells combined with various natural and synthetic immunoadjuvants. In many of these studies with improved modified cancer vaccines, a beneficial effect was reported for a proportion of patients, and the treatment was well tolerated. Various groups reported positive results with active immunization as a postsurgical adjuvant treatment (HOOVER et al. 1993; BERD et al. 1990; CASSEL et al. 1983; WUNDERLICH et al. 1985) or in patients with progressive advanced disease (MITCHELL et al. 1990; HAYASHI et al. 1992).

Because of its low neurotropism, the use of NDV as an antineoplastic agent was proposed in 1965 by CASSEL et al. (1983), who have performed more than 5000 vaccinations in melanoma patients with NDV melanoma oncolysates. This ASI protocol was well tolerated, and significant improvements of more than 50% in 5-year survival rates in comparison to historical controls without ASI were reported (CASSEL et al. 1983) and reproduced in a second unpublished study (CASSEL, personal communication). The authors also observed enhanced inflammatory cell responses in cerebral metastases during concurrent therapy will viral oncolysates (CASSEL et al. 1986). With its potency to induce IFN-α and β, TNF-α, adrenocorticotropic hormone (ACTH), TIMP (tissue inhibitor of metalloproteinases), and heat-shock proteins, NDV attached to tumor cell surfaces could be considered as a potent biological response modifier in a tumor vaccine.

At the site of vaccination (usually the skin), NDV-induced IFN-α and -β may be important initial mediators of lymphocyte migration. In the presence of immune T cells, IFN-γ might account for most of the lymphocyte recruitment related to delayed-type hypersensitivity (DTH). NDV-induced, locally produced cytokines may also facilitate delivery of a costimulatory signal in conjunction with the antigen receptor-mediated signal according to the two-signal pathway of T cell activation (SCHWARTZ 1990). In support of this assumption, we demonstrated that the NDV-mediated potentiation of tumor-specific CTL activity could be inhibited by anti-IFN antibodies. NDV also increases via its hemagglutinin-neuraminidase glycoprotein adhesive host–tumor cell interactions and thereby might increase the binding affinity and/or avidity of interacting antitumor immune lymphocytes (ERTEL et al. 1993). Antigen-specific effector cells activated through vaccination may recirculate and, upon recognition of antigen at the site of metastases, induce antitumor immune responses.

Secretion of IFN-γ seems to be important at this step, since the effectiveness of tumor-infiltrating lymphocytes when adoptively transferred to mice bearing micrometastasis correlated better with their ability to secrete IFN-γ than with their cytotoxicity in vitro.

On the basis of the ASI concept developed in the ESb tumor model, we have established a corresponding vaccination protocol for cancer patients. The vaccine consists of freshly isolated autologous intact tumor cells inactivated by irradiation and modified by coincubation with a small, predetermined amount of NDV (Ulster). In various types of cancer patients, new protocols for postoverative ASI treatment were worked out in either "adjuvant" (SCHIRRMACHER et al. 1989) or "advanced disease" (POMER et al. 1995) situations. Optimum procedures were first developed for the isolation and inactivation of live tumor cells from freshly operated specimens of primary tumor or metastases. This was done for malignant melanoma, hypernephroma, breast carcinoma, ovarian carcinoma, and colon cancer. For preparation of a vaccine, irradiated cells stored in liquid nitrogen were thawed and coincubated at 37°C for 1 h with live or inactivated NDV. The Virus-modified cells were characterized with immunological, biochemical, and electron-microscopical methods. Sterility and other quality controls were included (LIEBRICH et al. 1991). Results from phase I and II studies have been reported (LEHNER et al. 1991; SCHLAG et al. 1992). The first evaluation of clinical response to adjuvant ASI treatment with autologous tumor cell vaccines modified by NDV infection (ATV-NDV) was done in colorectal cancer patients following resection of liver metastases (SCHLAG et al. 1992).

We also treated patients with advanced renal cell carcinoma after nephrectomy with ATV-NDV vaccine, but used it in combination with subcutaneous rl-2 and IFN-α_{2b}. We now report on 4-year survival data of clinical responders (complete and partial remissions CR+PR), nonresponders, and those patients who experienced stable disease (SD) under therapy (POMER et al. 1995). More than 50% of all treated patients profited from this combination therapy by improvements in survival.

Several American groups are performing ASI studies in advanced malignant melanoma patients. Morton and colleagues (HAYASHI et al. 1992) used a combination of low-dose cyclophosphamide treatment plus irradiated, well-defined intact melanoma cells from cell lines. MITCHELL et al. (1990) administered an allogeneic melanoma cell lysate called Theraccine combined with the adjuvant Detox (Ribi, Immunchem Research Hamilton MT) and BERD et al. (1990) use a combination of low-dose cyclophosphamide treatment and irradiated autologous trinitrophenol (TNP)-modified melanoma cells. All three groups report about 15%–29% response rates in their treated patients.

Table 1 summarizes the rationale for using ATV-NDV in cancer patients. Variables of importance for any kind of cancer vaccine, such as specificity, immunogenicity, valency, tolerability, effectiveness, and practicability, are considered. We have already mentioned that in our hands viable tumor cells are superior to cell membranes with respect to activating tumor-specific CTLP. This is corroborated with a defined immunoglobulin idiotype, which served as tumor antigen. Immunization with idiotype-positive viable tumor cells induced protective immunity, whereas immunization with equivalent amounts of idiotype presented in soluble form gave no protection, but rather caused tumor enhancement (GOSH et al. 1990).

We would like to draw attention to possible limitations of cancer vaccines and their use in ASI studies. Limitations may exist with regard to responsiveness of the patient, amount of available material, residual tumor mass and/or recurrent tumor mass, antigenic heterogeneity, and immunoresistance. We are fully aware

Table 1. Rationale for using Newcastle disease virus (NDV)-infected autologous tumor cells (ATV-NDV) as vaccine for postoperative active specific immunotherapy (ASI) treatment

Variable	ATV-NDV	Limitations
Specificity	High: includes individually distinct MHC-associated tumor antigens; closest match to the patient's own cancer	Nonantigenic tumors; proportion of stroma cells and infiltrating host cells
Immunogenicity	High: intact cells, in particular virus-infected cells, superior to subcellular material for T cell responses	Immunoresistant tumor cell variants
Valency	As a polyvalent reagent, such a vaccine is better suited than monovalent reagents to treat heterogenoeous cellular disease such as cancer	Standardization
Tolerability/side effects	Both components well tolerated; side effects very few	Tumor enhancement, induction of suppression
Effectiveness	Local immune reactions: apparent synergistic interaction of ATV and NDV in skin reactivity to the vaccine	Nonresponding patients
	Systemic reactions: increased DTH reactivity to challenge with autologous tumor cells, reactions at sites of residual disease, reduced recurrence rate	Possibly tumor mass
Practicability	Can be prepared within 6–12 h from fresh specimens; feasible for a central laboratory with trained staff; mail service possible within 24 h	Amount of material; general availability

MHC, major histocompatibility complex; DTH, delayed-type hypersensitivity.

that this type of treatment must be combined with other modalities, but this is also true of other immunotherapeutic approaches. As a local form of treatment, ASI is better tolerated by cancer patients than systemic high-dose cytokine treatment or adoptive therapy with lymphokine-activated killer cells or tumor-infiltrating lymphocytes. Intensive research in this promising area of clinical tumor immunology is warranted to increase our strategies for treatment of metastatic disease.

The next generation of biotherapy approaches is likely to involve rational combinations of biologicals. Combination strategies are now "en vogue" and are supported by observations that tumors considered unresponsive even to high doses of individual factors can be responsive to certain combinations at lower doses. Rational combinations of biologicals with existing conventional treatment modalities should also be tried. After all, according to a recent analysis (Desai 1994), the current therapeutic advances that have accrued are due to an intense interaction between various therapeutic disciplines. Evidently what is needed now, at a time when we seem to have reached a plateau, is the inclusion of biological factors and treatment strategies and the integration of these without too much reluctances and resistance in clinical medicine.

A recent editorial (Kenney and Pagano 1994) spoke of a new age for "therapeutic" viruses for cancer treatment and claimed that, because of its relative safety, NDV has great promise for this purpose. It has already been sucessfully applied as a selective oncolytic agent (Lorence et al. 1994), as a nonspecific immunostimulant in advanced human cancer (Castary et al. 1993), and as a costimulatory biological in cancer vaccines for inducing systemic protective immunity against metastases (Ertel et al. 1993). Since NDV infection also leads to genetic modification, introduces a new adhesion molecule, induces locally cytokines, and has a costimulatory effect on T cells, ATV-NDV may be a feasible reagent for ASI induction. More defined reagents could possibly be developed in the future and used for ASI maintenance vaccination.

6 Cellular Therapies

Among the novel biological strategies that have emerged in recent years for the treatment of cancer, we should also mention the cell-based therapies involving transfer of activated tumor-reactive lymphocytes for adoptive cellular immunotherapy (ADI) (Schirrmacher et al. 1995). ADI with lymphokine-activated killer cells or with tumor-infiltrating lymphocytes was reported to have an overall response rate of about 25% in metastatic melanoma and renal cell carcinoma patients (Rosenberg et al. 1993; Osband et al. 1990). In this context, we must also mention the improvement of clinical leukemia treatment when using allogeneic human leukocyte antigen (HLA)-matched bone marrow in comparison to autologous bone marrow transplantation (BMT). This allogeneic graft versus leukemia effect (GvL) is clearly dependent on allogeneic T cells that are transferred together with bone marrow cells. This strategy of cellular immunotherapy has already been further improved by donor lymphocyte infusions with or without IFN-α in the treatment of recurrent leukemia after BMT (Slavin et al. 1992; Kumar et al. 1994).

In contrast to most other cellular therapies that use in vitro cultured lymphocytes stimulated with or without cytokine as effector cells, we have focused our attention on the use of in situ activated immune T lymphocytes for adoptive immunotherapy studies. Thus, peritoneal effector cells (PEC) generated from intrapinna tumor-immunized and intraperitoneally restimulated mice were able to transfer protective immunity in the absence of exogenous cytokines into normal, irradiated, or severe combined immunodeficient (SCID) mice which had been injected either subcutaneously or intravenously with increasing numbers of metastatic tumor cells. Expression of protective immunity was independent of host T cells, since it was similarly expressed in SCID mice and in sublethally irradiated mice. Preincubation in vitro for 24 h of the PEC before adoptive transfer led to a pronounced decrease of protective immunity, while the tumor-specific

CTL activity was maintained or even enhanced. These results demonstrated that protective immunity in vivo requires more than tumor-specific cytotoxic activity and that the in vitro culture of immune cells can change their in vivo functional properties (SCHIRRMACHER and GRIESBACH 1994).

We went on to improve the overall protective immunity of these immune cells by combining the ADI therapy with other treatment modalities. The cellular therapy effects seen with immune PEC could be augmented by host irradiation and by combination with ASI using previously established virus-modified tumor vaccines. In an attempt to study the possible role of cytokines in the system, we tested the effect of neutralizing antibodies to TNF-α and INF-γ on the therapeutic effect. Anti-TNF-α, but not anti-IFN-γ antibody treatment augmented the immunotherapeutic effectiveness of immune PEC-mediated ADI (SCHIRRMACHER 1995).

Using the ESb lymphoma as a model, we were able to describe experimental conditions for obtaining either synergistic or antagonistic effects between host irradiation and T cell-mediated antitumor immunity. Sublethal irradiation of mice during the afferent (induction) phase, but not during the efferent (effector) phase of the antitumor immune response had a strong suppressive effect on protective immunity. The immune system of syngeneic (DBA/2) and allogeneic (B10.D2) mice changed within 24 h after first contact with the ESb tumor cells from radio-sensitivity to radio-resistance. This possibly reflected an active cellular response associated with the change from virgin to an antigen-sensitized (primed) state of T lymphocyte differentiation. T cell depletion experiments revealed that the afferent phase was dependent on both CD4 and CD8 host T lymphocytes, while the efferent phase was mainly CD8 T cell dependent (SCHIRRMACHER and ZANGEMEISTER-WITTKE 1994).

Using in situ activated tumor immune T cells induced in allogeneic, tumor-resistant, MHC-identical but superantigen-different donor mice (B10.D2), we were able to show the specific eradication of micrometastases upon immune cell transfer into syngeneic tumor-bearing DBA/2 mice (SCHIRRMACHER et al. 1991). The B10.D2 antitumor immune Tcells were shown to contain both tumor-specific T cell clones and T-cell clones directed against minor DBA/2 histocompatibility antigens, so that GvL and graft versus host (GvH) effects could be expected. Nevertheless, the GvH reactivity was not lethal, and long-lasting tumor immunity could be established following migration, survival, and establishment of memory cells for tumor protectivity (ZANGEMEISTER-WITTKE and SCHIRRMACHER 1991).

Recently, we were able to demonstrate the unique efficiency of this model system in rejecting even late-stage disease of advanced metastasized cancer. Systemic immune cell transfer into 5-Gy-irradiated DBA/2 mice bearing syngeneic tumors established for up to 4 weeks and macrometastases in livers and kidneys led to massive infiltration of tumor tissues by CD4 and CD8 donor T lymphocytes. Upon interaction of immune CD4 donor T cells with host APC in synergy with immune CD8 donor T cells attacking the tumor cells directly, primary tumors (1.5 cm in diameter) were encapsulated and rejected from the skin, and liver metastases were eradicated (SCHIRRMACHER et al. 1995).

Further studies with separated and enriched CD4 or CD8 donor immune T cells alone or in combination revealed a strong synergistic effect of CD4 and CD8 immune T cells both in primary tumor rejection and in ADI-mediated prolongation of survival. Transfer of either CD4 or CD8 immune cells alone prevented further progressive tumor growth, but did not lead to regression of primary tumors. Combined transfer of CD8 and CD4 immune cells caused complete tumor regression, although somewhat delayed in comparison to immune spleen cell transfer (SCHIRRMACHER et al. 1995).

Synergistic effects of CD4 and CD8 immune T cells were also seen in the selective eradication of ESb-L *lac-Z* and ESb-MP liver metastases and in their targeting by the transferred immune cells. Quantitative data on tumor-infiltrating lymphocytes in liver metastases were derived from immunohistology by counting in representative liver sections the number of tumor cells and tumor-infiltrating lymphocytes at different time points under ADI. Independent of the size and cellularity of metastases, the percentage of tumor-infiltrating lymphocytes of both the CD8 and CD4 type was significantly higher in ADI-treated than in nontreated animals. Synergism between CD4 and CD8 immune T cells with regard to targeting to liver metastases was seen within the first week after treatment (SCHIRRMACHER et al. 1995).

These model studies using freshly in situ achieved antitumor immune cells from MHC congenic mice revealed strong GvL effects in the presence of tolerable GvH effects. These model studies may thus be of great relevance for further understanding and improving leukemia treatment following BMT. The reasons for the predominantly protective effects of this ADI protocol and a possible downregulated GvH reactivity are not yet entirely clear. Since we observed a late chronic GvH reaction in some experiments a contribution of minor histocompatibility antigen-specific T cells and thus of GvH reactivity in this anti tumor effect cannot be excluded. ADI by syngeneic in situ activated PEC also works in this model, but is not as effective as the allogeneic cells from MHC congenic mice. Preliminary findings indicate that mouse mammary tumor virus (MMTV)-derived viral superantigens which are present in DBA/2 but not B10.D2 mice and which delete T cells with certain v-β-chains from the repertoire of DBA/2 are associated with ESb lymphoma sensitivity of resistance of these strains.

Our studies on cellular therapies suggest that better GvL effects can be expected in clinical treatment if the donor T cells are presensitized against the tumor of the host. In principle this can be achieved either in vitro or in vivo. Since in vitro culture of T cells could affect their homing properties, possibilities for in vivo immunization of T cells for clinical ADI studies should not be excluded. All the experience that has been gained with tumor vaccines could be exploited for such application. It is to be expected that such immune T cells can be transferred into patients following allogeneic BMT because the latter procedure may have already established a lymphoid chimerism between donor and host.

7 Conclusions and Future Perspectives

Various approaches for immunotherapy of metastases have been described. Regardless of whether they are based on antibodies, antigens (tumor vaccines), or immune cells, they all have in common the aim of being more specific and selective than cytostatics or radiotherapy.

One of the main problems with immunotherapy is still the selective recognition of metastases and an efficient targeting. We were able to compare targeting of liver metastases by immune T cells with that of an mAb which reacts with a differentiation antigen (CD2) on ESb-MP lymphoma cells and which does not react with normal liver tissue. While the antibody targeted very well to the lymphoma liver metastases, especially after debulking of the primary tumor (SCHMID et al. 1993), a quantitative autoradiographic evaluation revealed significant heterogeneity in radioactivity retention with some metastases which remained nontargeted. Heterogeneity of uptake was even more pronounced in the primary tumor. The findings with immune T cells from tumor-resistant MHC-compatible strains clearly demonstrates the superiority of the T cell-targeting approach. This works for the primary tumor and for metastases of different sizes, macrometastases, foci, and small clusters down to single, disseminated cells. The superiority of T cell targeting in comparison to antibody targeting is likely due to the inherent active motility of the cells.

While many immunotherapy studies have reported on prophylactic immunization effects obtained with modified tumor vaccines, those that dealt with therapeutic applications usually reported on effects between 3 and 9 days after tumor implantation, while advanced diseased stages were usually not affected. Our ASI studies with whole cell vaccines showed that intact live cells are more immunogenic than cell lysates. Prophylactic immunization with live, nontumorigenic, IL-4-transfected Eb lymphoma cells induced long-term tumor-specific immunological protection. This correlated with the persistence of dormant tumor cells in the bone marrow (KHAZAIE et al. 1994), thus leading to persistent antigenic stimulation of the immune system. These findings appear to have a clinical correlate, because dormant tumor cells have also been detected in the bone marrow of many carcinoma patients (LINDEMANN et al. 1992). Future protocols may involve the use of live tumor cell vaccines incorporating a "suicide gene" which would allow their destruction if necessary.

Magnetic resonance spectroscopy data revealed that one of the early effects of ADI immune cell transfer was a change in the pH milieu in the primary tumor. Immune T cell transfer also influenced the microenvironment in the target organ liver. Both immunotherapy approaches, ASI and ADI, were able to activate liver sinusoidal endothelial cells and liver-derived Kupffer cells to become and remain activated and to produce NO and other substances and thus to contribute to antimetastatic resistance of the host (ROCHA et al. 1995a; UMANSKY et al., submitted). Further conclusions from the ADI studies were that CD4 immune T cells alone and CD8 immune T cells alone conferred only partial protection and resulted in stable disease, whereas when transferred together there was a strong synergistic interaction. CD4 plus CD8 immune T cells were able to reject large primary tumors from the skin and to completely eradicate metastases, independent of size and heterogeneity.

New tumor T cell targets will be discovered through the exploration of known genetic changes in tumors and their associated influences. New techniques have already been developed for the identification and cloning of genes for T cell targets and the identification of T cell epitopes by sequencing peptides eluted from HLA molecules on tumor cells. When combined with our increasing ability to manipulate the immune system, these advances must surely provide hope for effective immunotherapy and vaccination against cancer. As such approaches to T cell immunotherapy develop, the ability of a tumor cell to evade immunity by variation in HLA expression (or by other means) may prove to be a formidable barrier to effective cancer immunotherapy. It is thus extremly important that clinical experiments are well thought out and carefully designed. Immune monitoring should include measurements of the immune response and characterization of HLA expression on tumors and should be accompained by HLA and other immunogenic typing of the cancer patient. Only in this way will immunotherapy of metastases eventually receive the attention that it undoubtedly deserves.

References

Abrams PGA (1995) Antibody-based approaches in cancer therapy: monoclonal antibodies as pharmaceuticals. In: Eccles S (ed) The treatment of cancer: beyond chemotherapy. Documentation of conference in London, 13/14 March 1995

Ahlert T, Schirrmacher V (1990) Isolation of a human melanoma adapter Newcastle disease virus mutant with highly selective replication patterns. Cancer Res 50: 5962–5968

Ali M, Lemoine N R (1995) Gene therapy for cancer. Abstract at the 8th AEK Symposium, Heidelberg, 29–31 March 1995

Altevogt P, Leidig S, Heckl-Oestreicher B (1984) Resistance of metastatic tumor variants to tumor specific cytotoxic T lymphocytes not due to defects in expression of restricting MHC molecules in mice. Cancer Res 44: 5305–5313

Alvarez-Diez TM, Polhironis J, Reilly RM (1995) Pre-targeted imaging of LS174T colon cancer xenografts in nude mice with streptavidin (SA)-CC49 monoclonal antibody (Mab) and In-111-biotin. Proc Am Assoc Cancer Res 36: 616

Baskar S, Glimcher L, Nabavi N, Jones RT, Östrand-Rosenberg S (1995) Major histocompatibility complex class $II^+B7\text{-}1^+$ tumor cells are potent vaccines for stimulating tumor rejection in tumor-bearing mice. J Exp Med 181: 619–629

Berd D, Maguire HC, McCue P, Mastrangelo MJ (1990) Treatment of metastatic melanoma with an autologous tumor-cell vaccine: Clinical and immunologic results in 64 patients. J Clin Oncol 8: 1858–1867

Berd D, Takami S, Lattime EC, Maguire HC Jr, Hamilton R, Mastrangelo MJ (1995) Assessing the host response to haptenized melanoma vaccine. Proc Am Assoc Cancer Res 36: 677–678

Bosslet K, Schirrmacher V (1981) Escape of metastasizing clonal tumor cell variants from tumor-specific cytolytic T lymphocytes. J Exp Med 154: 557–561

Bosslet K, Schirrmacher V (1982) High-frequency generation of new immunoresistant tumor variants during metastasis of a cloned murine tumor line (ESb). Int J Cancer 29: 195–202

Bosslet K, Czech J, Opper M, Gerken M, Hoffmann D (1995) Tumor selective prodrug activation by b-glucuronidase-mediated catalysis. Abstract at the 8th AEK Symposium, Heidelberg, 29–31 March 1995

Cassel WA, Murray DR, Phillips HS (1983) A phase II study on the postsurgical management of stage II malignant melanoma with a Newcastle disease virus oncolysate. Cancer 57: 1302–1312
Cassel WA, Weidenheim KM, Campbell WG, Murray DR (1986) Malignant melanoma: inflammatory mononuclear cell infiltrates in cerebral metastases during concurrent therapy with viral oncolysate. Cancer 57: 1302–1312
Colombo M, Forni G (1994) Cytokine gene transfer in tumor inhibition and tumor therapy: where are we now? Immunol Today 15: 48–51
Csatary LK, Eckhard S et al. (1993) Attenuated veterinary virus vaccine for the treatment of cancer. Cancer Detec Prevent 17: 619–627
Desai PB (1994) Guest editorial. Understanding the biology of cancer: has this any impact on treatment? J Cancer Res Clin Oncol 120: 193–194
Ertel C, Millar NS, Emmerson PT, Schirrmacher V, von Hoegen P (1993) Viral hemagglutinin augments peptide specific cytotoxic T-cell responses. Eur J Immunol 23: 2592–2596
Fearon E, Pardoll D, Itaya T, Golumbek P, Levisky H, Simons J, Karasuyama H, Vogelstein B, Frost P (1990) IL-2 production by tumor cells bypasses T helper function in the generation of an antitumor response. Cell 60: 397–403
Ferber I, Scönrich G, Schenkel J, Mellor A, Hämmerling GJ, Arnold B (1994) Levels of peripheral T cell tolerance induced by different doses of tolerogen. Science 263: 674–676
Fidler IJ (1990) Critical factors in the biology of human cancer metastasis: 28th G.H.A. Clowes Memorial Award Lecture. Cancer Res 50: 6130–6138
Finke JH, Zea AH, Stanley J, Longo DL, Mizoguchi H, Tubbs RR, Wiltrout RH, O'Shea JJ, Kudoh S, Klein E, Bukowski RM, Ochoa AC (1993) Loss of T-cell receptor ζ chain and $p56^{lck}$ in T-cells infiltrating human renal cell carcinoma. Cancer Res 53: 5613–5616
Gosh P, Sica A, Young HA, Ye J, Franco JL, Wiltrout RH, Longo DL, Rice NR, Komschlies KL (1994) Alterations in NFkB/Rel family proteins in splenic T-cells from tumor-bearing mice and reversal following therapy. Cancer Res 54: 2969–2972
Gosh SK, White LM, Gosh R, Baukert RB (1990) Vaccination with membrane-associated idiotype provides greater and more prolonged protection of animals from tumor challenge than the soluble form of idiotype. J Immunol 145: 365
Gotter S, Kipriyanov S, Haas C, Dübel S, Breitling F, Khazaie K, Schirrmacher V, Little M (1995) A single chain antibody for coupling tumor cells treated with Newcastle disease virus to effector molecules. Tumor Targeting 1: 107–114
Gullino PM (1991) Microenvironment and angiogenic response. Angiogenesis, International Symposium St Gallen, 13–15 March, abstract 136
Hanna MG, Pollack VA, Peters LC, Hoover HC (1982) Active specific immunotherapy of established micrometastases with BCG plus tumor cell vaccines. Cancer 49: 659–664
Hayashi Y, Hoon DSB, Park MS, Terasaki PI, Foshag LJ, Morton DL (1992) Induction of CD4+ cytotoxic T cells by sensitization with allogeneic melanomas bearing shared or cross-reactive HLA-A. Cell Immunol 139: 411–425
Heicappel R, Schirrmacher V, von Hoegen P, Ahlert T, Appelhans B (1986) Prevention of metastatic spread by postoperative immunotherapy with virally modified autologous tumor cells. Int J Cancer 37: 569–577
Hellström I, Hellström KE (1993) Tumor immunology: an overview. Specific immunotherapy of cancer with vaccines. Ann NY Acad Sci 690: 6–18
Heppner GH (1989) Tumor cell societies. J Natl Cancer Inst 81: 648–649
Hilleman MR (1993) The promise and the reality of viral vaccines against cancer. In: Bystryn IC, Ferrone S, Livingstone P (eds) Specific immunotherapy of cancer with vaccines.
Hollinshead S, Stewart THM, Takita H, Dalbow M, Concannon J (1987) Adjuvant specific active lung cancer immunotherapy trials. Cancer 60: 1249–1262
Hoover HC, Brandhorst JS, Peters LC, Srudyke MG, Takehita Y, Madariagy I, Muenz LR, Hanna MG Jr (1993) Adjuvant active specific immunotherapy for human colorectal cancer. 6.5 year median follow-up of a phase III prospective randomized trial. J Clin Oncol 11: 390–399
Ingber D (1991) Extracellular matrix and cell shape: potential control points for inhibition of angiogenesis. J Cell Biol 47: 236–241
Ingber DE, Folkman J (1989) How does extracellular matrix control capillary morphogenesis? Cell 58: 803–805
Jain RK (1990) Vascular and interstitial barriers to delivery of therapeutic agents in tumors. Cancer Metastasis Rev 9: 253–266
Janeway CA(1989) A primitive immune system. Nature 341: 108
June CH, Gluesone JA, Nadler LM, Thompson CB (1994) The B7 and CD28 receptor families. Immunol Today 15: 321–331

Jurianz K, Haas C, Hubbe M, Enel C, Brunner G, Alterogt P, Schirrmacher V, von Hoegen P (1995) Adhesive function of Newcastle disease virus hemagglutinin in tumor-host interaction. Int J Oncol 7: 539–545

Kaido T, Maury C, Schirrmacher V, Gresser I (1994) Successful immunotherapy of the highly metastatic murine ESb lymphoma with sensitized $CD8^+$ T-cells and Ifn-α/β. Int J Cancer 57: 538–543

Kaminski M, Auerbach R (1988) Tumor cells are protected from NK-cell-mediated lysis by adhesion to endothelial cells. Int J Cancer 41: 847–849

Keller ET, Burkholder JK, Shi F, Pugh TD, McCabe D, Malter JS, MacEwen EG, Yang N, Ershler WB (1995) Particle-mediated cytokine gene transfer into canine oral mucosa and epidermis in vivo. Proc Am Assoc Cancer Res 36: 418

Kenney S, Pagano JS (1994) Editorial. Viruses as oncolytic agents: a new age for therapeutic viruses? J Natl Cancer Inst 86: 185

Kerbel RS (1990) Growth dominance of the metastatic cancer cell: cellular and molecular aspects. Adv Cancer Res 55: 87–132

Khazaie K, Prifti S, Beckhove P, Griesbach A, Russel S, Collins M, Schirrmacher V (1994) Persistence of dormant tumor-cells in the bone marrow of tumor-cell-vaccinated mice correlates with long term immunological protection. Proc Natl Acad Sci USA 91: 7430–7434

Kotera Y, Fontenot JD, Pecher G, Methgar RS, Finn OJ (1994) Humoral immunity against a tandem repeat epitope of human mucin MUC-1 in sera from breast, pancreatic and colon cancer patients. Cancer Res 54: 2856–2860

Krüger A, Schirrmacher V, von Hoegen P (1994a) Scattered micrometastasis visualized at the single cell level: detection and re-isolation of lac Z labeled metastasizing lymhoma cells. Int J Cancer 58: 275–284

Krüger A, Umansky V, Rocha M, Hacker JJ, Schirrmacher V, von Hoegen P (1994b) Pattern and load of spontaneous liver metastasis dependent on host immune status studied with lacZ transduced lymphoma. Blood 84(9): 3166–3174

Kumar L (1994) Leukemia: management of relapse after allogeneic bone marrow transplantation. J Clin Oncol 12: 1710–1717

Lehner B, Schlag P, Liebrich W, Schirrmacher V (1991) Postoperative active specific immunization in curatively respected colorectal cancer patients with virus-modified autologous tumor cell vaccine. Cancer Immunol Immunother 32: 173–178

Liebrich W, Schlag P, Manasterski M, Lehner B, Stöhr M, Möller P, Schirrmacher V (1991) In vitro and clinical characterization of a Newcastle-disease-virus-modified autologous tumor cell vaccine for treatment of colorectal cancer patients. Eur J Cancer 27: 703–710

Lindemann F, Schlinok G, Dwischedl P, Witte J, Riethmüller G (1992) Prognostic significance of micrometastatic tumor cells in bone marrow of colorectal cancer patients. Lancet 340: 685–689

Liotta LA, Steeg PS, Stetler-Stevenson WG (1991) Cancer metastasis and angiogenesis: an imbalance of positive and nagative regulation. Cell 64: 327–336

Lorence RM, Reichard KW, Katubig BB et al. (1994) Complete regression of human neuroblastoma xenografts in athymic mice after local Newcastle disease virus therapy. J Natl Cancer Inst 86: 1228–1233

Mitchell MS, Hare W, Kempf RA (1990) Active specific immunotherapy for melanoma. J Clin Oncol 8: 856–869

Mizoguchi H, O'Shea JJ, Longo DL, Loeffler CM, McVicar DW, Ochoa AC (1992) Alterations in signal transduction molecules in T lymphocytes from tumor-bearing mice. Science 258: 1795–1797

Mukherji B, Chakraborty NG, Ymasaki S, Okino T, Sporn J, Kurtzman S, Yamase H, Ergin MT (1995) In vivo immunogenicity of mage-1 nonapeptide-pulsed autologous macrophages in melanoma patients. Proc Am Assoc Cancer Res 36: 675–676

Nakagomi H, Petersson M, Magnusson I, Juhlin C, Matsuda M, Mellstedt H, Taupin J-L, Vivierr E, Anderson P, Kiessling R (1993) Decreased expression of the signal-transducting ζ chains in tumor-infiltrating T cells and NK cells of patients with colorectal carcinoma. Cancer Res 53: 5610–5612

Nicolson GL (1988) Organ specificity of tumor metastasis: role of preferential adhesion invasion and growth of malignant cells at specific secondary sites. Cancer Metastasis Rev 7: 143–188

Nicolson GL (1991) Gene expression, cellular diversification and tumor progression to the metastatic phenotype. Bioessays 13: 337–342

Östrand-Rosenberg S (1994) Tumor immunotherapy: the tumor cell as an antigen-presenting cell. Curr Opin Immunol 6: 722–727

Oliver RTD (1995) A new look at tumor immunology and immunotherapy. Biotherapy 9 (2): 119

Osband ME, Lavin PT, Babayan RK, Graham S, Lamm DL, Parker B, Sawaczyck I, Ross S, Krane RJ (1990) Effect of autolymphocyte therapy on survival and quality of life in patients with metastatic renal cell carcinoma. Lancet 335: 994–998

Paget S (1889) Lancet i: 571–573

Pekarek LA, Starr BA, Toledano A, Schreiber H (1995) inhibition of tumor growth by elimination of granulocytes. J Exp Med 181: 435–440

Plaksin D, Progador A, Vadai E, Feldman M, Schirrmacher V, Eisenbach L (1994) Effective anti-metastatic melanoma vaccination with tumor cells transfected with MHC genes and/or infected with Newcastle disease virus (NDV). Int J Cancer 59: 796–801

Pomer S, Schirrmacher V, Thiele R, Löhrke H, Staehler G (1995) Tumor response and 4 year survival data of patients with advanced renal cell carcinoma treated with autologous tumor vaccine and subcutaneous rr-IL-2 and IFN-alpha 2b. Int J Oncol (in press)

Rak JW, Hegmann EJ, Kerbel RS (1993) The role of angiogenesis in tumor progression and metastasis. Adv Mol Cell Biol 7: 205–251

Renner C, Jung W, Sahin U, Denfeld R, Pohl C, Trümper L, Hartmann F, Diehl V, van Lier R, Pfreundschuh M (1994) Cure of xenografted human tumors by bispecifics and human T cells. Science 264: 833

Renschler M, Dower WJ, Levy R (1995) Development of peptide therapy for lymphoma; random peptide libraries on phage for the identification of B cell receptor ligands. Proc Am Assoc Cancer Res 366: 669–670

Riethmüller G, Schneider-Gädicke E, Schlinok G et al. (1994) Randomized trial of monoclonal antibody for adjuvant therapy of resected Duke's C colorectal carcinoma. Lancet 343: 1177–1183

Rocha M, Krüger A, Van Rooijen N, Schirrmacher V, Umansky V (1995a) Liver endothelial cells participate in T cell dependent host resistance to lymphoma metastasis by production of nitric oxide in vivo. Int J Cancer 63: 405–411

Rocha M, Krüger A, Umansky V, von Hoegen P, Maor D, Schirrmacher V (1995b) Dynamic expression-changes in vivo of adhesion and costimulatory molecules determine load and pattern of lymphoma liver metastases (submitted)

Rosenberg SA, Lotze MT, Yang JC, Topalian SL, Chang AE, Schwartzentruber DJ, Aebersold P, Leitman S, Linehan WM, Seipp CA, White DE, Steinberg SM (1993) Prospective randomized trial of high-dose interleukin-2 alone or in conjuction with lymphokine-activated killer cells for the treatment of patients with advanced cancer. J Natl Cancer Inst 85: 622–632

Schirrmacher V (1980) Commentary. Shifts in tumor cell phenotypes induced by signals from the microenvironment: relevance for the immunobiology of cancer metastasis. Immunobiology 157: 89–98

Schirrmacher V (1985) Cancer metastasis: experimental approaches, theoretical concepts and impacts for treatment strategies. Adv Cancer Res 43: 1–73

Schirrmacher V (1992) Immunity and metastasis: in situ activation of protective T cells by virus modified cancer vaccines. Cancer Surv 13: 129–154

Schirrmacher V (1993) Active specific immunotherapy (ASI) – a new modality of cancer treatment involving the patients own immune system. Onkologie 16: 290–296

Schirrmacher V (1994) Tumor-Wirts Interaktionen und Signal-Austausch bei der Metastasierung. In: Macher E, Kolde G, Bröcker EB (eds) Jahrbuch der Dermatologie. Tumoren und Haut. Biermann, Zülpich, pp 53–65

Schirrmacher V (1995) Augmentation of protective immunotherapy against metastatic ESb lymphoma cells by combining immune cell transfer with host irradiation, antigen restimulation and/or anti-TNF-α antibody treatment. Int J Oncol 6: 17–25

Schirrmacher V, Barz D (1986) Characterization of cellular and extracellular plasmamembrane vesicles from a low metastatic lymphoma (Eb) and its high metastatic variant (ESb): inhibitory capacity in cell-cell interaction systems. Biochim Biophys Acta 860: 236–242

Schirrmacher V, Griesbach A (1994) Adoptive transfer of protective immunity against a high metastatic tumor cell variant by small numbers of tumor specific in situ activated peritoneal effector T-cells. Int J Oncol 5: 141–151

Schirrmacher V, von Hoegen P (1993) Importance of tumor cell membrane integrity and viability for CTL activation by cancer vaccines. Vaccine Res 2: 183–196

Schirrmacher V, Zangemeister-Wittke U (1994) γ-Irradiation suppresses T cell mediated protective immunity against a metastatic tumor in the afferent phase of the immune response but enhances it in the efferent phase if given before immune cell transfer. Int J Oncol 4: 335–346

Schirrmacher V, Landolfo S, Zawatzky R, Kirchner H (1981) Immunogenetic studies on the resistance of mice to highly metastatic DBA/2 tumor cell variants II. Influence of minor histocompatibility antigens on tumor resistance, gamma interferon induction and cytotoxic response. Invasion Metastasis 1: 175–194

Schirrmacher V, Fogel M, Rußmann E, Bosslet K, Altevogt P, Beck L (1982) Antigenic variation in cancer metastasis. Immune escape versus immune control. Cancer Metastasis Rev 1: 241–274

Schirrmacher V, Ahlert T, Heicappell R, Appelhans B, von Hoegen P (1986) Successful application of non-oncogenic viruses for anti-metastatic cancer immunotherapy. Cancer Rev 1: 241–272
Schirrmacher V, von Hoegen P, Schlag P, Liebrich W, Lehner B, Schumacher K, Ahlert T, Bastert G (1989) Active specific immunotherapy with autologous tumor cell vaccines modified by Newcastle disease virus: experimental and clinical studies. In: Schirrmacher V, Schwartz-Albiez R (eds) Cancer metastasis. Springer, Berlin Heidelberg New York, pp157–170
Schirrmacher V, von Hoegen P, Griesbach A, Zangemeister-Wittke U (1991) Specific eradication of micrometastases by transfer of tumor immune T-cells from MHC congenic mice. Cancer Immunol Immunother 32: 373–381
Schirrmacher V, Schild HJ, Gückel B, von Hoegen P (1992) Tumor specific CTL response requiring interactions of four different cell types and dual recognition of MHC class I and class II restricted tumor antigens. Immunol Cell Biol 71: 311–326
Schirrmacher V, Schlag P, Liebrich W, Patel B, Stoeck M (1993) Specific immunotherapy of colorectal carcinoma with Newcastle-disease-virus-modified autologous tumor cells prepared from resected liver metastases. Ann NY Acad Sci 690: 364–366
Schirrmacher V, Beckhove P, Krüger A, Rocha M, Umansky V, Fichtner KP, Hull WE, Zangemeister-Wittke U, Griesbach A, Jurianz K, von Hoegen P (1995) Effective immune rejection of advanced metastasized cancer. Int J Oncol 6: 505–521
Schlag P, Manasterski M, Gerneth T, Hohenberger P, Dueck M, Herfarth C, Liebrich W, Schirrmacher V (1992) Active specific immunotherapy with NDV modified autologous tumor cells following liver metastases resection in colorectal cancer. Cancer Immunol Immunother 35: 325–330
Schmid U, Schirrmacher V, Monburg F, Matzku S (1993) Preferential antibody targeting to small lymphoma metastases in the absence of the primary tumor. Eur J Cancer 29A: 217–225
Schönrich G, Kalinke U, Momburg F, Malissen M, Schmitt-Verhulst AM, Malissen B, Hämmerling GJ, Arnold B (1991) Downregulation of T cell receptors on self-reactive T cells as a novel mechanism for extrathymic tolerance induction. Cell 65: 293–304
Schreiber H (1993) Tumor immunology. In: Paul NE (ed) Fundamental immunology, 3rd edn. Raven, New York. 32: 1143–1178
Schwartz RH (1990) A cell culture model for T lymphocyte clonal anergy. Science 258: 1349–1356
Seung S, Urban JL, Schreiber H (1993) A tumor escape variant that has lost one major histocompatibility complex class I restriction element induces specific CD8+ T cells to an antigen that no longer serves as a target. J Exp Med 178: 933–940
Shrivastava PK, Udono H (1994) Heat-shock protein-peptide complexes in cancer immunotherapy. Curr Opin Immunol 6/5: 728–732
Slavin S, Ackerstein A, Weiss L, Nagler A, Reuven O, Naparstek E (1992) Immunotherapy of minimal residual disease by immunocompetent lymphocytes and their activation by cytokines. Cancer Invest 10: 221–227
Springer CJ (1995) Antibody-directed enzyme prodrug therapy (ADEPT) as an approach to cancer treatment. In: Eccles S (ed) The treatment of cancer: beyond chemotherapy. Documentation of conference in London, 13/14 March 1995
Suzuki Y, Suzuki T, Matsunaga M, Matsumoto M (1985) Gangliosides as paramyxovirus receptor. J Biochem 97: 1189–1199
Tallberg T, Tykkä H, Mahlberg K, Haltturen P, Lehtonen T, Kalima T, Sarna S (1985) Active-specific immunotherapy with supportive measures in the treatment of palliatively nephrectomized renal adenocarcinoma. A thrirteen year follow-up study. Eur Urol 811 (4): 233–234
Torre-Amione G, Beauchamp RD, Koeppen H, Park BH, Schreiber H, Moses HL, Rowley DA (1990) A highly immunogenic tumor transfected with a murine transforming growth factor type β1 cDNA escapes immune surveillance. Proc Natl Acad Sci USA 87: 1486–1490
Turk JL, Parker D (1982) Effect of cyclosphosphamide on immunologial control mechanisms. Immunol Rev 65: 99–113
Umansky V, Rocha M, Krüger A, von Hoegen P, Schirrmacher V (1995) In situ activated macrophages are involved in host resistance to lymphoma metastasis by production of nitricoxide. Int J Oncol 7: 33–40
Vaage J (1985) Natural and induced resistance to metastasis from mouse mammary carcinomas. In: Torisu M, Yoshida T (eds) Basic mechanisms and clinical treatment of tumor metastasis. Academic, New York, pp 163–189
Vitetta ES (1995) Monoclonal antibodies as carriers of toxins and as agonists in the therapy of lymphoma. Richard and Hinda Rosenthal Foundation Award Lecture. Proc Am Assoc Cancer Res 36: 646
Von Hoegen P, Altevogt P, Schirrmacher V (1987) New antigens presented on tumor cells can cause immune rejection without influencing the frequency of tumor-specific cytolytic T cells. Cell Immunol 109: 338–348

Von Hoegen P, Weber E, Schirrmacher V (1988) Modification of tumor cells by a low dose of Newcastle disease virus; augmentation of the tumor-specific T cell response in the absence of an anti-viral response. Eur J Immunology 18: 1159–1166

Von Hoegen P, Zawatzky R, Schirrmacher V (1990) Modification of tumor cells by a low dose of Newcastle disease virus III. Potentiation of tumor specific cytolytic T cell activity via induction of interferon alfa, beta. Cell Immunol 126: 80–90

Wunderlich M, Schiessel R, Rainer H, Rauhs R, Kovat E, Schemper M, Dittrich C, Mielesche M, Sedlacek HH (1985) Effect of adjuvant chemo- or immunotherapy on the prognosis of colorectal cancer operated for cure. Br J Surg 72 [Suppl]: 107–110

Ye QW, Mokyr MB (1984) Cyclophosphamide-induced appearance of immunopotentiating T cells in the spleens of mice bearing a large MOPC-315 tumor. Cancer Res 44: 3873–3879

Zangemeister-Wittke U, Schirrmacher V (1991) Transfer of long-lasting tumor immunity by immune T cells from MHC congenic mice: migration, survival and tumor protectivity of cytotoxic donor cells. Biotherapy 3: 319–329

Immunological and Biochemical Modulation in the Treatment of Advanced Colorectal Cancer: Update and Future Directions

R.U. HILGENFELD[1] and E.D. KREUSER[2]

1 Incidence, Epidemiology, and Clinical Relevance of Colorectal Carcinoma

Colorectal carcinoma (CRC) is among the most common cancer types in the western world. With the exception of Japan, it has a high incidence in the developed countries that is still rising (VOGEL and McPHERSON 1989). In the United

[1]Medical Department I, St. Joseph Hospital, Bäumerplan 24, 12101 Berlin, Germany
[2]Department of Hematology/Oncology, University Medical Center Benjamin Franklin, Free University of Berlin, Hindenburgdamm 30, 12200 Berlin, Germany

States of America, it accounts for approximately 150 000 new cases and about 50 000 fatalities per year (Boring et al. 1993). Despite several new treatment modalities, the prognosis depends largely on the extent of disease at the time of diagnosis. While surgery may be curative in earlier stages, about 50% of patients will suffer a relapse and die within 5 years (Fleischer et al. 1989). However, it has been demonstrated that adjuvant strategies can significantly improve survival in certain stages of CRC (Moertel et al. 1990, 1992).

2 Prognostic Factors in Advanced Disease

Over the years, several reports have correlated prognostic factors to response or survival. So far, the only reliable prognostic factor found by several authors to be statistically significant in CRC is the performance status (for a review, Graf et al. 1991). Kemeny and Braun (1983) detected no prognostic factor for response in 220 patients with advanced CRC. However, statistically significant for a poor prognosis were an elevated lactic dehydrogenase (LDH) level, an abnormal carcinoembryonic antigen (CEA) level, a white blood cell count (WBC) of over 10 000, a Karnofsky performance status of less than 60%, and liver metastases. Graf et al. (1991) examined 340 patients with advanced CRC and found the hemoglobin level, the disease-free interval, the number of symptoms, and the performance status to be statistically significant prognostic factors for survival, while the main independent determinant of response was the hemoglobin level. In another retrospective analysis of 324 patients included in four chemotherapy trials, the same group found only a response to chemotherapy of 4 months' duration to be associated with a statistically significant survival advantage (Graf et al. 1994a). When these prognostic factors were tested in an independent population for their relationship to survival by univariate and multivariate analyses to determine their predictive value, the group was able to confirm the importance of the hemoglobin level, the disease-free interval, and the Karnofsky performance status in both analyses. However, they found statistical significance for the number of symptoms only in the univariate analysis (Graf et al. 1994 b). Steinberg et al. (1992) reported baseline albumin and glutamic oxaloacetic transaminase (GOT) to be significant determinants of survival, while the performance status was prognostic for the response. In a recent prospective randomized trial in 142 patients with advanced CRC, a Karnofsky performance status of less than 80%, an elevated alkaline phosphatase and/or GOT, and no response to chemotherapy were all found to have a statistically significant negative impact on survival using multivariate analysis. In this study, patients with a Karnofsky performance status of less than 80%, liver metastases and no response to chemotherapy had only a 10% survival rate after 7 months, while those with a Karnofsky performance status greater than 80%, response to therapy within the first 4 months of treatment, and metastases to the lung only showed a survival rate of 75% after 7 months ($p = 0.0001$; Kreuser et al. 1995).

3 Biochemical Modulation

In recent years, biochemical modulators have been introduced in the treatment of cancer patients. There are noncytotoxic drugs which are supposed to enhance the efficacy of the main cytotoxic agent in various pathways by altering intracellular metabolism (Table 1).

3.1 5-Fluorouracil

Since the time it was first introduced in the treatment of cancer patients in the 1950s, 5-fluorouracil (FU) has remained the most active drug in CRC (Heidelberger et al. 1957; for a survey, see Haskell et al. 1990). While phase I/II trials disclosed a median response rate of 20% with a range of 8%–54% for treatment with FU alone in patients with CRC, this response rate was reduced to 12% in randomized trials (Haskell et al. 1990; Table 2) and to 11% in a recent meta-analysis (Advanced Colorectal Cancer Meta-Analysis Project 1992).

3.2 5-Fluorouracil and Folinic Acid

The hitherto most efficient modulator in CRC is folinic acid (FA). Its use in combination with FU is based on the ability of the FU metabolite fluorodeoxyuridine monophosphate (FdUMP) to inhibit the enzyme thymidilate synthase (TS). This results in the inhibition of thymidilate (dTMP) production and subsequently interferes with DNA synthesis. The ability of FdUMP to inhibit TS is affected by the intracellular level of 5,10-methylenetetrahydrofolate, a reduced folate. The coadministration of FA as an exogenous reduced folate has been shown to enhance the cytotoxicity of FU in several preclinical models (for review Sotos et al. 1994).

Phase II trials in previously untreated patients showed the average median response rate for the combination of FU/FA to be between 20% and 30% (for a

Table 1. Biochemical modulators of 5-fluorouracil (FU) and mechanisms involved

Modulator	Known or presumed mechanism of action
Folinic acid	Increased TS inhibition
Interferon-α	Increased FU metabolism?
Methotrexate	Purine synthesis inhibition
	Increased PRPP levels
	Increased FU incorporation into RNA
Trimetrexate	Purine synthesis inhibition
	Increased PRPP levels
	Increased FU incorporation into RNA
PALA	Pyrimidine synthesis inhibition
	Increased FU incorporation into RNA
Uridine	Selective decreased FU incorporation into RNA?

Modified from Sotos et al. (1994). TS, thymidilate synthase; PALA, *N*-(phosphonacetyl)-L-aspartate; PRPP, phosphoribosylpyrophospate.

Table 2. Randomized trials: 5-fluorouracil (Fu) versus 5-fluorouracil/folinic acid (FA) in colorectal (FU) carcinoma

Author	Year	Patients enrolled[a] (*n*)	Response rate (%)			Time to disease progression (months)			Median survival (months)		
			FU		FU + FA	FU		FU + FA	FU		FU + FA
PETRELLI et al.	1987	52	11	*	48		*				
ERLICHMAN et al.	1988	125	7	*	33	2.9	*	5.1	9.6	*	12.6
POON et al.	1989	429	10	*	43[b]		*		7.7	*	12.0
PETRELLI et al.	1989	343	12	*	30[c]	6.0		8.0	11.5		13.7
VALONE et al.	1989	245	17		19	4.6		5.5	11.5		11.8
CLENDENINN et al.	1990	218	23		33	4.4		5.5	12.7		10.6
DOROSHOW et al.	1990	79	13	*	44	4.0	*	5.5	12.8	(*)	14.4[d]
LABIANCA et al.	1991	182	10	*	22	6.0		6.0	11.0		11.0
DI COSTANZO et al.	1992	119	18		16	5.0		5.3	15.5		13.3
NOBILE et al.	1992	142	8	*	23	3.0		5.0	11.0		11.0
STEINKE et al.	1992	122	20		30	3.0	*	8.5	7.0		10.5
Medium			12.5		33	4.4		5.5	11.5		11.9

[a] For all arms.
[b] Low dose of FA.
[c] High dose of FA.
[d] Crossover study.
*Statistically significant difference.

review, see GREM et al. 1987). In seven out of 11 randomized trials conducted from 1987 to 1990, significantly higher response rates were seen for the combination of FU and FA than for FU alone (Table 2). An analysis of the time to disease progression yielded a statistical significance for five of the trials, but only two out of 11 studies showed a significantly improved survival in patients treated with the combination (Table 2). A meta-analysis of numerous randomized clinical trials comparing FU to a combination of FU/FA also revealed the latter to be associated with better response rates (23% versus 11%; ADVANCED COLORECTAL CANCER META-ANALYSIS PROJECT 1992). However, there was no clear evidence that the modulation of FU by FA also improved time to disease progression or survival.

While these results could not confirm the superiority of FU/FA, POON et al. (1989) reported a considerable weight gain as well as an improvement of the performance status and the symptoms of disease in patients treated with FU/FA, compared to those treated with FU only.

A randomized trial comparing two FU/FA protocols in advanced CRC was published in 1994. A 5-day FU regimen plus low-dose FA was not superior to weekly FU plus high-dose FA with regard to response rates and survival; however, the financial costs and side effects were lower (BUROKER et al. 1994). Several groups investigated the efficacy and toxicity of high-dose FA and high-dose FU as a continuous infusion in patients with advanced CRC.

Using 24-h infusion of 2600 mg FU/m^2 with 500 mg FA/m^2 over 24 h, ARDALAN et al. (1991) reported a response rate of 30% in ten patients. To confirm these results, WEH et al. (1994) conducted a multicenter trial. Patients received 2600 mg

Table 3. 5-Fluorouracil/methotrexate, in colorectal, carcinoma (phase II trials)

Author	Year	Regimen	Interval MTX/5-FU (h)	Patient (*n*)	Response rate (%)
CANTRELL et al.	1982	FU/MTX/FA	1	16	6
AJANI et al.	1985	FU/MTX	1	30	3
PANASCI et al.	1985	FU/MTX/FA	1	25	28
HANSEN et al.	1986	FU/MTX/FA	1	26	19
PYRHÖNEN and KOURI	1992	FU/MTX/FA/EPI	1	85	29
WEINERMAN et al.	1982	FU/MTX/FA	4	29	42
HERRMANN et al.	1984	FU/MTX/FA	7	20	40
LEONE et al.	1986	FU/MTX	20	24	46
LEONE et al.	1992	FU/MTX/FA	20	33	33
GLIMELIUS et al.	1986	FU/MTX/FA	3/23	50	50
KEMENY et al.	1984	FU/MTX	24	43	32
SINNIGE et al.	1990	FU/MTX/FA	24	20	20
ABAD et al.	1991	FU/MTX/FA	24	30	40

FU, 5-fluorouracil; MTX, methotrexate; FA, folinic acid; EPI, epirubicin.

FU/m^2 over 24 h weekly, preceded by 500 mg FA/m^2 as a 1-h infusion. All patients had been pretreated, mostly with FU/FA and/or FU/IFN. In patients who had a partial remission (PR) or stable disease (SD) with an improved clinical condition, therapy was continued until progression occurred. Five out of 57 patients (9%) achieved a PR, while 56% of patients (32 out of 57) remained stable. A total of 81% (30 out of 37) of those showing an overall response (PR/SD) under this regimen had a response before. The median duration of response was 3 months, and the median survival for all patients was 8 months (range, 3–17+ months). There was one toxic death; other side effects included mucositis, nausea, diarrhea, and hand-foot syndrome (WEH et al. 1994).

3.3 Methotrexate

Methotrexate (MTX), an antimetabolite, causes an increased intracellular concentration of phosphoribosylpyrophosphate due to MTX inhibition of purine metabolism, which results in an increased formation of fluorouridyl triphosphate (FUTP) (CADMAN et al. 1980). FUTP is then integrated into RNA instead of the physiologic uridine triphosphate (UTP).

The proper timing of sequential MTX and FU is an important issue. Phase II trials showed response rates of approximately 40% when MTX was applied 4–7 h before FU (Table 3). In randomized trials with FU and MTX versus FU alone, the combination was distinctly advantageous for the response and in some reports also for survival when the interval between MTX and FU application was longer (Table 4). For comparison, a MTX to FU interval of only 1 h resulted in a response rates of 3%–28%, while a 24-h interval yielded responses of 20%–32% of the patients. This trend toward better results with a longer interval was confirmed by a phase III trial comparing a 1-h to a 24-h interval. The arm with the 24-h interval was significantly better with respect to the overall response rate (29% versus

Table 4. 5-Fluorouracil/methotrexate versus 5-fluorouracil in advanced colorectal carcinoma (randomized trials)

Author	Year	Regimen	Interval MTX/5-FU (h)	No. Pat.	RR (%)	Survival (months)
Petrelli et al.	1987	FU/MTX	4	21	5	10
		FU		19	*11	11
		FU/FA		25	*48	12
O'Connell et al.	1990	FU/MTX	7	45	7	7
		FU/FA[a]		49	*33	*14
		FU/FA[b]		42	*33	*13
Herrmann et al.	1992	FU/MTX	7	83	25	13
		FU		76	18	15
Abad et al.	1992	FU/MTX/FA	10	60	25	13
		FU/FA		56	18	13
Delfino et al.	1992	FU/MTX/FA	18	28	*18	8
		FU		33	0	11
Macchiavelli et al.	1990	FU/MTX/FA	20	60	*28	N/A
		FU		58	12	8
NGTATG	1989	FU/MTX/FA	3 + 21	82	*24	*10
		FU		91	3	6
Kemeny et al.	1988	FU/MTX	24	17	6	10
		MOF/Strep		17	35	12
Poon et al.	1989	FU/MTX	24	39	*26	8
		FU/MTX/FA	7	39	21	8
		FU		39	10	8
Valone et al.	1989	FU/MTX	24	96	20	12
		FU		52	17	12
Marsh et al.	1991	FU/MTX/FA	24	76	*29	*15
		FU/MTX/FA	1	83	14.5	12

NGTATG, Nordic Gastrointestinal Tumor Adjuvant Therapy Group; RR, response rate; FU, 5-fluorouracil; MTX, methotrexate; FA, folinic acid.
[a] Low-dose FA.
[b] High-dose FA.
*Statistically significant.

14.5%), the time to disease progression (9.9 months versus 5.9 months), and the median survival (15.3 months versus 11.4 months; Marsh et al. 1991). A recent meta-analysis of phase III trials of sequential MTX and FU versus FU alone found a statistically significant doubling of the response rate as well as a significant survival advantage for the combination (Advanced Colorectal Cancer Meta-Analysis Project 1994).

3.4 Phosphonacetyl-L-Aspartate

In 1979, *N*-(phosphonacetyl)-L-aspartate (PALA) was introduced into clinical trials as a potent inhibitor of de novo pyrimidine biosynthesis. It inhibits the enzyme aspartate carbamyl transferase (ACTase), thus depleting cells of uridine nucleotide pools and subsequently causing an increased uptake of FU by RNA (Martin and Kemeny 1992). Martin et al. (1983) showed in preclinical models that PALA is more effective when given at a lower rather than a higher dose. Clinical

Table 5. 5-fluorouracil+*N*-(phosphonacetyl)-L-aspartate (PALA) in advanced colorectal cancer

Author	Year	PALA dose	Response rate (%)	Patients (*n*)	
				Responders	Total
ARDALAN et al.	1981	HD	11	2	18
WEISS et al.	1982	HD	10	2	21
ERLICHMAN et al.	1982	HD	4	1	28
PRESANT et al.	1983	HD	20	2	10
BUROKER et al.[a]	1985	HD	12	4	34
MUGGIA et al.	1987	HD	24	9	37
ARDALAN et al.	1988	LD	37	7	19
AHLGREN et al.	1990	LD	30	3	10
O'DWYER et al.	1990	LD	43	16	37
KEMENY et al.	1992a	LD	35	15	43

HD, high dose; LD, low dose.
[a]Randomized study.

trials with different dosages of PALA in combination with FU have shown response rates of 10%–43% (median, 26%; Table 5). However, randomized trials comparing FU/PALA to FU alone are still lacking.

4 Immunotherapy for Advanced Colorectal Cancer

Recent years have witnessed the exploration of many new immunotherapeutic strategies in CRC. Immunotherapy can be divided into an active and a passive category. Active immunotherapy should ideally stimulate host antitumor immunity, either cellular or humoral. Numerous specific tumor vaccines have been developed and tested for this purpose (for a review, see FOON 1989). Active specific immunotherapy (ASI) against tumor-associated antigens is thought to result in a host immune response with subsequent selective cytotoxicity for tumor cells (KENNETT et al. 1980). Passive immunotherapy relies on the administration of biologically active agents with innate antitumor properties such as reactivity to monoclonal antibodies and conjugates (FOON 1989).

4.1 Interleukin

Interleukin-2 (rIL-2) is a 15-kDa glycoprotein secreted by T-helper lymphocytes following activation by antigens or mitogens (MORGAN et al. 1976). It enhances natural cytotoxicity, mediated by natural killer (NK) and lymphokine-activated killer (LAK) cells (GRIMM et al. 1983; ORTALDO et al. 1984).

Earlier clinical studies showed response rates of up to 17% in CRC after treatment with rIL-2 alone or in combination with LAK cells (ROSENBERG et al. 1987, 1989). HAMBLIN et al. (1989) reported an enhanced response rate of 29% after combining rIL-2 and FU. YANG et al. (1993) reported a 44% response rate in patients treated with high doses of FU, FA, and rIL-2, but all of the patients had responded well to FU/FA before receiving IL-2. A recent randomized trial in which FU/FA was compared with a combination of FU, FA, and rIL-2 did not disclose significant differences between the two schedules. Response rates were 16% and 17%, respectively (HEYS et al. 1995).

4.2 Interferon

Interferons (IFN) are a group of glycoproteins with common antiviral, immunomodulatory, and antiproliferative functions. The three main types, α-, β-, and γ-interferon, are characterized by their acid stability, cell surface receptors, and primary sequences. While the exact mechanism of action is still unknown, antiproliferative effects against various tumor cell lines and immunostimulatory effects in vivo have been reported (for a review, see ITRI 1992). In the late 1980s, initial clinical trials with FU/IFN showed promising results, especially in patients with advanced CRC who had received no prior chemotherapy (WADLER et al. 1989). The excellent results reported in this initial study could not be confirmed in later trials. In 20 phase I/II trials conducted between 1987 and 1993 with a total of 453 patients, the median response rate of FU/IFN combined was 31% (range, 5%–63%; Table 6). Two randomized trials of FU/FA versus FU/IFN did not show a statistically significant difference of FU/IFN regarding response, time to disease progression, or survival (CORFU–A STUDY GROUP 1995, KREUSER et al. 1995; Table 7). Recently, a prospective randomized trial was performed to determine the efficacy, toxicity, and quality of life with FU and FA compared to FU and IFNα_{2b} (KREUSER et al. 1995). This study, in which a total of 142 patients were treated, revealed a significantly improved quality of life in the FU/FA treatment arm compared to the patients treated with FU/IFN. Quality of life in this study was assessed with the European Organization for Research and Treatment of Cancer (EORTC) QLQ C 30 questionnaire before and during the first 6 months of chemotherapy.

Clinical trials using double modulation of FU by FA and IFN did not show a clear benefit, the median response rate being 28% (Table 7). Several randomized trials comparing the triple combination (FU/FA/IFN) to FU/FA yielded contradictory results, the double modulation showing no clear advantage in the form of higher response rates or longer survival (Table 8).

5 Locoregional Therapy

Approximately 40% of patients with CRC will develop metastases to the liver as the primary organ during the course of their disease (WEISS et al. 1986). Therefore, several treatment approaches directed at liver metastases have been developed.

Table 6. Phase I/II trials with 5-fluorouracil and interferon in colorectal carcinoma

Author	Year	Response rate (%)	Patients (*n*)	
			Responders	Total
RIOS et al.	1987	19	3	16
CLARK et al.	1987	5	1	20
AJANI et al.	1989	7	2	29
DE VECCHIS et al.	1989	25	3	12
WADLER et al.	1989	63	20	32
KEMENY et al.	1990	26	9	34
PAZDUR et al.	1990	35	16	45
DOUILLARD et al.	1991	31	5	16
HUBERMAN et al.	1991	39	13	33
JAIYESIMI et al.	1991	16	4	23
WADLER et al.	1991	42	15	36
BANG et al.	1992	35	6	17
BOTTO et al.	1992	40	19	47
JOHN et al.	1992	33	16	48
PAVLICK et al.	1992	36	4	11
WEH et al.	1992	31	17	55
AIBA et al.	1993	24	6	17
MEEHAN et al.	1993	24	4	17
PAZDUR et al.	1993	31	12	39
Median		31		

Table 7. 5-fluorouracil, folinic acid, and interferon in colorectal carcinoma[a]

Author	Year	Response (*n*)	Rate (%)
LABIANCA et al.	1990	4/15	27
GREM et al.	1991	10/22	45
INOSHITA et al.	1991	14/46	30
KÖHNE-WÖMPNER et al.	1991	3/32	9
PIEDBOIS et al.	1991	7/10[b]	70
		1/6[c]	17
SEYMOUR et al.	1991	17/47	26
TAYLOR et al.	1991	3/14	21
BRUNETTI et al.	1992	3/17	18
CASCINU et al.	1992	23/45	51
VAN HAZEL et al.	1992	7/25	28
KREUSER et al.	1992c	18/62	29
LABIANCA et al.	1992	8/36	22
LEMBERSKY et al.	1992	9/11	82
MOORE et al.	1992	7/25	28
PUNT et al.	1992	5/19	26
SOBRERO et al.	1992	6/44	14
GREM et al.	1993	24/44	54
SINNIGE et al.	1993	17/30	57
AUBER et al.	1994	3/10	33
BUTER et al.	1994	26/49	53
PAZDUR et al.	1994	14/47	30
Median			28

[a]At least 10 patients included.
[b]Previously untreated.
[c]Pretreated.

Table 8. Randomized trials including 5-fluorouracil (FU), folinic acid (FA), and interferon (IFN) in colorectal carcinoma

Author/year	Patients (*n*)	Regimen	Response rate (%)	Time to disease progression (months)	Median survival (months)
DUFOUR et al. 1992	43	FU	18	5.0	N/A
		FU/IFN	24	6.9	N/A
YORK et al. 1993	146	FU	19	3.9+	13.6
		FU/IFN	31	4.6	12.4
CELLERINO et al. 1994	131	FU	14	4.0	12.0
		FU/IFN	6	4.0	12.0
RECCHIA et al. 1992	72	FU/FA	45	7.0	8.2
		FU/FA/IFN	22	6.6	8.2
KOSMIDIS et al. 1993	95	FU/FA	19	5.8	10.0
		FU/FA/IFN	6	4.0	7.0
PENSEL et al. 1993	48	FU/FA	33	8.3	11.1
		FU/FA/IFN	42	9.9	13.2
SEYMOUR et al. 1994	165	FU/FA	30	N/A	10.8
		FU/FA/IFN	31	N/A	10.0
Corfu – A Study Group 1995	449	FU/FA	18	6.2	11.0
		FU/IFN	21	7.3	11.0
KREUSER et al. 1995	142	FU/FA	8	3.7	7.4
		FU/IFN	8	2.2	9.3

N/A, not applicable.

5.1 Hepatic Artery Infusion

Since the 1960s, the delivery of chemotherapy directly to the liver via hepatic artery infusion (HAI) has aroused considerable interest and is now almost exclusively performed with percutaneous intra-arterial catheters and external pumps (SHEPARD et al. 1983). As in systemic therapy, fluoropyrimidines such as FU and fluorodeoxyuridine (FUdR) so far are the most potent cytotoxic drugs regarding response rates (PATT and MAVLIGIT 1991). Numerous phase II trials with FU/FUdR in patients with metastatic CRC have reported response rates of 5%–83% and a median response rate of 47% (Table 9). The survival in these studies varied between 9 and 31 months, and median survival was 14.5 months. Response rates and survival in pretreated patients was comparable. Toxicities of HAI included direct hepatic toxicity, biliary sclerosis, and gastric/duodenal irritation and ulceration. In a recent randomized trial, patients treated with dexamethasone in addition to HAI were able to achieve a greater dose intensity and had improved response rates while showing similar toxicity (KEMENY et al. 1992 b). To enhance the efficacy of HAI, several trials combined modulators such as FA and IFN with the fluoropyrimidines. While response rates in phase II trials were encouraging, the combination of FUdR and FA considerably increased hepatic toxicity(KEMENY et al. 1994; PATT et al. 1994). Five randomized trials conducted between 1987 and 1994 compared locoregional with systemic chemotherapy in patients with metastatic CRC (Table 10). While four of them showed a significantly higher response rate for intra-arterial treatment, only one found a significant improvement of survival for this treatment, and none showed a difference in time to disease progression.

Table 9. Locoregional chemotherapy in advanced colorectal cancer[a]

Author	Year	Drug	Patients enrolled (*n*)	Pretreated (%)	Response rate (%)	Median survival (months)
BALCH et al.	1983	FUdR	50	40	83[d]	26
COHEN et al.	1983	FUdR[c]	50	36	51	N/A
WEISS et al.	1983	FUdR	17	85	29	31
KEMENY et al.	1984b	FUdR	41	43	42	12
NIEDERHUBER et al.	1984	FUdR	70	45	83	25
JOHNSON and RIVKIN	1985	FUdR[c]	40	0	47	12
SCHWARTZ et al.	1985	FUdR	25	N/A	15	18
SHEPARD et al.	1985	FUdR[c]	53	42	32	17
COHEN et al.	1986	FUdR[c]	36	42	70	12
QUAGLIUOLO et al.	1987	FUdR	60	N/A	26	17
ROUGIER et al.	1987	FUdR	16	N/A	53	15
SCHEELE et al.	1991	FUdR[c]	72	0	54	13
LORENZ et al.	1992	FUdR	112		45	16
KEMENY et al.[b]	1993	FUdR	49	100	33	15
		FUdR[c]	45	100	47	19
STAGG et al.	1991	FUdR/5-FU	64	47	50	22
GRAGE et al.	1979	5-FU	31	0	34	10
DENCK et al.	1984	5-FU	24	0	44	14
THIRLWELL et al.	1986	5-FU	68	30	60	13
WILS et al.	1986	5-FU[c]	30	0	30	19
CORTESI et al.	1987	5-FU[c]	12	0	33	11
ARAI et al.	1988	5-FU[c]	33	N/A	64	12
SCHLAG et al.	1988	5-FU	36	0	72	14
GOLDBERG et al.	1990	5-FU	20	0	5	9
ROUGIER et al.	1991	5-FU[c]	48	0	60	14

RR=response rate; N/A=not applicable
[a]Nonrandomized trials.
[b]Randomized trial.
[c]Combination therapy.
[d]Evaluation of response by carcinoembryonic antigen (CEA) value.

Table 10. Locoregional versus systemic chemotherapy in advanced colorectal cancer[a]

Author	Year	Patients enrolled (*n*)	Response rate (%)			Time to disease progression (months)		Median Survival (months)	
			IA		IV	IA	IV	IA	IV
CHANG et al.	1987	64	62	*	17	7	9	18	12
KEMENY et al.	1987	99	50	*	20	9	5	17	12
HOHN et al.[b]	1989	143	42	*	10	13	7	17	16
MARTIN et al.	1990	74	48		21	5	6	12.5	10.5
ROUGIER et al.	1992	166	43	*	9	14.5	5.5	15	11

IA, intra-arterial (hepatic artery infusion, HAI); IV, intravenous, N/A, not applicable.
[a] Randomized trials.
[b] Crossover study.
*Statistically significant.

5.2 Chemoembolization

Chemoembolization is a new technique performed by placing a catheter in the hepatic artery followed by concomitant local delivery of chemotherapy and a vascular occlusion agent. In comparison with HAI, this should result in a more confined and prolonged delivery of chemotherapy. Experience with chemoembolization in metastatic CRC is still limited. Several phase II trials with different combinations of cytotoxic agents such as doxorubicin, mitomycin C, FU, and cisplatin have resulted in response rates of 17%–50%, and median survival was between 11 and 18 months (DANIELS et al. 1992; LANG and BROWN 1993; MARTINELLI et al. 1994).

6 New Drugs in the Treatment of Colorectal Cancer

Due to the unsatisfactory results obtained by treating advanced CRC with the agents described above, several new drugs are currently undergoing preclinical and clinical investigation.

6.1 Echinomycin

Echinomycin is a new bifunctional intercalating agent derived from *Streptomyces echinatus*. In a recent phase II trial in advanced CRC, it showed only modest activity with a response rate of 10% (WADLER et al. 1994). Since treatment was accompanied by massive anaphylactic reactions, premedication with dexamethasone is required.

6.2 Fludarabine

A number of phase I and II studies utilizing different dose levels and schedules of fludarabine phosphate, a new purine antimetabolite, have been reported for several malignancies. However, no activity was noted in advanced CRC (AJANI 1988).

6.3 Gemcitabine

Gemcitabine (2',2'-difluorodeoxycytidine) is a new deoxycytidine analogue. Despite structural and metabolic similarities to cytarabine, it exhibited a different schedule dependency: intermittent rather than daily drug administration resulted in higher cytotoxicity in vitro (HERTEL et al. 1990). Several phase II trials have found this drug to be effective in non-small-cell lung cancer (ANDERSON et al. 1994; NEGORO et al. 1994) and squamous cell cancer of the head and neck (CLAVEL et al.

1992), while in advanced CRC only one PR was reported among 26 patients in a recent phase II trial (ABBRUZZESE et al. 1991).

6.4 Iododoxorubicin

4'-Iodo-4'-deoxydoxorubicin (iododoxorubicin, IDX) is a halogenated doxorubicin analogue with increased lipophilic property and antitumor activity compared to doxorubicin in preclinical trials (MROSS et al. 1990). It also had a significant inhibitory effect on human colon cancer cell lines in vitro (VILLA et al. 1990). Despite this promising preclinical data, a recent German multicenter phase II trial did not find any objective response in 18 previously untreated patients with metastatic CRC (HERRMANN et al. 1995).

6.5 Trimetrexate

Trimetrexate (TMTX), like MTX, is an antifolate inhibitor of the enzyme dihydrofolate reductase (DHFR). In contrast to MTX, it does not compete with FA for cellular uptake and metabolism (ROMANINI et al. 1992) because of several different attributes (for a review, see O'DWYER et al. 1987). In vitro studies showed a cytotoxic effect of TMTX in combination with FU/FA (ROMANINI et al. 1992). The combination of FU, FA, and TMTX was also found to be more cytotoxic than MTX alone. A phase I trial in pretreated patients with colon carcinoma yielded a response rate of 20% (CONTI et al. 1994). Several phase II trials with a combination of TMTX and other modulators are presently under way.

6.6 Docetaxel

Taxol, a diterpene alkaloid isolated from the bark of *Taxus brevifolia*, has a unique mechanism of action. It promotes the formation of microtubule polymers in a cell by reversibly and specifically binding the β-subunit of tubulin. Taxol is administered intravenously by a 3- to 24-h infusion at 3-week intervals and has shown activity in ovarian and breast cancer, but not in CRC (CALDAS et al. 1993; SEIDMAN et al. 1993).

6.7 Tegafur

Tegafur, a prodrug to FU, is hydroxylated and converted to FU by hepatic microsomal enzymes, which may lead to a steadier level of FU in the tumor (AU et al. 1979; ANTTILA et al. 1983). Clinical development of tegafur was discontinued in the United States because of massive gastrointestinal and central nervous system (CNS) side effects without any apparent therapeutic advantage being gained over FU itself when given intravenously (FRIEDMAN and IGNOFFO 1980).

In a recent phase II trial at the M.D. Anderson Cancer Center, a response rate of 42% was achieved with UFT, a mixture of uracil and tegafur at a molar ratio of 4:1, combined with FA given orally (PAZDUR et al. 1994). Serious side effects were not seen with this regimen. There has not yet been any report on survival data.

6.8 Topoisomerase-I Inhibitors

Camptothecin (CPT) is a plant alkaloid obtained from *Camptotheca acuminata* that shows significant antitumor activity against several experimental tumors (GALLO et al. 1971). Reports of severe side effects such as leukopenia, gastrointestinal toxicity, and hemorrhagic cystitis in earlier trials led to abandonment of the drug for further clinical investigation. Camptothecin-11 (CPT-11), a CPT derivative, was first synthesized in 1983 and found to be less toxic than CPT (KUNIMOTO et al. 1987). Preclinical and phase II trials have demonstrated CPT-11 activity in several malignancies (for a review, see SHIMADA et al. 1993; MATSUOKA et al. 1994). A recent phase II study demonstrated some activity in CRC, with a response rate of 27% (17 out of 63 patients); however, no complete response was seen (SHIMADA et al. 1993).

7 Quality of Life

Since treatment of advanced CRC is still palliative, attention has recently shifted to the quality of life as a valid end point for clinical trials. So far several investigators have attempted to assess the quality of life in patients with CRC. SCHEITHAUER et al. (1993) reported that chemotherapy with FU, FA, and cisplatin improved both the quality of life and survival in a small number of patients compared to supportive care alone. The quality of life was measured by using a 22-item self-assessment scale. A Swedish group conducted interviews with patients treated for CRC using a questionnaire (GLIMELIUS et al. 1994). Comparison of two treatment modalities revealed a significant difference in the quality of life, but not in the response or survival.

A more widespread instrument for assessing quality of life over time is the QLQ C 30 questionnaire developed by EORTC, which consists of functional scales (physical state, ability to work, and cognitive, emotional, social, and global quality of life) as well as symptom scales (pain, nausea, vomiting, fatigue, dyspnea, loss of appetite, sleep disturbances, diarrhea, and constipation) (AARONSON et al. 1993). In addition to this validated "core questionnaire," there are several modular supplements specifically designed for certain types of cancer (BERGMAN et al. 1994). In a recent randomized clinical trial comparing FU/FA and FU/IFN in advanced CRC, quality of life was for the first time a major end point of the study. In a total of 142 patients, serial self-assessment revealed a significantly reduced quality of life in the FU/IFN treatment arm. This result was not expected from the physicians' rating conducted at the same time by using the National Cancer Institute (NCI) Common Toxicity Criteria. The compliance with the quality of life self-assessment by the patients was 92%. Evaluability of the EORTC questionnaire was also high (88%), thereby qualifying as a valuable instrument for assessing patients' quality of life in future randomized trials in CRC (KREUSER et al. 1995).

References

Aaronson NK, Ahmedzai S, Bullinger M (1993) The European Organisation for Research and Treatment of Cancer QLQ-C-30: a quality-of-life instrument for use in international clinical trials in oncology. J Natl Cancer Inst 85: 333–338

Abad A, Rosell R, Pastor MC, Barnadas A, Carles J, Ribelles N, Oller B, Armengol M, Salva R, Ribas M (1991) Sequential combination of methotrexate (MTX), 5-fluorouracil (FU), and high-dose folinic acid (FA) in advanced colorectal cancer: double biochemical modulation? Am J Clin Oncol 14: 393–396

Abad A, Garcia P, Grávalos C, Tusquets I, Rosell R, Perez-Manga G, Cortés–Funes H, Fabregat X, Barnadas A (1992) Phase III trial with methotrexate, 5-FU and high-dose leucovorin (MFL) vs 5-FU, leucovorin (FL) vs 5-FU (F) in advanced and metastatic colorectal cancer. Proc Am Soc Clin Oncol 11: 161 (abstr)

Abbruzzese JL, Grunewald R, Weeks EA, Gravel D, Adams T, Nowak B, Mineishi S, Tarassoff P, Satterlee W, Raber MN, Plunkett W (1991) A phase I clinical, plasma, and cellular pharmacology study of gemcitabine. J Clin Oncol 9: 491–498

Advanced Colorectal Cancer Meta-Analysis Project (1992) Modulation of fluorouracil by leucovorin in patients with advanced colorectal cancer: evidence in terms of response rates. J Clin Oncol 10: 896–903

Advanced Colorectal Cancer Meta-Analysis Project (1994) Meta-analysis of randomized trials testing the biochemical modulation of fluorouracil by methotrexate in metastatic colorectal cancer. J Clin Oncol 12: 960–969

Ahlgren JD, Lokich J, Schulz J, McCulloch W, Lynn J, Schein PS (1990) Protracted infusion of 5-fluorouracil (5-FU) with weekly PALA: pilot study in metastatic colorectal carcinoma. Proc Am Soc Clin Oncol 9: 119 (abstr)

Aiba K, Ogawa M, Kuraishi Y, Akazawa S, Sasaki T, Sakata Y, Esaki K (1993) A combination phase II study of 5-fluorouracil and natural type alpha interferon in patients with advanced colorectal adenocarcinoma. Proc Am Soc Clin Oncol 12: 208 (abstr)

Ajani JA, Kanojia MD, Bedikian AY, Korinek JK, Stein SH, Espinoza EG, Bodey GP (1985) Sequential methotrexate and 5-fluorouracil in the primary treatment of metastatic colorectal carcinoma. Am J Clin Oncol 8: 69–71

Ajani JA, Abbruzzese JL, Faintuch JS, Blackburn R, Levin B, Boman BM (1988) Phase II study of fludarabine phosphate in patients with advanced colorectal carcinoma. Invest New Drugs 6: 47–50

Ajani JA, Rios AA, Ende K, Abbruzzese JL, Edwards C, Faintuch JS, Saks S, Gutterman JU, Levin B (1989) Phase I and II studies of the combination of recombinant human interferon-γ and 5-fluorouracil in patients with advanced colorectal carcinoma. J Biol Response Mod 8: 140–146

Anderson H, Lund B, Bach F, Thatcher N, Walling J, Hansen HH (1994) Single-agent activity of weekly gemcitabine in advanced non-small-cell lung cancer: a phase II study. J Clin Oncol 12: 1821–1826

Anttila MI, Slotaniemi EA, Kairaluoma MI, Mokka RE, Sundquist AT (1983) Pharmacokinetics of ftorafur after intravenous and oral administration. Cancer Chemother Pharmacol 10: 150–153

Arai Y, Kido C, Endo T, Miyake Y, Ariyoshi Y, Fukushima M (1988) Hepatic arterial infusion (HAI) chemotherapy for liver metastasis from colon, gastric and breast cancer (ca). Proc Am Soc Clin Oncol 7: 102 (abstr)

Ardalan B, Glazer RI, Kensler TW, Jayaram HN, Pham TV, Macdonald JS, Cooney DA (1981) Synergistic effect of 5-fluorouracil and N-(phosphonacetyl)-L-asparate on cell growth and ribonucleic acid synthesis in a hyman mammary carcinoma. Biochem Pharmacol 30: 2045–2049

Ardalan B, Singh G, Silberman H (1988) A randomized phase I and II study of short-term infusion of high-dose fluorourcacil with or without N-(phosphonacetyl)-L-aspartic acid in patients with advanced pancreatic and colorectal cancers. J Clin Oncol 6: 1053–1058

Ardalan B, Chua L, Tian E, Reddy R, Sridhar K, Benedetto P, Richman S, Legaspi A, Waldman S, Morrell L, Feun L, Savaraj N, Livingstone A (1991) A phase II study of weekly 24-h infusion with high-dose fluorouracil with leucovorin in colorectal carcinoma. J Clin Oncol 9: 625–630

Au JL, Freidman MA, Sadeed W (1979) Pharmacokinetics and metabolism of ftorafur in man. Cancer Treat Rep 63: 343–350

Auber M, Lynch J, High G, Romero P, DeVore R, Crowell E, Rogers J, Raich P (1994) Phase I study of 5-FU, folinic acid (FA), and recombinant interferon alpha-2 (IFN) in patients (pts) with metastatic colorectal carcinoma. Proc Am Soc Clin Oncol 13: 209 (abstr)

Balch CM, Urist MM, McGregor ML (1983) Continuous regional chemotherapy for metastatic colorectal cancer using a totally implantable infusion pump. Am J Surg 145: 285–290

Bang YJ, Kim HT, Suh CI, Kim SY, Heo DS, Kim NK (1992) Phase II study of 5-fluorouracil and recombinant interferon-gamma in patients with metastatic colorectal cancer. Proc Am Soc Clin Oncol 11: 179 (abstr)

Bergman B, Aaronson NK, Ahmedzai S, Kaasa S, Sullivan M (1994) The EORTC-QLQ-LC 13: a modular supplement to the EORTC core quality of life questionnaire (QLQ-C30) for use in lung cancer clinical trials. Eur J Cancer 30: 635–642

Boring CC, Squires TS, Tong T (1993) Cancer statistics 1993. CA Cancer J Clin 43: 7–26

Botto HG, Galvez C, Palao Marco F, Bonamassa M, Fabeiro MA, Marin F, Botto IS (1992) 5-Fluorouracil (5-FU) and interferon alpha 2 b (INF) in advanced colorectal cancer: results in 47 patients. Proc Am Soc Clin Oncol 11: 176 (abstr)

Brunetti I, Falcone A, Bertuccelli M, Giannessi PG, Conte PF (1992) Double 5-fluorouracil (5-FU) modulation with folinic acid (FA) and recombinant alfa 2-B interferon (IFN) in patients (PTS) with metastatic colorectal carcinoma (MCRC). Proc Am Soc Clin Oncol 11: 175 (abstr)

Buroker TR, Moertel CG, Fleming TR, Everson LK, Cullinan SA, Krook JE, Mailliard JA, Marschke RF, Klassen DL, Laurie JA, Moon MD (1985) A controlled evaluation of recent approaches to biochemical modulation on enhancement of 5-fluorouracil therapy in colorectal carcinoma. J Clin Oncol 3: 1624–1631

Buroker TR, O'Connell MJ, Wieand HS, Krook JE, Gerstner JB, Mailliard JA, Schaefer PL, Levitt R, Kardinal CG, Gesme DH Jr (1994) Randomized comparison of two schedules of fluorouracil and leucovorin in the treatment of advanced colorectal cancer. J Clin Oncol 12: 14–20

Buter J, Sinnige HAm, Cats A, Sleijfer DTh, Verschueren RCJ, Mulder NH (1994) Maintenance of remission on 5-fluorouracil (5-FU), leucovorin (LV) and α-2a interferon (IFNα) in advanced colorectal cancer. Proc Am Soc Clin Oncol 13: 193 (abstr)

Cadman E, Heimer R, Banz C (1980) The influence of methotrexate pretreatment on 5-fluorouracil metabolism in L1210 cells. J Biol Chem 256: 1695–1704

Caldas C, McGuire WP 3rd (1993) Praclitaxel (Taxol) therapy in ovarian carcinoma. Semin Oncol 20 [suppl 3]: 50–55

Cantrell JE Jr, Brunet R, Lagarde C, Schein PS, Smith FP (1982) Phase II study of sequential methotrexate-5-FU therapy in advanced measurable colorectal cancer. Cancer Treat Rep 66: 1563–1565

Cascinu S, Fedeli A, Luzi Fedeli L, Catalano G (1992) Double biochemical modulation of 5-fluorouracil by leucovorin and cyclic low dose interferon alpha 2b in advanced colorectal cancer patients. Ann Oncol 3: 489–491

Cellerino R, Antognoli S, Giustini L, Acito L, Marcellini M, Bavosi M, Trivisonne R, Francini G, Del Papa M, Rossi G, Testa E, Latini L, Piga A (1994) A randomised study of fluorouracil (5FU) with or without α-interferon (IFN) in advanced colorectal cancer. Proc Am Soc Clin Oncol 13: 217 (abstr)

Chang AE, Schneider PD, Sugarbaker PH, Simpson C, Culnane M, Steinberg SM (1987) A prospective randomized trial of regional *versus* systemic continuous 5-fluorodeoxyuridine chemotherapy in the treatment of colorectal liver metastases. Ann Surg 206: 685–693

Clark PI, Slevin MK, Reznek RH, Niederle N, Kurschel E, Lundell G, Cedermark B, Fallenius A, Blomgren H, Ohman U, Silfersward C, Theve NO, Wrigley PFM (1987) Two randomized phase II trails of intermittent intravenous versus subcutaneous α-2 interferon alone (trial 1) and in combination with 5-fluorouracil (trial 2) in advanced colorectal cancer. Int J Colorect Dis 2: 26–29

Clavel M, Vermorken JB, Judson I, Wagener T, Hatty S, Wanders J (1992) Gemcitabine is an active drug in patients with squamous cell carcinoma of the head and neck. Proc Am Soc Clin Oncol 11: 249 (abstr)

Clendeninn NJ, Guaspari A, Laufman LR, Brenckman WD Jr and the BW Colorectal Cancer Cooperative Group (1990) A randomized, double-bind trial of fluorouracil (FU) plus placebo (P) vs FU plus high dose oral leucovorin (LV) in colorectal cancer. J Cancer Res Clin Oncol 116(Suppl, Part I): 661 (abstr)
Cohen AM, Kaufman SD, Wood WC, Greenfield AJ (1983) Regional hepatic chemotherapy using an implantable drug infusion pump. Am J Surg 145: 529–533
Cohen AM, Schaeffer N, Higgins J (1986) Treatment of metastatic colorectal cancer with hepatic artery combination chemotherapy. Cancer 57: 1115–1117
Collins KD, Stark GR (1971) Aspartate transcarbamylase. Interaction with the transition state analogue N-(phosphonacetyl)-L-asparate. J Biol Chem 246: 6599–6605
Conti JA, Kemeny N, Seiter K, Goker E, Tong W, McKenzie A, Ragusa K, Bertino JR (1994) Trial of sequential trimetrexate, fluorouracil and high-dose leucovorin in previously treated patients with gastrointestinal carcinoma. J Clin Oncol 12: 695–700
Corfu-A Study Group (1995) Phase III randomized study of two fluorouracil combinations with either interferon alfa-2a or leucovorin in advanced colorectal cancer. J Clin Oncol 13: 921–928
Cortesi E, Aschelter AM, Marchetti P, Padovani A, Pellegrini A (1987) Three drugs treatment for regional intra-arterial therapy liver metastases from colorectal cancer. Proc Am Soc Clin Oncol 6: 95 (abstr)
Daniels J, Pentecost M, Teitelbaum G, Daniels A, Paylunian C, Sternlicht M (1992) Hepatic artery chemoembolization (HAE) for carcinoma of colon using angiostat collagen and cisplatin, mitomycin, and doxorubicin: response, survival, and serum levels. Proc Am Soc Clin Oncol 11: 171 (abstr)
Delfino C, Gaccia G, Manigo O (1992) 5-Fluorouracil (FU) alone versus methotrexate (M)+leucovorin (L)+(FU) in patients (pts) with metastatic colorectal cancer (MCC). Proc Am Soc Clin Oncol 11: 177 (abstr)
Denck H (1984) Ergebnisse einer intraarteriellen intermittierenden Chemotherapie mit 5-FU bei Metastasenleber sowie Tumoren des Gastrointestinal- und Urogenitaltrakts. Oncologie 7: 167–176
De Vecchis L, Nunziata C, Ricci F, Cucchiara G, Picconi A, Bonmassar G, Lagomarsino Caprino M, Bonmassar E (1989) Immunotherapy of advanced colorectal cancer with alpha-interferon and 5-fluorouracil. J Chemother 1: 128–135
Di Costanzo F, Bartolucci R, Calabresi F, Sofra M, Marzola M, Belsanti V, Boni C, Bacchi M (1992) Fluorouracil-alone versus high-dose folinic acid and fluorouracil in advanced colorectal cancer: a randomized trial of the Italian Oncology Group for Clinical Research (GOIRC). Ann Oncol 3: 371–376
Doroshow JH, Bertrand M, Newman E, Mullhauf P, Leong L, Blayney D, Goldberg D, Margolin K, Carr B, Metter G (1987) Preliminary analysis of a randomized comparison of 5-fluorouracil versus 5-fluorouracil and high-dose continuous-infusion folinic acid in disseminated colorectal cancer. NCI Monogr 5: 171–174
Douillard JY, Leborgne J, Danielou JY, Lehur PA, Pottron L, Le Neel JC, Bolze JL, Jutel P (1991) Phase II trial of 5 fluorouracil (5 FU) and recombinant alpha interferon (R IFN) (Intron A^{R}) in metastatic, previously untreated colorectal cancer (CRC). Proc Am Soc Clin Oncol 10: 139 (abstr)
Dufour P, Husseini F, Cure P, Olivier JP, Dumas F, Dreyfuss F, Prevot G, Cholet Ph, Audhuy B, Oberling F (1992) Randomized study of 5-fluorouracil (5-FU) versus 5-FU + alpha 2A interferon (IFN) for metastatic colorectal carcinoma. Proc Am Soc Clin Oncol 11: 192 (abstr)
Erlichman C, Donehower RC, Speyer JL, Klecker R, Chabner BA (1982) Phase I-phase II trial of N-phosphonacetyl-L-aspartic acid given by intravenous infusion and 5-fluorouracil given by bolus injection. J Natl Cancer Inst 68: 227–231
Erlichman C, Fine S, Wang A, Elkahim T (1988) A randomized trial of fluorouracil and folinic acid in patients with metastatic colorectal carcinoma. J Clin Oncol 6: 469–475
Fleischer DE, Goldberg SB, Browning TH, Cooper JN, Friedman E, Goldner FH, Keeffe EB, Smith LE (1989) Detection and surveillance of colorectal cancer. JAMA 261: 580–585
Foon KA (1989) Biological response modifiers: the new immunotherapy. Cancer Res 49: 1621–1639
Friedman MA, Ignoffo RJ (1980) A review of the United States clinical experience of the fluoropyrimidine, Ftorafur (NSC-148958). Cancer Treat Rev 7: 205–213
Gallo RC, Whang-Peng J, Adamson RH (1971) Studies on the antitumor activity, mechanism of action, and cell cycle effects of camptothecin. J Natl Cancer Inst 46: 789–795
Glimelius B, Ginman C, Graffman S, Påhlman L, Ståhle E (1986) Sequential methotrexate-5-FU-leucovorin (MFL) in advanced colorectal cancer. Eur J Cancer Clin Oncol 22: 295–306
Glimelius B, Hoffmann K, Graf W, Påhlman L, Sjöden PO (1994) Quality of life during chemotherapy in patients with symptomatic advanced colorectal cancer. Cancer 73: 556–562
Goldberg JA, Kerr DJ, Wilmott N, McKillop JH, McArdle CS (1990) Regional chemotherapy for colorectal liver metastases: A phase II evaluation of targeted hepatic arterial 5-fluorouracil colorectal liver metastases. Br J Surg 77: 1238–1240

Graf W, Glimelius B, Påhlman L, Bergström R (1991) Determinants of prognosis in advanced colorectal cancer. Eur J Cancer 27: 1119–1123

Graf W, Påhlman L, Bergström R, Glimelius B (1994a) The relationship between an objective response to chemotherapy and survival in advanced colorectal cancer. Br J Cancer 70: 559–563

Graf W, Bergström R, Påhlman L, Glimelius B (1994b) Appraisal of a model for prediction of prognosis in advanced colorectal cancer. Eur J Cancer 30A: 453–457

Grage TB, Vassilopoulos PP, Shingleton WW, Jubert AV, Elias EG, Aust JB, Moss SE (1979) Results of a prospective randomized study of hepatic arterial infusion with 5-fluorouracil versus intravenous 5-fluorouracil in patients with hepatic metastases from colorectal cancer: A Central Oncology Group study. Surgery 86: 550–555

Grem JL, Hoth DF, Hamilton JM, King SA, Leyland-Jones B (1987) Overview of current status and future direction of clinical trials with 5-fluorouracil in combination with folinic acid. Cancer Treat Rep 71: 1249–1264

Grem JL, King SA, O'Dwyer PJ, Leyland-Jones B (1988) Biochemistry and clinical activity of N-(phosphonacetyl)-L-aspartate: a review. Cancer Res 48: 4441–4454

Grem JL, McAtee N, Murphy RF, Balis FM, Steinberg SM, Hamilton JM, Sorensen JM, Sartor O, Kramer BS, Goldstein LJ, Gay LM, Caubo KM, Goldspiel B, Allerga CJ (1991) A pilot study of interferon alfa-2a in combination with fluorouracil plus high-dose leucovorin in metastatic gastrointestinal carcinoma. J Clin Oncol 9: 1811–1820

Grem JL, Jordan E, Robson ME, Binder RA, Hamilton JM, Steinberg SM, Arbuck SG, Beveridge RA, Kales AN, Miller JA, Weiss RB, McAtee N, Chen A, Goldspiel B, Saver E, Allegra CJ (1993) Phase II study of fluorouracil, leucovorin, and interferon alfa-2a in metastatic colorectal carcinoma. J Clin Oncol 11: 1737–1745

Grimm EA, Robb RJ, Roth JA, Neckers LM, Lachman LB, Wilson DJ, Rosenberg SA (1983) Lymphokine-activated killer cell phenomenon III. Evidence that IL2 is sufficient for direct activation of peripheral blood lymphocytes into lymphokine-activated killer cells. J Exp Med 158: 1356–1361

Hamblin TJ, Inzani V, Sadullah S, Stevenson FK, Williamson P, Van der Plas J, Palmer P, Franks CR (1989) A phase-II trial of recombinant interleukin-2 and 5-FU chemotherapy in patients with metastatic colorectal carcinoma. Cancer Treat Rev 16[suppl A]: 163–167

Hansen R, Ritch P, Anderson T (1986) Sequential methotrexate, 5-fluorouracil, and calcium leucovorin in colorectal carcinoma. Am J Clin Oncol 9: 352–354

Haskell CM, Selch MT, Ramming KP (1990) Colon and rectum. In: Haskell DM (ed) Cancer treatment. Saunders, Philadelphia, pp 232–254

Heidelberger C, Chaudhari NK, Danenberg P, Mooren D, Griesbach L, Duschinsky R, Schnitzer RJ, Pleven E, Scheiner J (1957) Fluorinated pyrimidines, a new class of tumor-inhibitory compounds. Nature 179: 663–666

Herrmann R, Spehn J, Beyer JH, von Franqué U, Schmieder A, Holzmann K, Abel U (1984) Sequential methotrexate and 5-fluorouracil: improved response rate in metastatic colorectal cancer. J Clin Oncol 2: 591–594

Herrmann R, Knuth A, Kleeberg U, Middeke H, Korsten FW, Trux FA, Schmieder A, Evers C (1992) Sequential methotrexate and 5-fluorouracil (5-FU) vs 5-FU alone in metastatic colorectal cancer. Ann Oncol 3: 539–543

Herrmann R, Mross K, Kleeberg U, Kreuser ED, Becher R, Edler L, Queißer W, Liati P, Kerpel-Fronius S for the Phase II Study Group of the Association of Medical Oncology of the German Cancer Society (1995) 4'-Iodo-4'-deoxydoxorubicin (IDX) in colorectal cancer – a multicenter phase II study. Onkologie 18: 28–30

Hertel LW, Boder GB, Kroin JS, Rinzel SM, Poore GA, Todd GC, Grindey JB (1990) Evaluation of the antitumor activity of gemcitabine (2',2'-difluoro-2'deoxycytidine). Cancer Res 50: 4417–4422

Heys SD, Eremin O, Ruggeri EM, Pein F, Rainer H, Oskam R, de Peuter RA, Palmer PA, Franks CR (1995) A phase III study of recombinant interleukin-2, 5-fluorouracil and leucovorin versus 5-fluorouracil and leucovorin in patients with unresectable or metastatic colorectal carcinoma. Eur J Cancer 31A: 19–25

Hohn DC, Stagg RJ, Friedman MA, Hannigan JF Jr, Rayner A, Ignoffo RJ, Acord P, Lewis BJ (1989) A randomized trial of continuous intravenous versus hepatic intra-arterial floxuridine in patients with colorectal cancer metastatic to the liver: the Northern California Oncology Group trial. J Clin Oncol 7: 1646–1654

Huberman M, McClay E, Atkins M, Goldsweig H, Bonanno D, Layton M, Teissitore J (1991) Phase II trial of 5 fluorouracil (5FU) and recombinant interferon-α-2a (IFN) in advanced colorectal cancer. Proc Am Soc Clin Oncol 10: 153 (abstr)

Inoshita G, Yalavarthi P, Murthy S, Budd GT, Andresen S, Netaji B, Gibson V, Sergi J, Bukowski RM (1991) Phase I trial of 5-FU, leucovorin (LV) and rHuIFN-alpha2a in metastatic colorectal cancer (CR Ca). Proc Am Soc Clin Oncol 10: 152 (abstr)

Itri LM (1992) The interferons. Cancer 70: 945–950

Jaiyesimi I, Pazdur R, Patt YZ, Ajani J, Abbruzzese JL, Daugherty K, Levin B (1991) Phase II study of recombinant interferon alpha 2a (rIFN) and continuous infusion 5-fluorouracil (5FU) in metastatic colorectal carcinoma. Proc Am Soc Clin Oncol 10: 147 (abstr)

John W, Neefe J, Smith M, Massie M, Cantrell J (1992) 5-Fluorouracil (5-FU) and interferon-α (IFN) in advanced colon cancer: the University of Kentucky experience. Proc Am Soc Clin Oncol 11: 189 (abstr)

Johnson LP, Rivkin SE (1985) The implanted pump in metastatic colorectal cancer of the liver. Risk versus benefit. Am J Surg 149: 595–598

Kemeny NE (1992) Is hepatic infusion of chemotherapy effective treatment for liver metastases? Yes. In: DeVita VT Jr, Hellman S, Rosenberg SA (eds) Important advances in oncology. Lippincott, Philadelphia, pp 207–227

Kemeny N, Braun DW (1983) Prognostic factors in advanced colorectal carcinoma. Am J Med 74: 786–794

Kemeny NE, Ahmed T, Michaelson RA, Harper HD, Yip LC (1984a) Activity of sequential low-dose methotrexate and fluorouracil in advanced colorectal carcinoma: attempt at correlation with tissue and blood levels of phophoribosylpyrophosphate. J Clin Oncol 2: 311–315

Kemeny N, Daly J, Oderman P, Shike M, Chun H, Petroni G, Geller N (1984b) Hepatic artery pump infusion: toxicity and results in patients with metastatic colorectal carcinoma. J Clin Oncol 2: 595–600

Kemeny N, Daly J, Reichman B, Geller N, Botet J, Oderman P (1987) Intrahepatic or systemic infusion of fluorodeoxyridine in patients with liver metastases from colorectal carcinoma. A randomized trial. Ann Intern Med 107: 459–465

Kemeny N, Reichman B, Geller N, Hollander P (1988) Implementation of the group sequential methodology in a randomized trial in metastatic colorectal carcinoma. Am J Clin Oncol 11: 66–72

Kemeny N, Younes A, Seiter K, Kelsen D, Sammarco P, Adams L, Derby S, Nurray P, Houston C (1990) Interferon alpha-2a and 5-fluorouracil for advanced colorectal carcinoma. Assessment of activity and toxicity. Cancer 66: 2470–2475

Kemeny N, Costa JA, Seiter K, Neidzwiecki D, Botet J, Martin D, Costa P, Wiseberg J, McCulloch W (1992a) Biochemical modulation of bolus fluorouracil by PALA in patients with advanced colorectal cancer. J Clin Oncol 10: 747–752

Kemeny N, Seiter K, Niedzwiecki D, Chapman D, Sigurdson E, Cohen A, Botet J, Oderman P (1992b) A randomized trial of intrahepatic infusion of fluorodeoxyuridine with dexamethasome versus fluorodeoxyuridine alone in the treatment of metastic colorectal cancer. Cancer 69: 327–334

Kemeny N, Cohen A, Seiter K, Conti JA, Sigurdson ER, Tao Y, Niedzwiecki D, Botet J, Budd A (1993) Randomized trial of hepatic arterial floxuridine, mitomycin, and carmustine versus floxuridine alone in previously treated patients with liver metastases from colorectal cancer. J Clin Oncol 11: 330–335

Kemeny N, Conti JA, Cohen A, Campana P, Huang Y, Shi WJ, Botet J, Pulliam S, Bertino JR (1994) Phase II study of the hepatic artery floxuridine, leucovorin, and dexamethasone for unresectable liver metastases from colorectal carcinoma. J Clin Oncol 12: 2288–2295

Kennett RH, Jonak ZL, Bechtol (1980) Monoclonal antibodies against human tumor-associated antigens. In: Kennett RH, McKearn TJ, Bechtol KB (eds) Monoclonal antibodies: hybridomas: a new dimension in biological analyses. Plenum, New York, pp 55–168

Köhne-Wömpner CH, Schmoll HJ, Hiddeman W, Knipp H, Wilke H, Bodenstein H, Lohrmann HP, Preiss J, Käufer C, Fischer JT, Balleisen L, Urbanitz D, Schwark J, Schöffski P, Kreuser ED, Graubner M (1991) 5-Fluorouracil (5-FU), leucovorin (LV), alpha-2b interferon (IFN) in advanced colorectal cancer (CC): a phase I/II study. Proc Am Soc Clin Oncol 10: 159 (abstr)

Kosmidis P, Tsavaris N, Skarlos D, Pavlidis N, Beer M, Theocharis D, Samanta D, Briasoulis E, Fountzilas G (1993) Fluorouracil (5-FU) and folinic acid (FA) with or without a 2b-interferon (IFN) in advanced colorectal cancer (ACC). A prospective randomized trial. Proc Am Soc Clin Oncol 12: 211 (abstr)

Kreuser ED, Hilgenfeld RU, Matthias M, Hoksch B, Boewer C, Oldenkott B, Knauf WU, Boese-Landgraf J, Schalhorn A, Zeitz M, Thiel E (1992a) A phase I trial interferon α-2b with folinic acid and 5-fluorouracil administered by 4-hour infusion in metastatic colorectal carcinoma. Semin Oncol 19 [suppl 3]: 197–203

Kreuser ED, Wadler S, Thiel E (1992b) Interactions between cytokines and cytotoxic drugs: putative molecular mechanisms in experimental hematology and oncology. Semin Oncol 19 [suppl 4]: 1–7

Kreuser ED, Hilgenfeld RU, Matthias M, Hoksch B, Boewer C, Oldenkott B, Lenk H, Wiener N, Boese-Landgraf J, Schalhorn A, Thiel E (1992c) A phase II trial of interferon α-2b with folinic acid and 5-fluorouracil administered by 4-h infusion in metastic colorectal carcinoma. Semin Oncol 19 [suppl 4]: 57–62

Kreuser ED, Küchler T, Zwingers T, Streit M, Hilgenfeld RU, Ridwelski K, Matthias M, Boewer Ch, Zillig D, Kairies M, Pasold R, Thiel E (1995) A multicenter prospective randomized trial with quality-of-life assessment comparing interferon α-2b and folinic acid as modulators of 5-fluorouracil in patients with metastatic colorectal carcinoma. Springer, Berlin Heidelberg New York (Recent Results in Cancer Research)

Kunimoto T, Nitta K, Tanaka T, Uehara N, Baba H, Takeuchi M, Yokokura T, Sawada S, Miyasaka T, Mutai M (1987) Antitumor activity of 7-ethyl-10-[4-(1-piperidino)-1-piperidino] carbonyloxy-camptothecin, a novel water-soluble derivative of camptothecin, against murine tumors. Cancer Res 47: 5944–5947

Labianica R, Pancera G, Luporini A, Dallavalle G, Locatelli MC, Fraschini P, Luporini G (1990) Double modulation of 5-fluorouracil (5-FU) with α2b interferon (IFN) and folinic acid (FA) in advanced colorectal cancer. Ann Oncol 1 [suppl]: 92 (abstr)

Labianca R, Pancera G, Aitini E, Barni S, Beretta A, Beretta GD, Cesana B, Comella G, Cozzaglio L, Cristoni M, Spagnolli P, Frontini L, Gottardi O, Martignoni G, Scapaticci R, Smerieri F, Vinci M, Zadro A, Zaniboni A, Luporini G (1991) Folinic acid + 5-fluorouracil (5-FU) versus euqidose 5-FU in advanced colorectal cancer. Phase III study of 'GISGAD' (Italian Group for the Study of Digestive Tract Cancer). Ann Oncol 2: 673–679

Labianca R, Giaccon G, Pancera G, Ambrosini G, Amichetti G, Barni S, Cascinu S, Martignoni G, Pavananto G, Scapticci R, Zaniboni A, Luporini G (1992) Low dose α2b interferon (IFN) + folinic acid (FA) + 5-fluorouracil (5FU) in advanced colorectal cancer (ACC): A phase II study of GISGAD (Italian Group for the Study of Gastrointestinal Cancer). Proc Am Soc Clin Oncol 11: 177 (abstr)

Lang EK, Brown CL (1993) Colorectal metastases to the liver: selective chemoembolization. Radiology 189: 417–422

Lembersky B, Koff H, Miketic L, Posner M, Jacobs S, Carr B (1992) Phase II trial of 5-fluorouracil (FU), leucovorin (LV) and inteferon alfa-2b (IFN) in the treatment of metastatic colorectal canceer (CRC). Proc Am Soc Clin Oncol 11: 186 (abstr)

Leone BA, Romero A, Rabinovich MG, Perez JE, Macchiavelli M, Strauss E (1986) Sequential therapy with methotrexate and 5-fluorouracil in the treatment of advanced colorectal carcinoma. J Clin Oncol 4: 23–27

Leone B, Romero A, Rabinovich M, Vallejo C, Bianco A, Perez J, Rodriguez R, Cuevas A, Macchiavelli M, Paris A, Lacava J (1992) Double modulation of 5-fluorouracil (5-FU) by methotrexate (MTX) and leucovorin (LV) and in metastatic colorectal carcinoma. Proc Am Soc Clin Oncol 11: 174 (abstr)

Lorenz M, Maier P, Reimann M, Inglis R, Hottenrott C (1992) Regionale Chemotherapie von 160 Patienten mit Lebermetastasen kolorektaler Karzinome. In: Schmoll HJ, Meyer HJ, Wilke H, Pichlmayer R (eds) Aktuelle Therapie gastrointestinaler Tumoren, 1st edn. Springer, Berlin Heidelberg New York, pp 372–381

Macchiavelli M, Leone B, Romero A, Rabinovich M, Vallejo C, Bianco A, Pérez J, Rodriguez R, Cuevas M, Alvarez L, Chacon R, Hannois A (1990) Advanced colorectal carcinoma (CRC): a prospective randomized trial of sequential methotrexate (MTX)-5-fluorouracil (5FU) vs. 5FU alone. Proc Am Soc Clin Oncol 9: 102 (abstr)

Marsh JC, Bertino JR, Katz KH, Davis CA, Durivage HJ, Rome LS, Richards F 2nd, Capizzi RL, Farber LR, Pasquale DN, Stuart R, Koletsky AJ, Makuch R, O'Hollaren K (1991) The influence of drug interval on the effect of methotrexate and fluorouracil in the treatment of advanced colorecta cancer. J Clin Oncol 9: 371–380

Martin DS, Kemeny NE (1992) Modulation of fluorouracil by N-(phosphonacetyl)-L-asparate: a review. Semin Oncol 19: 49–55

Martin DS, Stolfi RL, Sawyer RC, Spiegelman S, Casper ES, Young CW (1983) Therapeutic utility of utilizing low-doses N-(phosphonacetyl)-L- aspartic acid in combination with 5-fluorouracil: a murine study with clinical relevance. Cancer Res 43: 2317–2321

Martin JK Jr, O'Connell MJ, Wieand HS, Fitzgibbons RJ Jr, Mailliard JA, Rubin J, Nagorney DM, Tschetter LK, Krook JE (1990) Intra-arterial floxuridine vs systemic fluorouracil for hepatic metastases from colorectal cancer. A randomized trial. Arch Surg 125: 1022–1027

Martinelli DJ, Wadler S, Bakal CW, Cynamon J, Rozenblit A, Haynes H, Kaleya R, Wiernik PH (1994) Utility of embolization or chemoembolization as second line treatment in patients with advanced or recurrent colorectal carcinoma. Cancer 74: 1706–1712

Matsuoka H, Furasawa M, Tomoda H, Seo Y, Kono A, Takeda S, Sato K (1994) Cytotoxicity of CPT-11 and SN-38 for gastrointestinal and recurrent carcinomas cultured on contact-sensitive plates. Anticancer Res 14: 405–409

Meehan K, Seidler C, Gehr G, Maurer LM (1993) An outpatient regimen of 5-fluorouracil (5-FU) and recombinant interferon alpha-2b (rIFNA-2B) in metastatic adenocarcinoma of the colon (MAC). Proc Am Soc Clin Oncol 12: 222 (abstr)

Moertel CG, Fleming TR, Macdonald JS, Haller DG, Laurie JA, Goodman PJ, Ungerleider JS, Emerson WA, Tormey DC, Glick JH, Veeder MH, Mailliard JA (1990) Levamisole and fluorouracil for adjuvant therapy of resected colon carcinoma. N Engl J Med 322: 352–358

Moertel C, Fleming T, Macdonald J, Haller D, Laurie J (1992) The intergroup study of fluorouracil (5-FU) plus levamisole (LEV) and levamisole alone as adjuvant therapy for stage C colon cancer. Proc Am Soc Clin Oncol 11: 161 (abstr)

Moore MJ, Kaizer L, Erlichman C, Bunting P, Feld R, Fine S, Myers R, Thiessen JJ, Yuan S (1992) A phase II clinical and pharmacologic study of 5-fluorouracil (5FU) + leucovorin + interferon in advanced colorectal cancer. Proc Am Soc Clin Oncol 11: 183 (abstr)

Morgan DA, Ruscetti FW, Gallo R (1976) Selective in vitro growth of T lymphocytes from normal human bone marrows. Science 193: 1007–1010

Mross K, Mayer U, Hamm K, Burk K, Hossfeld DK (1990) Pharmacokinetics and metabolism of iodo-doxorubicin and doxorubicin in humans. Eur J Clin Pharmacol 39: 507–513

Muggia FM, Camacho FJ, Kaplan BH, Green MD, Greenwald ES, Wernz JC, Engstrom PF (1987) Weekly 5-fluorouracil combined with PALA: toxic and therapeutic effects in colorectal cancer. Cancer Treat Rep 71: 253–256

Negro S, Fukuoka M, Kurita Y, Niitani H (1994) Results of phase II studies of gemcitabine in non-small cell lung cancer (NSCLC) Proc Am Soc Clin Oncol 13: 367 (abstr)

Niederhuber JE, Ensminger W, Gyves J, Thrall J, Walker S, Cozzi E (1984) Regional chemotherapy of colorectal cancer metastatic to the liver. Cancer 53: 1336–1343

Nobile MT, Rosso R, Sertoli MR, Rubagotti A, Vidili MG, Guglielmi A, Venturini M, Canobbio L, Fassio T, Gallo L, Galligioni E, Galotti P, Bruzzi P, Sobrero A (1992) Randomised comparison of weekly bolus 5-fluorouracil with or without leucovorin in metastatic colorectal carcinoma. Eur J Cancer 28A: 1823–1827

Nordic Gastrointestinal Tumor Adjuvant Therapy Group (1989) Superiority of sequential methotrexate, fluorouracil, and leucovorin to fluorouracil alone in advanced symptomatic colorectal carcinoma: a randomized trial. J Clin Oncol 7: 1437–1446

O'Connell M, Poon MA, Wieand H, Moertel C, Krook J, Gerstner J (1990) Biochemical modulation of 5-fluorouracil (5FU) with leucovorin (LV): confirmatory evidence of improved therapeutic efficacy in the treatment of advanced colorectal cancer. Proc Am Soc Clin Oncol 9: 106 (abstr)

O'Dwyer PJ, DeLap RJ, King SA, et al (1987) Trimetrexate: clinical development of a nonclassical antifolate. NCI Monogr 5: 105–109

O'Dwyer PJ, Paul AR, Walczak J, Weiner LM, Litwin S, Comis RL (1990) Phase II study of biochemical modulation of fluorouracil by low-dose PALA in patients with colorectal cancer. J Clin Oncol 8: 1497–1503

Ortaldo JR, Mason AT, Gerard JP, Henderson LE, Farrar W, Hopkins RF 3d, Herberman RB, Rabin H (1984) Effects of natural and recombinant IL2 on regulation of IFN gamma production and natural killer cell activity: lack of involvement of the TAC antigen for these immunoregulatory effects. J Immunol 133: 779–783

Panasci L, Ford J, Margolese R (1985) A phase II study of sequential methotrexate and fluorouracil in advanced colorectal cancer. Cancer Chemother Pharmacol 15: 164–166

Patt YZ, Mavligit GM (1991) Arterial chemotherpy in the management of colorectal cancer: an overview. Semin Oncol 18: 478–490

Patt YZ, Jones D, Chase J, Charnsangavej C, Chuang V, Carraso H, Lawrence D, Roh M (1994) Hepatic arterial infusion (HAI) of recombinant interferon-alpha (rIFNα) & 5-fluorouracil (5FU) for colorectal cancer (CRC) liver metastases refractory to intravenous (IV) 5FU & leucovorin (LCV). Proc Am Soc Clin Oncol 13: 201 (abstr)

Pavlick AC, DelGaudio D, Harper H, Rosenbluth R, Pecora AL (1992) Salvage therapy with continuous infusion 5-fluorouracil and alpha interferon in patients with colorectal cancer. Proc Am Soc Clin Oncol 11: 188 (abstr)

Pazdur R, Ajani JA, Patt YZ, Winn R, Jackson D, Shepard B, DuBrow R, Campos L, Quaraishi M, Faintuch J, Abbruzzese JL, Gutterman J, Levin B (1990) Phase II study of fluorouracil and recombinant interferon alfa-2a in previously untreated advanced colorectal carcinoma. J Clin Oncol 8: 2027–2031

Pazdur R, Ajani JA, Patt YZ, Gomez J, Bready B, Levin B (1993) Phase II evaluation of recombinant alpha-2a-interferon and continuous infusion fluorouracil in previously untreated metastatic colorectal adenocarcinoma. Cancer 71: 1214–1218

Pazdur R, Lassere Y, Rhodes V, Ajani JA, Sugarman SM, Patt YZ, Jones DV, Markowitz AB, Abbruzzese JL, Bready B, Levin B (1994) Phase II trial of uracil and tegafur plus oral leucovorin: an effective oral regimen in the treatment of metastatic colorectal carcinoma. J Clin Oncol 12: 2296–2300

Pazdur R, Roach R, Braud E, Osmon C, Verschragen C, Kolbye S, Bready B, Winn R (1994) Phase II study of folinic acid, 5-fluorouracil, and interferon alfa-2a for patients with advanced colorectal carcinoma. Proc Am Soc Clin Oncol 13: 224 (abstr)

Pensel R (1993) Advanced colon cancer (ACC): A randomized trial of fluorouracil (5-FU) + folinic acid (FA) and 5-FU + FA + interferon alpha 2b (IFN). Proc Am Soc Clin Oncol 12: 203 (abstr)

Petrelli N, Herrera L, Rustum Y, Burke P, Creaven P, Stulc J, Emrich LJ, Mittelman A (1987) A prospective randomized trial of 5-fluorouracil versus 5-fluorouracil and high-dose leucovorin versus 5-fluorouracil and methotrexate in previously untreated patients with advanced colorectal carcinoma. J Clin Oncol 5: 1559–1565

Petrelli N, Douglass HO Jr, Herrera L, Russell D, Stablein DM, Bruckner HW, Mayeer RJ, Schinella R, Green MD, Muggia FM, Megibow A, Greenwald ES, Bukowski RM, Harris J, Levin B, Gaynor E, Loutfi A, Kalser MH, Barkin JS, Benedetto P, Woolley PV, Nauta R, Weaver DW, Leichman LP (1989) The modulation of fluorouracil with leucovorin in metastatic colorectal carcinoma: a prospective randomized phase III trial. J Clin Oncol 7: 1419–1426

Piedbois P, Gimonet JF, Feuilhade F, Le Péchoux C, Lévy C, Brun B, Rotman N, Delchier JC, Blasquez M, Thill L, Fagniez PL, Bader JP, Julien M, Le Bourgeois JP (1991) 5FU, folinic acid and alpha-2a-interferon combination in advanced gastrointestinal cancer. Proc Am Soc Clin Oncol 10: 141 (abstr)

Poon MA, O'Connell MJ, Moertel CG, Wieand HS, Cullinan SA, Everson LK, Krook JE, Mailliard JA, Laurie JA, Tschetter LK, Wiesenfeld M (1989) Biochemical modulation of fluorouracil: evidence of significant improvement of survival and quality of life in patients with advanced colorectal carcinoma. J Clin Oncol 7: 1407–1418

Presant CA, Ardalan B, Multhauf P, Chan C, Staples R, Green L, Browning S, Carr BI, Chang FF, Thayer W (1983) Continuous five-day infusion of PALA and 5FU: a pilot phase II trial. Med Pediatr Oncol 11: 162–163

Punt CJA, de Mulder PHM, Burghouts JTM, Wagener DJT (1992) Fluorouracil continuous infusion plus alfa interferon plus oral folinic acid in advanced colorectal cancer. Semin Oncol 19 [suppl 3]: 208–210

Pyrhönen SO, Kouri MO (1992) Phase II study of epirubicin sequential methotrexate and 5-fluorouracil for advanced colorectal cancer. Eur J Cancer 28A: 1828–1832

Quagliuolo V, Bignami P, Doci R, Civalleri D, Cosimelli M (1987) Continuous artery infusion of floxuridine for metastatic colorectal cancer. Proc ECCO 4: 179 (abstr)

Recchia F, Nuzzo A, Lombardo M, Lalli A, Fabiani F, Fanini R, Di Lullo L, Venturoni L, Peretti G for the IORAM (Istituto Oncologico Regione Abruzzo e Molise) (1992) 5-Fluorouracil (5FU) and high dose folinic acid (HDFA) chemotherapy (CT) with or without alfa 2b interferon (IFN) in advanced colorectal cancer (ACC). Preliminary report of a randomized trial. Proc Am Soc Clin Oncol 11: 162 (abstr)

Romanini A, Li WW, Colofiore JR, Bertino JR (1992) Leucovorin enhances cytotoxicity of trimetrexate/fluorouracil, but not methotrexate/fluorouracil in CCRF-CEM cells. J Natl Cancer Inst 84: 1033–1038

Rosenberg SA, Lotze MT, Muul LM, Chang AE, Avis FP, Leitman S, Linehan WM, Robertson GN, Lee RE, Rubin JT, Seipp CA, Simpson CG, White DE (1987) A progress report on the treatment of 157 patients with advanced cancer using lymphokine-activated killer cells and interleukin-2 or high-dose interleukin-2 alone. N Engl J Med 316: 889–897

Rosenberg SA, Lotze MT, Yang JC, Aebersold PM, Linehan WM, Seipp CA, White DE (1989) Experience with the use of high-dose interleukin-2 in the treatment of 652 cancer patients. Ann Surg 210: 474–485

Rougier P, Lasser P, Elias D, Ghosen M, Droz JP, Sidibe S, Theodore C, Lumbroso J (1987) Intra-arterial hepatic chemotherapy (IAHC) for liver metastases (LM) from colorectal (CR) origin. Proc Am Soc Clin Oncol 6: 94 (abstr)

Rougier P, Ducreux M, Pignon JP, Elias D, Tigaud J, Lumbroso J, Ruffie P, Lassere PH (1991) Prognostic factors in patients with liver metastases from colorectal carcinoma treated with discontinuous intra-arterial hepatic chemotherapy. Eur J Cancer 27: 1226–1230

Rougier P, Laplanche A, Huguier M, Hay JM, Ollivier JM, Escat J, Salmon R, Julien M, Roullet Audy JC, Gallot D, Gouzi JL, Pailler JL, Elisa D, Lacaine F, Roos S, Rotman N, Luboinski M, Lasser P (1992) Hepatic arterial infusion of floxuridine in patients with liver metastases from colorectal carcinoma: Long-term results of a prospective randomized trial. J Clin Oncol 10: 1112–1118

Scheele J, Stangl R, Altendorf-Hofmann A, Gall FP (1991) Regionale Chemotherapie bei diffusen Lebermetastasen des kolorektalen Karzinoms. Zentralbl Chir 116: 1215–1235

Scheithauer W, Rosen H, Kornek GV, Sebesta C, Depisch D (1993) Randomised comparison of combination chemotherapy plus supportive care with supportive care alone in patients with metastatic colorectal cancer. Br Med J 306: 752–755

Schlag P, Hohenberger P, Schwarz V, Herfarth CH (1988) Intra-arterielle 5-Fluorouracil-Chemotherapie bei Lebermetastasen kolorektaler Karzinome. Med Klin 83: 705–709

Schwartz SI, Jones LS, McCune CS (1985) Assessment of treatment of intrahepatic malignancies using chemotherapy via an implantable pump. Ann Surg 201: 560–567

Seidman AD, Norton L, Reichman BS, Crown JP, Yao TJ, Heelan R, Hakes TB, Lebwohl DE, Gilewski TA, Surbone A (1993) Preliminary exprience with paclitaxel (Taxol) plus recombinant human granulocyte colony-stimulating factor in the treatment of breast cancer. Semin Oncol 20 [suppl 3]: 40–45

Seymour MT, Johnson PWM, Wrigley PFM, Slevin ML (1991) 5-Fluorouracil (5FU), leukovorin (LV) and α-interferon (IFNα): a new regimen suitable for randomised trial. Proc Am Soc Clin Oncol 10: 160 (abstr)

Seymour MT, Slevin M, Cunningham D, Kerr D, James R, Lederman R, Perren T, McAdam W, Duffy A, Stenning S, Taylor I (1994) A randomized assessment of interferon-α2a (IFNα) as a modulator of 5-fluorouracil (5FU) and leucovorin (LV) in advanced colorectal cancer. Proc Am Soc Clin Oncol 13: 209 (abstr)

Shepard KV, Levin B, Karl RC, Faintuch J, DuBrow RA, Hagle M, Cooper RM, Beschorner J, Stablein D (1985) Therapy for metastatic colorectal cancer with hepatic artery infusion chemotherapy using a subcutaneous implanted pump. J Clin Oncol 3: 161–169

Shimada Y, Yoshino M, Wakui A, Nakao I, Futatsuki K, Sakata Y, Kambe M, Taguchi T, Ogawa N, and the CPT-11 Gastrointestinal Cancer Study Group (1993) Phase II study of CPT-11, a new camptothecin derivative, in metastatic colorectal cancer. J Clin Oncol 11: 909–913

Sinnige HAM, Sleijfer DT, de Vries EGE, Willemse PHB, Mulder NH (1990) Modification of 5-fluorouracil activity by high-dose methotrexate or leucovorin in advanced colorectal carcinoma. Eur J Cancer 26: 625–628.

Sinnige HAM, Buter J, Sleijfer DTh, Verschueren RCJ, Mulder NH (1993) Update of a phase II study of 5-fluorouracil (5-FU), leucovorin (LV) and α-2a interferon (IF) in advanced colorectal cancer. Proc Am Soc Clin Oncol 12: 206 (abstr)

Sobrero A, Nobile MT, Guglielmi A, Mori A, Aschele C, Bolli E, Tixi L, Gallo L, Parodi GC, Bruzzi P, Brema F, Rosso R (1992) Phase II study of 5-fluorouracil plus leucovorin and interferon alpha 2b in advanced colorectal cancer. Eur J Cancer 28A: 850–852

Sotos GA, Grogan L, Allegra CJ (1994) Preclinical and clinical aspects of biomodulation of 5-fluorouracil. Cancer Treat Rev 20: 11–49

Staag RJ, Venook AP, Chase JL, Lewis BJ, Warren RS, Roh M, Mulvihill SJ, Grobman BJ, Rayner AA, Hohn DC (1991) Alternating hepatic intra-arterial floxuridine and fluorouracil: A less toxic regimen for treatment of liver metastases from colorectal cancer. J Natl Cancer Inst 83: 423–428

Steinberg J, Erlichman C, Gadalla T, Fine S, Wong A (1992) Prognostic factors in patients with metastatic colorectal cancer receiving 5-fluorouracil and folinic acid. Eur J Cancer 28A: 1817–1820

Steinke B, Günther E, Hirschmann WD, Sondern W, Koniczek KH, Wander HE, Natt F, Wagner T, Hinrichs HF, Werdier D (1992) Fluorouracill versus folinic acid/fluorouracil in advanced colorectal cancer – preliminary results of a randomized trial. Semin Oncol 19 [suppl 3]: 141–147

Taylor C, Modiano M, Alberts D, Marcus S, Hersh E (1991) Combination chemotherapy with 5-fluorouracil (5-FU), leucovorin (LCV) and interferon-alpha (IFN-α) in patients with adenocarcinoma. Proc Am Soc Clin Oncol 10: 141 (abstr)

Thirlwell MP, Hollingsworth LM, Herba MJ, Boileau G, Boos G, MacFarlane JK (1986) Ambulatory hepatic artery infusion chemotherapy for cancer of the liver. Am J Surg 151: 585–589

Valone FH, Friedman MA, Wittlinger PS, Drakes T, Eisenberg PD, Malec M, Hannigan JF, Brown BW Jr (1989) Treatment of patients with advanced colorectal carcinomas with fluorouracil alone, high-dose leucovorin plus fluorouracil, or sequential methotrexate fluorouracil, and leucovorin: a randomized trial of the Northern California Oncology Group. J Clin Oncol 7: 1427–1436

Van Hazel GA, Buck M, Byrne MJ (1992) Phase II study of the combination of 5-fluorouracil (5FU), folinic acid (FA) and alpha interferon (IFN) in metastatic colorectal adenocarcinoma (CC). Proc Am Soc Clin Oncol 11: 195 (abstr)

Villa R, Zaffaroni N, Giulani FC, Colella G, Sanfilippo O, Sivestrini R (1990) Absolute and relative activities of 4'-iodo-4'-deoxydoxorubicin against human colorectal tumors, as evaluated by a short-term in vitro assay. Anticancer Res 10: 105–108

Vogel VG, McPherson RS (1989) Dietary epidemiology of colon cancer. Hematol Oncol Clin North Am 3: 35–63

Wadler S, Schwartz EL, Goldman M, Lyver A, Rader M, Zimmerman M, Itri L, Weinberg V, Wiernik PH (1989) Fluorouracil and recombinant alpha-2a-interferon: an active regimen against advanced colorectal carcinoma. J Clin Oncol 7: 1769–1775

Wadler S, Lembersky B, Atkins M, Kirkwood J, Petrelli N (1991) Phase II trial of fluorouracil and recombinant interferon alfa-2a in patients with advanced colorectal carcinoma: an Eastern Cooperative Oncology Group study. J Clin Oncol 9: 1806–1810

Wadler S, Tenteromano L, Gazenave L, Sparano JA, Greenwald ES, Rozenblit A, Kaleya R, Wiernik PH (1994) Phase II trial of echinomycin in patients with advanced or recurrent colorectal cancer. Cancer Chemother Pharmacol 34: 266–269

Weh HJ, Platz D, Braumann D, Buggisch P, Eckardt N, Schmiegel WH, Drescher S, Kleeberg UR, Müllerleile U, Crone-Munzebrock W, Hoffmann R, Müller P, Klapdor R, Pompecki R, Erdmann H, Reichel L, Jungbluth M, Hoffmann L, Mainzer K, Hossfeld DK (1992) Phase II trial of 5-fluorouracil and recombinant interferon alfa-2b in metastatic colorectal carcinoma. Eur J Cancer 28A: 1820–1823

Weh HJ, Wilke HJ, Dierlamm J, Klaassen U, Siegmund R, Illiger HJ, Schalhorn A, Kreuser ED, Hilgenfeld RU, Steinke B, Weber W, Burkhard O, Zoller A, Pfitzner J, Suber R, Kribel R, Hossfeld DK (1994) Weekly therapy with folinic acid (FA) and high-dose 5-fluorouracil (5-FU) 24-h infusion in pretreated patients with metastatic colorectal carcinoma. Ann Oncol 5: 233–237

Weinerman B, Schacter B, Schipper H, Bowman D, Levitt M (1982) Sequential methotrexate and 5-FU in the treatment of colorectal cancer. Cancer Treat Rep 66: 1553–1555

Weiss GR, Ervin TJ, Meshad MW, Kufe DW (1982) Phase II trial of combination therapy with continuous-infusion PALA and bolus-injection 5-FU. Cancer Treat Rep 66: 299–303

Weiss GR, Garnick MB, Osteen RT, Steele GD Jr, Wilson RE, Schode D, Kaplan WD, Boxt LM, Kandarpa K, Mayer RJ, Frei ET III (1983) Long-term hepatic arterial infusion of 5-fluorodeoxyuridine for liver metastases using an implantable infusion pump. J Clin Oncol 1: 337–344

Weiss L, Grundmann E, Torhorst J, Hartveit F, Moberg I, Eder M, Fenoglio-Preiser CM, Napier J, Horne CH, Lopez MJ (1986) Haematogenous metastatic patterns in colonic carcinoma: an analysis of 1541 necropsies. J Pathol 150: 195–203

Wils J, Schlangen J, Naus A (1986) Phase II study of hepatic artery infusion with 5-fluorouracil, adriamycin, and mitomycin C (FAM) in liver metastases from colorectal carcinoma. Recent Results Cancer Res 100: 157–162

Yang JC, Shlasko E, Ritchey JL, Landry JG, Write DE, Rosenberg SA (1993) Combination chemotherapy for metastatic colorectal cancer using 5-fluorouracil, leucovorin and interleukin-2. Eur J Cancer 29A: 355–359

York M, Greco FA, Figlin RA, Einhorn L, Man T, Cockey L, Mott D, Light SE (1993) A randomized phase III trail comparing 5-FU with or without interferon alfa 2a for advanced colorectal cancer. Proc Am Soc Clin Oncol 12: 200 (abstr)

Young JC, Shlasko E, Ritchey JL, Landry JG, White DE, Rosenberg SA (1993) Combination chemotherapy for metastatic colorectal cancer using 5-fluorouracil, leucovorin and interleukin-2. Eur J Cancer 29A: 355–359

Surgical and Multimodality Approaches to Liver Metastases

P.M. Schlag, M. Hünerbein, and P. Hohenberger

1 Introduction

Although liver metastases may provide the first evidence of cancer progression, they almost always indicate general dissemination of malignancy. Treatment options for hepatic metastases include surgical resection, administration of regional chemotherapy, systemic chemotherapy, radiotherapy, and various innovative strategies to achieve specific drug targeting. In spite of all advances, surgery is currently the only therapy that prolongs survival or even provides the chance of cure. Technical standards constitute no longer a limiting factor in the use of hepatic resection for malignant disease. Although hepatic resection remains a major procedure, current mortality rates of 0%–6% justify widespread use of this type of surgery (Foster and Lundy 1981; Gayowski et al. 1994). Consequently, it is now even more important to establish accurate guidelines defining the indications of hepatic resection for malignant disease.

Virchow Hospital, Robert Rössle Hospital and Tumor Institute, Humboldt University, 13122 Berlin, Germany

Due to its high incidence, most of the available information regarding hepatic resection for metastatic disease relates to metastases arising from colorectal primaries. About 80% of hepatic resections are performed for colorectal secondaries, while 20% are undertaken for non-colorectal metastases (IWATSUKI et al. 1989). In spite of the poor prognosis of patients with liver matastases, occasional success of surgical resection resulting in long-term survival has been reported. This observation suggests the existence of a subset of patients who benefit from surgical therapy.

Improved understanding of tumor biology has been incorporated in treatment strategies and has been used to refine techniques in surgery and adjuvant therapies, which offer promising prospects for future therapy strategies.

2 Clinical Diagnosis

The basic step in the diagnostic evaluation of liver lesions involves confirming the presence of malignancy, which may be difficult without biopsy. While liver function tests provide no specific information, determination of the serum level of a range of tumor markers can be helpful in establishing the correct diagnosis (LUCAROTTI et al. 1991; AL JARALLAH et al. 1993). If high levels of α-fetoprotein (AFP) (greater than 100 ng/l) are detected, presence of a primary hepatocellular carcinoma is very likely. Colorectal metastases may be associated with increased levels of carcinoembryonic antigen (CEA) and CA19-9. An increased level of CEA (more than 5 ng/ml) is found in about 90% of all patients with hepatic metastases of colorectal carcinoma. However, the specifity and sensitivity for CA19-9 is low (60%–65%) (HOHENBERGER et al. 1994). Pathologic liver function tests and increased levels of CA15-3 are indicative of hepatic metastases of breast carcinoma (GERAGHTY et al. 1992).

Ultrasonography is generally the initial imaging method for detection of suspected liver metastases. The accuracy of this method is substantially influenced by the presence of independent hepatic pathology such as liver cirrhosis. Consequently, the sensitivity and specifity of ultrasonography may be as low as 50% for the detection of hepatic tumors, and similar results have been reported for conventional computed tomography (CT) (WERNECK et al. 1991). Special CT techniques, e.g., angiographically enhanced CT and incremental dynamic bolus CT, have been demonstrated to be more sensitive for the detection of small liver tumors (SOYER et al. 1994; COLLIER and FOLEY 1993). Recently, comparable results were obtained for magnetic resonance imaging (MRI), and further development of contrast media such as gadopentate dimeglumine may induce improved imaging of hepatic tumors by this technique (MITCHELL et al. 1994; SEMELKA et al. 1994).

At present a reasonable degree of accuracy in the differentiation of benign and malignant liver tumors can be achieved by a combination of several imaging methods. Dynamic bolus CT is reasonably specific in the diagnosis of *hemangioma*. Characteristically, hemangiomas appear as a hypodense lesion followed by peripheral enhancement during infusion of bolus of contrast and gradual filling of the lesion. However, these criteria are typical in no more than 55% of the cases (FREENY and MARKS 1986). Other radiologic investigation techniques such as angiography, scintigraphy, and single photon emission CT (SPECT) can contribute significantly to increasing the accuracy of the diagnosis. In a recent study on 83 patients with hepatic lesions, MRI was used to discriminate between hemangiomas and malignant lesions with a 97% accuracy (McFARLAND et al. 1994). TC 99m sulfur colloid scintigraphy is very useful for differentiating focal nodullar hyperplasia (FNH) from other hepatic masses, because uptake of this compound is extremely uncommon in other lesions (WELCH et al. 1985). Promising results were also reported for MRI in FNH and hepatic adenoma (FERRUCCI 1990).

Immunoscintigraphy using monoclonal anti-CEA antibodies has proved valuable for the detection of liver metastases of colorectal carcinoma. A sensitivity ranging from 50% to 80% and a specifity of up to 85% has been obtained (HASEMANN et al. 1992; IMDAHL et al. 1994). The results of this technique depend on the specifity of the antibodies as well as on the stability of the labeling and the imaging technique used (HÖLTING et al. 1990).

3 General Considerations for Surgical Resection

The indication for hepatic resection is limited to patients who are potentially curable by surgery. The presence of hepatic or celiac node involvement as well as other extrahepatic spread implies the impossibilty of eradicating the disease completely by liver resection. Unfortunately, discontinuous extrahepatic spread, including lymph node involvement, local recurrence, and peritoneal carcinosis, is described in 10%–25% of patients with hepatic metastases, and infiltration beyond the liver in adjacent organs is found in up to 30% of cases (HUGHES et al. 1988; WAGNER et al. 1984). However, preoperative imaging is often not sensitive enough to identify patients with extrahepatic spread. In up to 40% of patients, involvement of extrahepatic structures is demonstrated on laparotomy (STEELE et al. 1991). Although infiltration per continuitatem may be treatable by mutivisceral resection, the prognosis in these patients is poor. Laparoscpy and laparoscopic ultrasound can be helpful in diagnosing dissemiated malignancy, thus avoiding futile laparotomy (HÜNERBEIN et al. 1994). In the future, the analysis of CK-18-positive cells in the bone marrow may become a further important factor in decision-making regarding liver resection. These cells probably indicate disseminated disease with a poor prognosis not suitable for surgical, i.e., local measures.

4 Surgical Results in the Treatment of Liver Metastases

4.1 Colorectal Cancer

At the time of diagnosis, liver metastases are found in 15%–25% of patients with colorectal cancer, while a further 25%–30% will develop metachronous metastases after resection of the primary tumor (NIH 1990 Asbun et al. 1994; Fowler et al. 1993; Elias et al. 1993). This means that ultimately about 50% of the patients will show liver metastases in the course of their disease. Progression of liver involvement is the most limiting factor for survival in colorectal malignancy. Untreated liver metastases from colorectal cancer are associated with a surivival time of between 5 and 25 months (Table 1). The median survival time is somewhat less than 10 months (Wagner et al. 1984; Lahr 1983; Steele et al.1991).

A multitude of therapeutic approaches have been evaluated to improve the outcome of patients with hepatic metastases. Presently the only treatment option providing a realistic chance for cure remains surgical resection. Surgery has been increasingly used as primary therapy. This has mainly been caused by the decrease in morbidity and mortality of the procedure due to increased experience and technical advances.

However, as a result of advanced nonresectable or disseminated disease or poor general condition only about 25% of patients are suitable for surgery (Ballantyne and Quin 1993). Despite preoperative staging in patients considered for liver resection, a considerable proportion have extensive disease that is discovered by laparotomy. In a recent study of the Gastrointestinal Tumor Study Group (GITSG), 69 of 150 patients (42%) with potentially resectable liver metastases on the basis of preoperative staging were found to have nonresectable disease at surgery (Steele et al. 1991). Curative resection was possible in 46%, while 12% of the patients underwent noncurative resection. The median survival time of patients who underwent a curative procedure was significantly longer than in those who underwent noncurative resection (37 versus 21 months). However, there was no survival benefit for the noncurative resection group compared to untreated patients (16 months). Consequently, noncurative resection is best avoided in patients with liver metastases from colorectal cancer.

Table 1. Survival of patients with untreated colorectal secondaries to the liver

Reference	Patients (*n*)	Median survival time (months)	Five-year survival rate(%)
Goslin et al. 1982	125	10–24	0
Lahr et al. 1983	147	4.5–12	1
Wagner et al. 1984	252	11 –21	<5
Scheele et al. 1990	983	6.9–14.2	0
Steele et al. 1991	47	16.5	n.s.

n.s., not stated.

The role of surgical resection of colorectal metastases to the liver has been investigated extensively. Surgery has improved overall survival and has been curative in selected patients. Actuarial 5-year survival rates of 23%–40% have consistently been reported in major studies (Table 2). The 10 year survival rate has been recorded to be 12%–25%. The operative mortality rate ranged from 0% to 6%.

These data indicate that hepatic resection for colorectal metastases is a safe and efficient procedure. Nonetheless there is ongoing controversy regarding patient selection for resective surgery, although a number of clinical and pathological prognostic factors have been identified. Numerous studies have suggested that the number and distribution of metastases significantly affect survial. Although there is no general consensus, most authors agree that hepatic resection should be limited to patients with less than four metastases (HUGHES et al. 1988; ASBUN et al. 1994). While patients with one to three metastases have a similar prognosis resulting in 5-year survival rates of about 30%, patients with four or more metastases have a very poor prognosis, with a 5-year survival rate of 3% (GAYOWSKI et al. 1994). Interestingly, in this recent report bilobular distribution of the metastases had adverse effects on overall and median survival, which is in contrast to the results of most other studies (STEELE et al. 1991; FOSTER 1990).

Although the specific type of operation (e.g. anatomic versus wedge resection) apparently has no influence on survival, it is crucial to obtain a tumor-free resection margin of at least 1 cm. Consequently, for larger lesions anatomic lobectomy may be more appropriate even if a non-anatomic procedure is technically feasible. A diameter of the metastastic lesion of more than 5 cm has been considered an important determinant by some authors, but this has not reached general acceptance. A multicenter retrospective review by the Registry of Hepatic Metastases demonstrates that a disease-free interval of less than 1 year was associated with less favorable survival rates (HUGHES et al. 1988). Patients with a disease-free interval of more than 1 year achieved a significantly improved 5-year survival rate (26%) compared to those with a shorter interval (16%). This may reflect the biological properties of the primary tumor and favors resection of metachronous secondaries that develop more than 1 year after resection of the primary tumor. Nonetheless, resection of a synchronous liver metastases may

Table 2. Survival after resection of hepatic metastasis from colorectal cancer; studies with more than 100 patients

Reference	Patients (*n*)	Five-year survival rate (%)	Perioperative mortality rate (%)
GAYOWSKI et al.1994	204	32	0
SCHEELE et al. 1990	183	40	5.5
HUGHES et al. 1988	859	33	n.s.
ADSON et al. 1984	141	25	4
FOSTER and LUNDY 1981	231	23	6

n.s., not stated.

be performed with good results if a solitary lesion can be removed by limited resection with a minimal risk of bleeding and contamination (VOGT et al. 1991).

Unfortunately, recurrence of liver metastases after seemingly curative resection is observed in about 30%–50% of cases (HUGHES et al. 1988; NORDLINGER et al. 1994). Only a small percentage of recurrent metastases appear to be resectable. In a French multicenter study, 116 patients underwent second liver resection, resulting in an overall 3-year survival rate of 33% (NORDLINGER et al. 1994). In contrast, only 26% of the patients with unresected recurrence confined to the liver survived 3 years. The operative mortality due to a second liver resection was below 1%. In our experience in 151 patients with hepatic recurrence, a second resection was performed in 34 patients, with a postoperative mortality rate of 5.8% and a median survival time of 18 months. These results demonstrate that a second resection of recurrent liver metastases can be beneficial in a carefully selected group of patients with limited disease. Therefore, in the absence of extrahepatic disease we recommend surgery as the therapy of choice even for a second recurrence in the liver, using the same indication criteria as for primary hepatic metastases.

The role of orthotopic liver transplantation in the treatment of colorectal liver metastases has been examined in several studies. A review of data from seven European centers and 30 patients undergoing liver transplantation for colorectal secondaries showed a 1- and 2-year survival rate of 28% and 14%, respectively (KONERU et al. 1988). Most patients develop recurrence within a short interval after transplantation due to immunosuppression. At present, hepatic transplantation is not justified for the treatment of colorectal liver metastases.

4.2 Gastric Cancer

The prognosis of advanced gastric cancer is extremely poor, and surgery is the only therapeutic approach that offers the prospect of cure. In the presence of metastatic spread to the liver, the 1-year and 5-year survial rates are less than 20% and 10%, respectively (BABA et al. 1992). Survival beyond 1 year of patients with untreated hepatic metastases is rare.

With respect to the different biological behavior and prognostic significance, it is mandatory to distinguish between involvement of the liver by direct infiltration and discontinuous metastases.

Current opinion maintains that resection of gastric cancer including en bloc resection of the left lobe of the liver should be performed if the tumor can be removed completely (ADAM and EFRON 1989). HABU et al. recently reported the results of surgery in 196 patients with gastric cancer extending to adjacent organs (HABU et al. 1990). Of 12 patients who underwent combined resection for infiltration of the liver, four (33%) survived more than 12 months, while all seven patients without combined resection died within 6 months. Combined resection showed a slight but significant effect on the mean survival of patients with liver involvement. Consequently, en bloc resection may be considered beneficial in selected patients with respect to median survival, although 5-year survival cannot be guaranteed (KORENAGA et al. 1988; TABUCHI et al. 1977). Little has been published about hepatic resection for discontinuous metastases arising from primary gastric carcinoma. Most reports are collective reviews with pooled data of liver resections for metastasis of various primaries. These studies provide only sparse information on the therapeutic significance of liver resection for gastric cancer (Table 3). The figures for 5-year survival range between 5% and 12% after liver resection for gastric cancer (FOSTER 1978; MORROW et al. 1982; WOLF et al. 1991). Recently more comprehensive data have been presented by BINES et al. (1993). Liver resection for metastatic spread in ten patients resulted in long-term survival in one patient with adenocarcinoma of the stomach. The collective 1- and 5-year survival rates were 45% and 30%, respectively. However, a rather high hospital mortality rate of 25% was observed. Other groups confirmed a short-term improvement in survival for patients undergoing synchronous liver resection at the time of gastrectomy if there was limited spread to the liver (KOGA et al. 1980; OKUYAMA et al. 1985). Metachronous resections should only be considered for those patients who can potentially be rendered disease free.

4.3 Renal Cell Carcinoma

In contrast to colorectal cancer with its portal venous route of hematogenous spread, isolated liver metastases of renal cell carcinoma are rare. Renal cell cancer shows a caval type of dissemination, which favors metastatic spread to sites other than the liver. Consequently, even in specialized centers there is only a low incidence of liver metastases originating from renal cell cancer. In a series of patients with metastatic renal cell cancer evaluated for survival, only 11% developed liver metastases (SWANSON et al. 1983).

Reviewing the data of 13 studies reporting the results of 1805 liver resections for metastatic disease, only 39 patients underwent resection for renal cell carcinoma (Table 4). These cases represent 2.2% of all liver resections. Unfortunately, only seven of the authors state their treatment results. To our knowledge only five patients survived 5 years or longer.

Table 3. Results of combined hepatic resection and gastrectomy in patients with advanced gastric cancer

Reference	Patients (*n*)	Five-year survival rate (%)	Perioperative mortality rate (%)
FOSTER 1978	7	0	
MORROW et al. 1982	11	12	
HABU et al. 1990	12	–[a]	6
WOLF et al. 1991	19	5	
BINES et al. 1993	12	30	8.3

[a]Mean survival time, 10 months.

Table 4. Results of hepatic resection for renal cell cancer

Reference	Resections (*n*)	Survivors (*n*)	Total number of patients
RAMMING et al. 1977	8	2[a]	123
FOSTER and LUNDY 1981	3	2[b]	335
MORROW et al. 1982	1	n.s.	64
THOMPSON et al. 1983	1	1[a]	138
TOMAS-DE LA VEGA et al. 1984	3	2[a]	29
KORTZ et al. 1984	1	n.s.	21
FUNOVICS et al. 1986	4	n.s.	245
STIMPSON et al.1987	1	n.s.	87
COBURN et al. 1987	1	n.s.	56
STEHLIN et al. 1988	4	1[a]	111
IWATSUKI et al. 1989	3	0[a]	411
RINGE et al. 1990	5	n.s.	151
STURM et al. 1991	2	1[a]	34
Total	39		1805

n.s., not stated.
[a] At 5 years.
[b] 2 years.

4.4 Breast Cancer

Metastatic spread to the liver is a common observation in patients with advanced carcinoma of the breast when autopsy is performed. However, conventional preoperative staging often fails to detect these metastases. In a study of 43 patients undergoing laparotomy for staging of advanced carcinoma of the breast, MEIRION et al. found intra-abdominal disease not detected by preoperative staging in 37% of the patients (MEIRION-THOMAS et al. 1978).

In spite of the high incidence, there are only sporadic reports on hepatic resection for liver metastases in breast cancer. The presence of liver metastases is frequently synonymous with generalized disease not curable by surgery. FOSTER et al. reported that none of five patients treated by hepatectomy for metastatic breast cancer survived 5 years (FOSTER 1978). ELIAS et al. have evaluated the role of hepatic resection in a multimodal treatment protocol including pre- and postoperative chemotherapy (ELIAS et al. 1991). In a series of 22 patients, surgical exploration revealed benign lesions in four patients, while nonresectable intra-abdominal spread was present in six patients. Consequently, hepatectomy was performed in only 12 patients (54%). The median survival time for these patients was 37 months, which was twice as long as that of control patients (13–17 months). However, a high rate of recurrence occurred in the survivors, with only two patients remaining free of disease at 46 and 29 months, respectively.

These data demonstrate that resection of liver metastases from breast cancer can prolong survival in highly selected patients if a multimodal approach is used. The results of hepatic resection for secondary involvement of breast cancer might be improved by using more sensitive staging techniques such as laparoscopy. However, the main limiting factor in cytoreductive surgery for metastatic breast cancer remains the efficacy of systemic treatment regimens.

4.5 Miscellaneous Sites

Although most of the data regarding hepatic resection for metastatic disease relate to colorectal cancer, a significant proportion (10%–15%) of secondary hepatic neoplasms originate from various other sites. The overall 5-year survival rate after resection of noncolorectal metastases may be as high as 20% (WOLF et al. 1991). The primaries include pancreatic carcinoma, malignant melanoma, sarcoma, and endocrine tumors such as carcinoid, gastrinoma, and gynecologic neoplasms. Occasionally, patients with these primaries present with metastatic disease that seems to be limited to resectable sites of the liver. Due to technical advances that allow liver resections to be performed safely, surgical treatment may be considered in selected patients. However, the results of hepatic resection must be compared with the natural course of the disease. Presently no general guidelines for the indications for surgery can be provided because of the lack of representative data.

To date the only information available consists of reports from single institutions on the treatment of small numbers of patients. The prognosis of patients undergoing liver resection for non-colorectal liver metastases is generally poor, with a mean 5-year survival rate of 17% and a mean survival time of 23 months (WOLF et al. 1991). Long-term survival is not usually observed in patients with pancreatic cancer and gynecologic malignancies, but liver resection may be beneficial for some patients with hepatic metastases of carcinoid tumors or sarcomas. STEHLIN et al. reported that liver resection for metastases resulted in 5 year survival of two of three patients with carcinoid tumors (STEHLIN et al. 1988). Even if cure is not possible, significant symptom palliation can be achieved by liver resection in some patients with endocrine tumors. Especially in patients with carcinoid tumors or gastrinoma, cytoreduction can reduce systemic tumor effects and may therefore facilitate management of the disease. As an exception to the rule, patients with hormone-active endocrine tumors may benefit from liver transplantation, as even in a palliative situation alleviation of severe symptoms may justify transplantation.

5 Adjuvant Therapy After Liver Surgery

Recurrence after curative resection of hepatic metastases may be due to persistent microsopic disease in the liver or to homing of circulating malignant cells following liver resection. Preoperative or adjuvant treatment strategies are aimed at the eradication of such minimal disease. Adjuvant chemotherapy following liver surgery has been administered intra-arterially via the hepatic artery, interaperitoneally, or systemically. None of the studies showed a significant benefit in comparison to surgery alone (Table 5). More recently, active, specific immunotherapy was tested. In our experience, at 18 months after R0 resection of

Table 5. Results of selected trials on adjuvant regional chemotherapy of the liver after resection of hepatic metastases from colorectal cancer

Reference	Route	Patients (*n*)	Drugs	Median survival time (months)	Survival rate (%)
O'CONNELL et al. 1985	Intravenous	26	–	–	–
O'CONNELL et al. 1985	Intra-arterial	26	5-FU, Me-CCNU	–	15[a]
CURLEY et al. 1993	Intra-arterial	18	5-FU	39	–
LORENZ et al. 1995	Intra-arterial	60	5-FU, FUDR	44	–
DONATO et al. 1994	Intra-arterial	102	5-FU	–	–
TSUJITANI et al. 1991	Intraportal	17	5-FU, MMC	–	–
LORZEN et al. 1992	Intraportal	61	FUDR, 5-FU, LV	36	–
AUGUST et al. 1985	Intraperitoneal	21	5-FU	–	53[a]
FORTNER et al. 1984	Intravenous	6	5-FU	24	–
O'CONELL et al. 1985	Intravenous	26	Me-CCNU	34	–

5-FU, 5-fluorouracil; FUDR, 2-fluoro-2'-deoxyuridine; Me-CCNU, methyl-CCNU (semustine); MMC, mitomycin; LV, leucovorin.
[a] At 5 years.

colorectal liver metastases, 61% of patients had developed recurrence compared to 87% of a historically matched group (SCHLAG et al. 1992). The definitive value of this treatment strategy cannot yet be determined, and further investigation in a prospective randomized trial is now underway. Progress in tumor immunology and vaccines designed using molecular biologic and genetic methods will further improve adjuvant therapy in the future.

6 Regional Chemotherapy for Liver Metastases

Response rates for systemic chemotherapy in most patients with liver metastases are limited and do not contribute to an increased survival. Polychemotherapy and high-dose regimens are accompanied by severe side effects and unacceptable discomfort to the patient. Therefore, regional chemotherapy is of special interest for the treatment of nonresectable liver metastases, because high local tumor concentrations of chemotherapeutic agents are achieved with low systemic toxicity. 2-fluoro-2-deoxyuridine. (FUDR) and 5-fluorouracil (5-Fu) have mainly been used for hepatic artery infusion (HAI), as these drugs have a high hepatic excretion rate. Response rates are consistently better for intra-arterial infusion (40%–60%) than for systemic chemotherapy (10%–20%). However, in most of the reported trials the overall survival was not significantly different, and extrahepatic disease progression was more frequent in patients treated by intra-arterial therapy (ROUGIER et al. 1990). To overcome this drawback, HAI has been studied in combination with systemic chemotherapy (KEMENEY et al. 1992). Again, this has not resulted in better survival when compared to HAI alone (MARTIN et al. 1990). With tumor necrosis factor-α administered intra-arterially, sporadic antitumor effects were observed (ZUKIWSKI et al. 1989). In a randomized trial comparing splenic artery infusion of interleukin-2 (IL-2) for hepatic metastases of various malignancies, the splenic route was more advantageous (MAVLIGIT et al. 1988). The blood supply of hepatic metastases relies predominantly on the arterial route, although the periphery of the metastasis is mainly perfused by the portal system. Sudden interruption of the arterial blood supply is followed by partial tumor necrosis. Various techniques have been investigated to achieve such interruption including hepatic artery ligation, complete desarterialization of the liver, hepatic artery embolization using degradable or nondegradable microspheres, Gelfoam (absorbable gelation) sponges, Lipiodol (iodized oil) or lyophilized dura, indwelling ballon catheters, and extra-arterial inflatable balloons, allowing repeated interruption of arterial blood flow (GERARD et al. 1991; HUNT et al. 1990). Finally, isolated hepatic perfusion has been performed for different types of hepatic metastases to obtain higher local drug concentrations (BARONE et al. 1990; MORIYA et al. 1991). Unfortunately, it has not been established whether any of these techniques improve survival in patients with hepatic metastases. Other treatment modalities include induction of interstitial hyperthermia by microvaves, electromagnetic waves, or laser application. Radiation treatment has been performed by the external or internal route with radioisotopes coupled to microspheres or antibodies. The latter have also been coupled to cytotoxic drugs. Local treatment of liver metastases has been attempted by cryotherapy, alcohol injection and interstitial radiation with iridium-192 alone or in combination with chemotherapy. The primary goal of these treatments is palliation, and their effect on survival is questionable.

7 Conclusion

The liver is a common site of metastastic spread arising from various cancers, especially from gastrointestinal primaries. Various treatment options, including regional and systemic chemotherapy, have been used for metastatic disease to the liver. However, it has not yet been shown, that these methods prolong survival or lead to long-term survival. In contrast, surgical resection of metastases can result in cure for selected patients.

Hepatic resection can now be performed with an acceptable risk. The mortality rate has decreased significantly in the last few decades (GAYOWSKI et al. 1994; FOSTER and LUNDY 1981). Since there are no longer any restrictions on hepatic surgery from a technical point of view, accurate selection of patients who will benefit from aggressive surgery has become even more important. Numerous studies have clearly demonstrated that hepatic resection of secondary lesions of colorectal cancer can imporve survival (HUGHES et al. 1988; STEELE et al. 1991; GAYOWSKI et al. 1994). Five-year survival rates as high as 40% have been reported when appropriate patient selection criteria were used (SCHEELE et al. 1990). Although various studies were performed to identify these criteria, there are currently only a few generally accepted prognostic factors to guide surgical decisions. The presence of widespread disease, i.e., portal/celiac lymph nodes, extrahepatic metastases, and multiple liver metastases, must be considered clearly as a contraindication for resection of hepatic metastases of colorectal cancer. In most institutions, three to four metastases are considered to be the upper limit for resection. Patients with longer interval between resection of the primary and diagnosis of liver metastases have a significantly improved prognosis after resection of the metastases. However, a short disease-free interval or the presence of synchronous metastases is not an absolute contraindication to surgery. Nevertheless, surgical resection of liver metastases should be abandoned in the presence of extrahepatic disease or if a complete removal of the metastatic lesions (R0 resection) is unlikely to be achieved. The prognostic factors for resection of noncolorectal metastases to the liver are contradictory at the moment, and the indications for resection are not consistent. The results of hepatic resection for secondary involvement must be judged with respect to the natural course of untreated disease. However, resective treatment should not be denied to patients with noncolorectal liver metastases if no other treatment is available and curative resection may be possible.

Adjuvant chemotherapy following surgery of liver metastases did not improve survival rates. Tumor cell vaccination after resection of liver metastases is an attractive concept. Early reports are promising, and randomized controlled trials have been initiated.

References

Adam YG, Efron G (1989) Trends and controversies in the management of carcinoma of the stomach. Surg Gynecol Obstet 169: 371–385

Adson MA, vanHeerden JA, Adson MH, Wagner JS, Ilstrup DM (1984) Resection of hepatic metastases from colorectal cancer. Arch Surg 119: 647–651

Al Jarallah MA, Behbehani AE, el Nass SA, Temim L, Ebraheem AK, Ali MA, Szymendera JJ (1993) Serum CA-15.3 and CEA patterns in postsurgical follow-up, and in monitoring clinical course of metastatic cancer in patients with breast carcinoma. Eur J Surg Oncol 19: 74–79

Asbum HJ, Tsao JI, Hughes KS (1994) Resection of hepatic metastases from colorectal carcinoma. The registry data. Cancer Tract Res 69: 33–41

August DA, Sugarbaker PH, Ottow RT, Gianola FJ, Schneider PD (1985) Hepatic resection of colorectal metastases. Influence of clinical factors and adjuvant interaperitoneal 5-fluorouracil via Tenckhoff catheter on survival. Ann Surg 201: 210–218

Baba H, Okuyama T, Hiroyuki O, Anai H, Korenaga D, Maehara Y, Akazawa K, Sugimachi K (1992) Prognostic factors for noncurative gastric cancer: univariate and multivariate analyses. J Surg Oncol 51: 104–108

Ballantyne GH, Quin J (1993) Surgical treatment of liver metastases in patients with colorectal cancer. Cancer 71: 4252–4266

Barone R, Goldfarb P, Sahleh F, Kogler J, Shiftan T, Weiss R (1990) Liver resection and adjuvant intraarterial FUDR therapy for the treatment of metastatic colon carcinoma. Proc Am Soc Clin Oncol 9: 443

Bines SD, England G, Deziel DJ, Witt TR, Doolas A, Roseman DL (1993) Synchronous, metachronous, and multiple hepatic resections of liver tumors originating from primary gastric tumors. Surgery 114: 799–805

Cobourn CS, Makowka L, Langer B, Taylor BR, Falk RE (1987) Examination of patient selection and outcome for hepatic resection for metastatic disease. Surg Gynecol Obstet 165: 239–246
Collier BD, Foley WD (1993) Current imaging strategies for colorectal cancer. J Nucl Med 34: 537–540
Curley SA, Roh MS, Chase IL, Hohn DC (1993) Adjuvant hepatic infusion chemotherapy after curative resection of colorectal liver metastases. Am J Surg 166: 746–748
Donato N, Dario C, Giovanni S, Virgilio B, Paolo DP, Roberto L, Gianfranco P, Mario L, Daniela P, Angelo T et al. (1994) Retrospective study on adjuvant chemotherapy after surgical resection of colorectal cancer metastatic to the liver. Eur J Surg Oncol 20: 454–460
Elias D, Lasser P, Spielmann M, May Levin F, el Malt O, Thomas H, Mouriess H (1991) Surgical and chemotherapeutic treatment of hepatic metastases from carcinoma of the breast. Surg Gynecol Obstet 172: 461–464
Elias D, Lasser P, Hoang JM, Lecler J, Debaene B, Bognel C, Spencer A, Rougier P (1993) Repeat hepatectomy for cancer. Br J Surg 80: 1557–1562
Ferrucci JT (1990) Liver tumor imaging: current concepts. AJR 155: 473–483
Fortner JG, Silva JS, Cox EB, Maclean JB (1984) Multivariate analysis of a personal series of 247 consecutive patients with liver metastases from colorectal cancer. Ann Surg 199: 306–316
Foster JH (1978) Survival after liver resection for secondary tumors. Am J Surg 135: 389–394
Foster JH (1990) Surgical treatment of metastatic liver tumors. Hepatogastroenterology 37: 182–187
Foster JH, Lundy J (1981) Pathology of liver metastases. Curr Probl Surg 18: 157–160
Fowler WC, Hoffman JP, Eisenberg BL (1993) Redo hepatic resection for metastatic colorectal carcinoma. World J Surg 17: 658–661
Freeny PC, Marks WM (1986) Hepatic hemangioma: dynamic bolus CT. AJR 147: 713–719
Funovics JM, Wenzel E, Függer R, Schemper M (1986) Leberresektionen wegen hämatogener oder infiltrierender Metastasen. Wien Klin Wochenschr 98: 813–820
Gayowski TJ, Iwatsuki S, Madaraga JR, Selby R, Todo S, Irish W, Starzl TE (1994) Experience in hepatic resection for metastatic colorectal cancer: analysis of clinical and pathologic risk factors. Surgery 116: 703–711
Geraghty JG, Coveney EC, Sherry F, O'Higgins NJ, Duffy MJ (1992). CA 15-3 in patients with locoregional and metastatic breast carcinoma. Cancer 70: 2831–2834
Gerard A, Buyse M, Pector JC, Bleiberg H, Arnaud JP, Willoms G, Delvaux G, Nitti D, Depadt G et al. (1991) Hepatic artery ligation with and without portal infusion of 5-FU. A randomized study in patients with unresectable liver metastases from colorectal carcinoma. The E.O.R.T.C. Gastrointestinal Cancer Cooperative Group (G.I. Group). Eur J Surg Oncol 17: 289–294
Goslin R, Steele G, Zamscheck N (1982) Factors influencing survival in patients with hepatic metastases from adenocarcinoma of the colon and rectum. Dis Colon Rectum 25: 249–754
Habu H, Saito N, Sato Y, Takerchita K, Sunagawa M, Endo M (1990) Results of surgery in patients with gastric cancer extending to the adjacent organs. Hepatogastroenterology 37: 417–420
Hasemann MK, Brown DW, Keeling CA, Reed NL (1992). Radioimmunodetection of occult carcinoembryonic antigen producing cancer. J Nucl Med 33: 1750–1756
Hohenberger P, Schlag PM, Gerneth T, Herfarth C (1994) Pre- and postoperative carcinoembryonic antigen determinations in hepatic resection for colorectal metastases. Predictive value and implications for adjuvant treatment based on multivariate analysis. Ann Surg 219: 135–143
Hölting T, Schlag P, Georgi P (1990). Current status of immunoscintigraphy in colorectal cancer – results of 5 years clinical experience. Eur J Surg Oncol 16: 312–318
Hughes KS, Simon RM, Songhorabodi S (1988) Resection of the liver for colorectal carcinoma metastases: a multiinstitutional study of indications for resection. Surgery 103: 278–288.
Hünerbein M, Rau B, Schlag PM (1994) Laparoscopy and laparoscopic ultrasound for staging of gastrointestinal tumors. Eur J Surg Oncol 21: 50–55
Hunt TM, Flowerdew ADS, Birch SJ, Williams JD, Mulle MA, Taylor I (1990) Prospective randomized controlled trial of hepatic arterial embolization of infusion chemotherapy with 5-fluorouracil and degradable starch microspheres for colorectal liver metastastases. Br J Surg 77: 779–783
Imdahl A, Bräutigam P, Hauenstein KH, Eggstein S, Waninger J, Farthmann EH (1994) The value of CEA immunoscintigraphy for diagnosis of colorectal cancer and its metastases: results of a prospective study. Zentralbl Chir 119: 17–22
Iwatsuki S, Sheahan DG, Starzl TE (1989) The changing face of hepatic resection. Curr Probl Surg 26: 283–379
Kemeny N, Conti J, Seiter K (1992) A phase I trial intrahepatic fluorodesoxyuridine (FUDR) and systemic 5-fluorouracil (5-FU) + leucovorin in metastatic colorectal cancer: a useful adjuvant program after resection of colorectal metastases? Proc Am Soc Clin Oncol 11: 194

Koga, S, Kawaguchi H, Kishimoto H (1980) Therapeutic significance of non-curative gastrectomy for gastric cancer with liver metastasis. Am J Surg 140: 356–359

Koneru, B, Bassavilla A, Bowman J (1988) Liver transplantation for malignant tumors. Gastroenterol Clin North Am 17: 177–182

Korenaga, D, Okamura T, Baba H, Saito A, Sugimachi K (1988) Results of resection of gastric cancer extending to adjacent organs. Br J Surg 75: 12–15

Kortz WJ, Myers WC, Hanks JB, Schirmer B, Jones S (1984) Hepatic resection for metastatic cancer. Ann Surg 199: 182–184

Lahr CJ, Scong SJ, Cloud G (1983) A multifactorial analysis of prognostic factors in patients with liver metastases from colorectal carcinoma. J Clin Oncol 1: 720–726

Lorenz M, Hottenroth C, Maier P, Reimann M, Inglis R, Encke A (1992) Continuous regional treatment with fluoropyrimidines for metastases from colorectal carcinomas: influence of modulation with leucovorin. Semin Oncol 19: 163–170

Lorenz M, Stabe-Schler E, Rossion I, Koch R, Gag C, Encke A (1995) Ergebnisse der Resektion und adjuvanten Therapie von Lebermetastasen. Zentralbl Chir 120: 279–289

Lucarotti ME, Rothnie ND, Kelly SB, Hershman MJ, Johanssen C, Nilsson O, Lindholm L, Wood CB, Williamson R.C,Habib NA (1991) Clinical evaluation of tumour marker combinations in the differential diagnosis of benign and malignant liver disease. HPB Surg 5: 23–28

Martin JK Jr, O'Connell MJ, Wieand HS, Fitzgibbons RJ Jr, Mailliard JA, Nagorney DM, Tschetter LK, Krook JE (1990). Intra-arterial floxuridine vs systemic fluorouracil for hepatic metastases from colorectal cancer. A randomized trial. Arch Surg 125: 1022–1027

Mavligit G, Charnsangavej C, Talpaz M, Gutterman J, Wallace S (1988) Splenic versus hepatic artery infusion of recombinant interleukin-2 for patients with liver metastases. Proc Am Soc Clin Oncol 7: 107

McFarland EG, Mayo Smith WW, Saini S, Hahn PF, Goldberg MA, Lee MJ (1994). Hepatic hemangiomas and malignant tumors: improved differentiation with heavily T2-weighted conventional spin-echo MR imaging. Radiology 193: 43–47

Meirion-Thomas J, Redding WH, Coombes RC (1978) Failure to detect intraabdominal metastases from breast cancer: a case for staging laparotomy. Br J Med 2: 157–159

Mitchell DG, Saini S, Weinreb J, De Lange EE, Runge VM, Kuhlman JE, Parisky Y, Johnson CD, Brown JJ, Schnall M et al. (1994) Hepatic metastases and cavernous hemangiomas: distinction with standard- and triple-dose gadoteridol-enhanced MR imaging. Radiology 193: 49–57

Moriya Y, Sugihara K, Hojo K, Makuuchi M (1991) Adjuvant hepatic intra-arterial chemotherapy after potentially curative hepatectomy for liver metastases from colorectal cancer. Eur J Surg Oncol 17: 519

Morrow CE, Grage TB, Sutherland DER, Najarian JS (1982) Hepatic resection for secondary neoplasms. Surgery 92: 610–614

NIH (1990) NIH consensus conference. Adjuvant therapy for patients with colon and rectal cancer. JAMA 264 (11): 1444–1450

Nordlinger B, Vaillant, JC, Guiguet M, Balladur P, Paris F, Bachellier P, Jaeck D. (1994) Survival benefit of repeat liver resections for recurrent colorectal metastases: 143 cases. Association Francaise de Chirurgie. J Clin Oncol 12: 1491–1496.

O'Connell, MJ, Adson, MA, Schutt, AJ, Rubin, J, Moertel, CG, Ilstrup. DM (1985) Clinical trial of adjuvant chemotherapy after surgical resection of colorectal cancer metastatic to the liver. Mavo Clin Proc 60: 517–520

Okuyama K, Isono K, Juan IK (1985) Evaluation of treatment for gastric cancer with liver metastases. Cancer 55: 2498–2505

Ramming KP, Sparks FC, Eilber F, Morton D (1977) Management of hepatic metastases. Semin Oncol 4: 71–80

Ringe B, Bechstein WO, Raab R, Meyer HJ, Pichlmayr R (1990) Leberresektionen bei 157 Patienten mit colorektalen Metastasen. Chirurg 61: 272–279

Rougier P, Hay JM, Olliver JM (1990) A controlled multicentric trial of intrahepatic chemotherapy (IHC) vs standard palliative treatment for colorectal liver metastases. Proc Am Soc Clin Oncol 9: 104

Scheele J, Stangl R, Altendorf Hofmann A (1990) Hepatic metastases from colorectal carcinoma: impact of surgical resection on the natural history. Br J Surg 77(11): 1241–1246.

Schlag P, Manasterski M, Gerneth T, Hohenberger P, Dueck M, Herfarth C, Liebrich W, Schirrmacher V (1992) Active specific immunotherapy with Newcastle disease virus modified autologous cancer: first evaluation of clinical response of a phase II trial. Cancer Immunol Immunother 35: 325–329

Semelka RC, Shoenut JP, Ascher SM, Kroeker MA, Greenberg HM, Yaffe CS, Micflikier AB (1994) Solitary hepatic metastasis: comparison of dynamic contrast-enhanced CT and MR imaging with fat-suppressed T2-weighted breath-hold T1-weighted FLASH, and dynamic gadolinium-enhanced FLASH sequences. J Magn Reson Imaging 4: 319–323

Soyer P, Bluemke DA, Hruban RH, Sitzmann JV, Fishman EK (1994) Hepatic metastases from colorectal cancer: detection and false-positive findings with helical CT during arterial portography. Radiology 193: 71–74

Steele G Jr, Bleday R, Mayer RJ, Lindblad A, Petrelli N, Weaver D (1991) A prospective evaluation of hepatic resection for colorectal carcinoma metastases to the liver: Gastrointestinal Tumor Study Group Protocol 6584. J Clin Oncol 9: 1105–1112

Stehlin JS, Delpoly PD, Greef PJ et al. (1988) Twenty years experience with infusion and resection in 414 patients. Ann Surg 208: 23–25

Stimpson RE, Pellegrini CA, Way IW (1987) Factors affecting the morbidity of elective liver resection. Am J Surg 153: 189–196

Sturm J, Raute M, Trede M (1991) Operationsindikation bei Lebermetastasen nicht kolorektaler Karzinome. In: Herfarth C, Schlag PU (eds) Neue Entwicklungen in der Therapie von Lebertumoren Springer, Berlin Heidelberg New York, pp 337–344

Swanson DA, Johnson DE, von Eschenbach A, Chuang VP, Wallace S (1983) Angioinfarction plus nephrectomy for metastatic renal cell carcinoma – an update. J Urol 130: 4498–4451

Tabuchi Y, Boku S, Kato M, Takita Y, Kawaguch K (1977) Clinicopathological and prognostic studies of gastric cancer macroscopically invading neighboring organs. Jpn J Cancer Clin 22: 964–973

Thompson H, Tompkins R, Longmire WP (1983) Major hepatic resection. Ann Surg 197: 375–388

Tomas-de la Vega JE, Donahue EJ, Doolas A (1984) A ten year experience with hepatic resection. Surg Gynecol Obstet 159: 223–228

Tsujitani S, Watanabe A, Kakeji Y, Machara Y, Machara Y, Tomoda H, Furusawa M (1991) Hepatic recurrence not prevented with low-dosage long-term intraportal 5-FU infusion after resection of colorectal liver metastasis. Eur J Surg Oncol 17: 526–529

Vogt P, Raab R, Ringe B, Pichlmayr R (1991) Resection of synchronous liver metastases from colorectal cancer. World J Surg 15: 62–67

Wagner JS, Adson MA, van Heerden JA (1984) The natural history of hepatic metastases from colorectal cancer. Ann Surg 199: 199–502

Welch TJ, Sheedy PF, Johnson CM (1985) Focal nodular hyperplasia and hepatic adenoma: comparision of angiography, CT,US and scintigraphy. Radiology 156: 593–595

Wernecke K, Rummeney E, Bongartz G, Vassallo P, Kivelitz D, Wiesmann W, Peters PE, Reiser M, Pircher W (1991) Detection of hepatic masses in patients with carcinoma: comparative sensitivities of sonography, CT, and MR imaging. AJR 157: 731–739

Wolf RF, Goodnight JE, Krag DE, Schneider PD (1991) Results of resection and proposed guidelines for patient selection in instances of non-colorectal hepatic metastases. Surg Gynecol Obstect 173: 454–460.

Zukiwski A, Wallac S, Guterman J, Saks S, Mavligit G (1989) Hepatic arterial infusion of recombinant tumor necrosis factor in patients with metastatic carcinoma to the liver. Proc Am Soc Clin Oncol 8: 122

Subject Index

Current Topics in Microbiology and Immunology

Volumes published since 1989 (and still available)

Vol. 173: **Pfeffer, Klaus; Heeg, Klaus; Wagner, Hermann; Riethmüller, Gert (Eds.):** Function and Specificity of γ/δ TCells. 1991. 41 figs. XII, 296 pp. ISBN 3-540-53781-3

Vol. 174: **Fleischer, Bernhard; Sjögren, Hans Olov (Eds.):** Superantigens. 1991. 13 figs. IX, 137 pp. ISBN 3-540-54205-1

Vol. 175: **Aktories, Klaus (Ed.):** ADP-Ribosylating Toxins. 1992. 23 figs. IX, 148 pp. ISBN 3-540-54598-0

Vol. 176: **Holland, John J. (Ed.):** Genetic Diversity of RNA Viruses. 1992. 34 figs. IX, 226 pp. ISBN 3-540-54652-9

Vol. 177: **Müller-Sieburg, Christa; Torok-Storb, Beverly; Visser, Jan; Storb, Rainer (Eds.):** Hematopoietic Stem Cells. 1992. 18 figs. XIII, 143 pp. ISBN 3-540-54531-X

Vol. 178: **Parker, Charles J. (Ed.):** Membrane Defenses Against Attack by Complement and Perforins. 1992. 26 figs. VIII, 188 pp. ISBN 3-540-54653-7

Vol. 179: **Rouse, Barry T. (Ed.):** Herpes Simplex Virus. 1992. 9 figs. X, 180 pp. ISBN 3-540-55066-6

Vol. 180: **Sansonetti, P. J. (Ed.):** Pathogenesis of Shigellosis. 1992. 15 figs. X, 143 pp. ISBN 3-540-55058-5

Vol. 181: **Russell, Stephen W.; Gordon, Siamon (Eds.):** Macrophage Biology and Activation. 1992. 42 figs. IX, 299 pp. ISBN 3-540-55293-6

Vol. 182: **Potter, Michael; Melchers, Fritz (Eds.):** Mechanisms in B-Cell Neoplasia. 1992. 188 figs. XX, 499 pp. ISBN 3-540-55658-3

Vol. 183: **Dimmock, Nigel J.:** Neutralization of Animal Viruses. 1993. 10 figs. VII, 149 pp. ISBN 3-540-56030-0

Vol. 184: **Dunon, Dominique; Mackay, Charles R.; Imhof, Beat A. (Eds.):** Adhesion in Leukocyte Homing and Differentiation. 1993. 37 figs. IX, 260 pp. ISBN 3-540-56756-9

Vol. 185: **Ramig, Robert F. (Ed.):** Rotaviruses. 1994. 37 figs. X, 380 pp. ISBN 3-540-56761-5

Vol. 186: **zur Hausen, Harald (Ed.):** Human Pathogenic Papillomaviruses. 1994. 37 figs. XIII, 274 pp. ISBN 3-540-57193-0

Vol. 187: **Rupprecht, Charles E.; Dietzschold, Bernhard; Koprowski, Hilary (Eds.):** Lyssaviruses. 1994. 50 figs. IX, 352 pp. ISBN 3-540-57194-9

Vol. 188: **Letvin, Norman L.; Desrosiers, Ronald C. (Eds.):** Simian Immunodeficiency Virus. 1994. 37 figs. X, 240 pp. ISBN 3-540-57274-0

Vol. 189: **Oldstone, Michael B. A. (Ed.):** Cytotoxic T-Lymphocytes in Human Viral and Malaria Infections. 1994. 37 figs. IX, 210 pp. ISBN 3-540-57259-7

Vol. 190: **Koprowski, Hilary; Lipkin, W. Ian (Eds.):** Borna Disease. 1995. 33 figs. IX, 134 pp. ISBN 3-540-57388-7

Vol. 191: **ter Meulen, Volker; Billeter, Martin A. (Eds.):** Measles Virus. 1995. 23 figs. IX, 196 pp. ISBN 3-540-57389-5

Vol. 192: **Dangl, Jeffrey L. (Ed.):** Bacterial Pathogenesis of Plants and Animals. 1994. 41 figs. IX, 343 pp. ISBN 3-540-57391-7

Vol. 193: **Chen, Irvin S. Y.; Koprowski, Hilary; Srinivasan, Alagarsamy; Vogt, Peter K. (Eds.):** Transacting Functions of Human Retroviruses. 1995. 49 figs. IX, 240 pp. ISBN 3-540-57901-X

Vol. 194: **Potter, Michael; Melchers, Fritz (Eds.):** Mechanisms in B-cell Neoplasia. 1995. 152 figs. XXV, 458 pp. ISBN 3-540-58447-1

Vol. 195: **Montecucco, Cesare (Ed.):** Clostridial Neurotoxins. 1995. 28 figs. XI., 278 pp. ISBN 3-540-58452-8

Vol. 196: **Koprowski, Hilary; Maeda, Hiroshi (Eds.):** The Role of Nitric Oxide in Physiology and Pathophysiology. 1995. 21 figs. IX, 90 pp. ISBN 3-540-58214-2

Vol. 197: **Meyer, Peter (Ed.):** Gene Silencing in Higher Plants and Related Phenomena in Other Eukaryotes. 1995. 17 figs. IX, 232 pp. ISBN 3-540-58236-3

Vol. 198: **Griffiths, Gillian M.; Tschopp, Jürg (Eds.):** Pathways for Cytolysis. 1995. 45 figs. IX, 224 pp. ISBN 3-540-58725-X

Vol. 199/I: **Doerfler, Walter; Böhm, Petra (Eds.):** The Molecular Repertoire of Adenoviruses I. 1995. 51 figs. XIII, 280 pp. ISBN 3-540-58828-0

Vol. 199/II: **Doerfler, Walter; Böhm, Petra (Eds.):** The Molecular Repertoire of Adenoviruses II. 1995. 36 figs. XIII, 278 pp. ISBN 3-540-58829-9

Vol. 199/III: **Doerfler, Walter; Böhm, Petra (Eds.):** The Molecular Repertoire of Adenoviruses III. 1995. 51 figs. XIII, 310 pp. ISBN 3-540-58987-2

Vol. 200: **Kroemer, Guido; Martinez-A., Carlos (Eds.):** Apoptosis in Immunology. 1995. 14 figs. XI, 242 pp. ISBN 3-540-58756-X

Vol. 201: **Kosco-Vilbois, Marie H. (Ed.):** An Antigen Depository of the Immune System: Follicular Dendritic Cells. 1995. 39 figs. IX, 209 pp. ISBN 3-540-59013-7

Vol. 202: **Oldstone, Michael B. A.; Vitković, Ljubiša (Eds.):** HIV and Dementia. 1995. 40 figs. XIII, 279 pp. ISBN 3-540-59117-6

Vol. 203: **Sarnow, Peter (Ed.):** Cap-Independent Translation. 1995. 31 figs. XI, 183 pp. ISBN 3-540-59121-4

Vol. 204: **Saedler, Heinz; Gierl, Alfons (Eds.):** Transposable Elements. 1995. 42 figs. IX, 234 pp. ISBN 3-540-59342-X

Vol. 205: **Littman, Dan. R. (Ed.):** The CD4 Molecule. 1995. 29 figs. XIII, 182 pp. ISBN 3-540-59344-6

Vol. 206: **Chisari, Francis V.; Oldstone, Michael B. A. (Eds.):** Transgenic Models of Human Viral and Immunological Disease. 1995. 53 figs. XI, 345 pp. ISBN 3-540-59341-1

Vol. 207: **Prusiner, Stanley B. (Ed.):** Prions Prions Prions. 1995. 42 figs. VII, 163 pp. ISBN 3-540-59343-8

Vol. 208: **Farnham, Peggy J. (Ed.):** Transcriptional Control of Cell Growth. 1995. 17 figs. IX, 141 pp. ISBN 3-540-60113-9

Vol. 209: **Miller, Virginia L. (Ed.):** Bacterial Invasiveness. 1996. 16 figs. IX, 115 pp. ISBN 3-540-60065-5

Vol. 210: **Potter, M.; Rose, N. R. (Eds.):** Immunology of Silicones. 1996. 136 figs. XX, 430 pp. ISBN 3-540-60272-0

Vol. 211: **Wolff, L.; Perkins, A. S. (Eds.):** Molecular Aspects of Myeloid Stem Cell Development. 1996. 98 figs. XIV, 298 pp. ISBN 3-540-60414-6

Vol. 212: **Vainio, O.; Imhof, B. A. (Eds.):** Immunology and Developmental Biology of the Chicken. 1996. 43 figs. IX, 281 pp. ISBN 3-540-60585-1

Vol. 213/I: **Günthert, U.; Birchmeier, W. (Eds.):** Attempts to Understand Metastasis Formation I. 1996. 35 figs. XV, 293 pp. ISBN 3-540-60680-7

Vol. 213/II: **Günthert, U.; Birchmeier, W. (Eds.):** Attempts to Understand Metastasis Formation II. 1996. 33 figs. XV, 288 pp. ISBN 3-540-60681-5

MIX
Papier aus verantwortungsvollen Quellen
Paper from responsible sources
FSC® C105338

If you have any concerns about our products,
you can contact us on
ProductSafety@springernature.com

In case Publisher is established outside the EU,
the EU authorized representative is:
Springer Nature Customer Service Center GmbH
Europaplatz 3, 69115 Heidelberg, Germany

Printed by Libri Plureos GmbH
in Hamburg, Germany